3D INTEGRATION for VLSI SYSTEMS

3D
INTEGRATION
for VLSI
SYSTEMS

edited by
Chuan Seng Tan
Kuan-Neng Chen
Steven J. Koester

PAN STANFORD PUBLISHING

Published by
Pan Stanford Publishing Pte. Ltd.
Penthouse Level, Suntec Tower 3
8 Temasek Boulevard
Singapore 038988

E-mail: editorial@panstanford.com
Web: www.panstanford.com

British Library Cataloguing-in-Publication Data
A catalogue record for this book is available from the British Library.

3D Integration for VLSI Systems

ISBN: 978-981-4303-81-1 (Hardcover)
ISBN: 978-981-4303-82-8 (eBook)

Printed in the USA

Preface

Three-dimensional (3D) integration has emerged as a critical performance enabler for integrated circuits, at a time when the microelectornics industry is faced with unprecedented scaling barriers, which have arisen both due to fundamental physics and economic constraints. 3D integration provides a mechanism for space transformation of the traditional planar implementation of integrated circuits into three-dimensional space. It therefore provides a pathway to extend geometrical scaling for further performance enhancement ("More Moore"), as well as provide functional diversification ("More than Moore") to improve higher-level system operation. At its core, 3D integration is simply the process of vertically stacking of circuits and forming electrical connections between them. Despite this seemingly simple concept, however, 3D integration involves significant development of many new technologies, from the basic processes and materials issues involved, to new approaches to system architectures, and it is the status and progress in these new areas that provide the focus of this book.

The advent of 3D integration is a direct result of relentless research in academia, research laboratories, and industry over the last 10 years. Today, 3D integration exists as a diverse set of stacking and vertical interconnection technologies that can take a multitude of forms, with the precise implementation depending on the applications. At the time of this writing, commercial 3D products already exist, including small form factor image sensors that include through silicon vias (TSV), and several announcements of future products, including 3D memory chips, have been made.

The concept of this book on 3D technology dates back to more than two years ago. At that time, an increasing number of publications and conferences had started to focus on 3D integration. At the end of an IEEE-sponsored International Workshop on Next Generation Electronics in Tainan, Taiwan, in November 2008, Pan Stanford Publishing (PSP) identified 3D integration as an important topic for book publication. While a few reference books on this emerging field already existsed, there was an urgent need to highlight more recent developments in a new book and the idea of this book was formed. Given the many varieties of 3D integration technologies and its large span in the semiconductor supply chain, we decided to edit a book with contributions from experts in academia, research laboratories, and industry. After careful planning, we identified and invited chapter contribution from an impressive line-up of highly qualified researchers. It took more than one full year for planning, writing, and editing.

The objective of this book is to present novel ideas in pre-packaging wafer-level 3D integration technologies. The book covers process technologies such as wafer bonding, through silicon via (TSV), wafer thinning and handling, infrastructures, integration schemes, design as well as providing a succinct outlook. All process technologies are carefully described and potential applications are listed. Technical challenges are also highlighted. This book is particularly beneficial to researchers or engineers who are already working or are beginning to work on 3D technology.

This book would not have been possible without a team of highly qualified and dedicated people. We are particularly grateful to Stanford Chong of PSP for initiating this undertaking and for providing his support. Rhamie Wahap and dedicated editorial staff at PSP worked alongside with us and provided us with the necessary editorial support. The three co-editors were funded for many years through the MARCO and DARPA funded Interconnect Focus Center (IFC) as well as the DARPA funded 3D IC Program; our 3D technology platform research, and this book, would not have been possible without this extended research support.

C.S. Tan is immensely grateful for the unfailing love and support from his wife, Lee Peng, without whom this book would not have been possible. He is currently supported by a Nanyang Assistant Professorship at the Nanyang Technological University. His research is generously supported by a grant from the Semiconductor Research Corporation (SRC), USA, through a subcontract from the Interconnect and Packaging Center at Georgia Institute of Technology, a seedling fund from Defence Advanced Research Program Agency (DARPA), USA, as well as the Defence Science and Technology Agency (DSTA) in Singapore. K.N. Chen would like to acknowledge funding supports from the National Science Council, Taiwan, as well as 3D IC research supports from the National Chiao Tung University, IBM T.J. Watson Research Center, and Massachusetts Institute of Technology. S. J. Koester is gratefully with support from the University of Minnesota.

Last but not least, we are extremely thankful to authors who accepted our invitation and contributed chapters to this book. We hope that the readers will find this book useful in their pursuit of 3D technology. Please do not hesitate to contact us if you have any comments or suggestions.

Chuan Seng Tan
tancs@ntu.edu.sg
USA

Kuan-Neng Chen
knchen@mail.nctu.edu.tw
Taiwan

Steven J. Koester
skoester@umn.edu
USA
May 2011

Contents

Chapter 1

3D INTEGRATION TECHNOLOGY – INTRODUCTION AND OVERVIEW

Chuan Seng Tan
Nanyang Technological University

Steven J. Koester
University of Minnesota

Kuan-Neng Chen
National Chiao Tung University

1.1 INTRODUCTION

The past decade has seen three-dimensional (3D) integration technology mature rapidly from a hypothetical concept to a technology that is on the cusp widespread commercial implementation. This rapid trend towards acceptance of 3D integration has been both a result of key demonstrations of the technical feasibility of the process, as well as a growing consensus that 3D integration will be necessary to continue current computational system performance trends. 3D technology also offers an abundance of opportunities for new applications and functionality. In this introduction, we provide an overview of the system needs that are driving 3D integration development, the recent advances in the underlying technology that have been key to its recent acceptance, and new opportunities for additional functionality that 3D has the potential to provide.

3D Integration for VLSI Systems
Edited by Chuan Seng Tan, Kuan-Neng Chen and Steven J. Koester
Copyright © 2012 by Pan Stanford Publishing Pte. Ltd.
www.panstanford.com

1.2 DRIVERS AND APPLICATIONS

The driving force for microprocessors in the last several decades has been the scaling of Si metal-oxide semiconductor field-effect transistors (MOSFETs),[1] which has allowed transistor dimensions to shrink from fractions of millimeters in the 1960's, to the 10's of nanometers in present-day technologies. As Gordon Moore predicted in his seminal paper,[1] reducing the feature size also allows chip area to be decreased, improving production thereby reducing cost per function. The scaling laws original put in place by Dennard[2] also showed that improved device and ultimately processor speed could be achieved through dimensional scaling. The impact of these trends on society has been nothing short of remarkable, as computational systems have become an indispensible aspect of nearly every facet of society. However, all trends ultimately have limits, and Moore's Law is no exception. The limits to Moore's Law scaling have come simultaneously from many directions. Lithographic limits have made it extremely difficult to pack more features onto a semiconductor chip, and the most advanced lithographic techniques needed to scale are becoming prohibitively expensive for most fabs.[3] Furthermore, short-channel effects and random fluctuations are making conventional planar device geometries obsolete.[4] Finally, the fact that scaling has proceeded without appreciable voltage reduction over the past decade has increased power densities to the precipice of cooling and reliability limits.[5] This latter fact has essentially ended clock frequency enhancement as the primary driver of micro-processor performance improvement.

Amidst this landscape, multi-core architectures have become the new engine driving system-level performance enhancement.[6] In multi-core architectures, computation proceeds in parallel, thereby lessening the frequency requirements for the individual cores. Instead of increasing clock frequency, performance enhancements can be achieved by adding cores. Reducing the frequency of individual cores is particularly beneficial from a power point of view, since it allows the supply voltage to be reduced, which in turn improves computational efficiency. However, multi-core architectures place new requirements on the technology, and as we will show, these requirements are ideally met by utilizing 3D integration. The main requirement for multi-core architectures is increased memory bandwidth (either main memory or cache). This condition derives from the well-known law[7] that states that the cache miss rate goes as the inverse square root of the cache size for a single core, and as the number of cores is increased, the cache capacity must be increased geometrically for each additional core added.

3D offers the opportunity to increase the amount of available memory to multi-core processors by stacking additional primary or cache memory on the chip, as opposed to relying upon off-chip access. 3D integration can also increase the aggregate bandwidth to this memory since it provides much greater density of interconnects than traditional C4s. Finally, the power required to access memory through 3D interconnects is reduced compared to off-chip memory since smaller, less power-hungry interconnect drive circuits are needed due to the reduced interconnect capacitance. A conceptual diagram showing the benefits of 3D integration for multi-core memory design is shown in Fig. 1.1.[8]

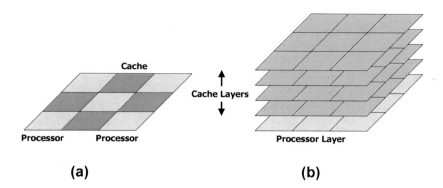

(a)   **(b)**

Fig. 1.1 Conceptual diagram showing core and memory partitioning for (a) 2D and (b) 3D chip designs (adapted from Ref. 8).

If memory expansion for multi-core architectures were the only application of 3D technology, enthusiasm for 3D technology development might have been substantially more muted. However, 3D technology has potential to benefit computational systems in a vast number of new ways. For instance, voltage regulation is another aspect of multi-core architectures where 3D could provide tremendous benefits.[9-10] The advantages of a 3D chip with a dedicated voltage regulation layer compared to a 2D system with off-chip regulation are substantial.[9] First, 3D reduces the resistance between the voltage source and chip, minimizing supply voltage droop. 3D also allows the use of multiple voltage planes, potentially one of every core. This would allow the supply voltage of the cores to be precisely tuned for optimal performance, and even allow cores to be turned off if needed. It has also been proposed that if switching converters are utilized, I^2R losses through the package could be reduced since the power would enter the package at a higher voltage than the chip supply voltage.[10]

3D technology is also useful for a variety of advanced memory applications. For flash technology, form factor is the overriding concern, and therefore 3D integration can provide increased density within a given footprint. On the other hand, the benefits of 3D stacking for SRAM and DRAM technology result both from increased areal density and increased performance.[11] With sufficient interconnect density, a 3D DRAM can partitioned with the memory controller on one layer, and the arrays on the remaining layers, resulting in significant improvement in access time.[11]

Perhaps some of the most intriguing aspects of 3D integration technology are the new capabilities that could be brought to bear by bringing very dissimilar technologies into intimate contact. One interesting possibility is the integration of 3D photonic interconnects with multi-core processors, which have the potential to dramatically reduce the power of interconnects particularly in the regime where the interconnect distance is long, and the data rates are high. A conceptual 3D optical interconnect scheme is shown in Fig. 1.3.[12] Here a complete network-on-a-chip is shown, which utilizes a bottom multi-core processor layer, an intermediate memory layer, and an optical interconnect layer on top. In such a system, local interconnects could be provided by standard metal wiring, whereas global connections would be made using a photonic interconnect fabric consisting of silicon waveguides. The interconnect power has been calculated for such a scheme and is found to be nearly 2 orders of magnitude lower compared to electrical interconnects.[12] The photonic network provides the additional advantage that off-chip I/O is achieved at the same bandwidth with little additional power.

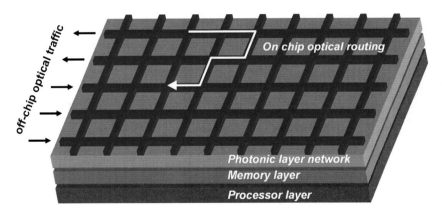

Fig. 1.2 Conceptual diagram of a 3D integrated photonic interconnect network (adapted from Ref. 12).

Co-integration of III-V components with CMOS is another possible application of 3D integration that could have a profound impact on high-performance systems. The optical properties of III-V are the most compelling reason for doing this, as the integration of lasers, optical amplifiers and detectors[13] could provide additional performance enhancement and design flexibility for photonic interconnects.[14] The work in[13] is particularly intriguing as the heterogeneous integration is performed at the device level, creating an entirely new device, in this case by integrating a III-V optical gain or absorbing region with a Si waveguide. Integration of high-speed III-V electronic devices has also be proposed to improve performance of mixed-signal chips by integrating either high-speed or high-power III-V FETs or HBTs with CMOS digital circuitry.[15] Future challenges for heterogeneous 3D integration will certainly include yield, reliability and cost. However, in the end, if basic manufacturability issues can be addressed, heterogeneous 3D technology could allow new applications for Si-based systems by allowing each sub-component to be optimized according to its specific function without the trade-offs associated with monolithic integration.

1.3 CLASSIFICATION

System Integration Landscape

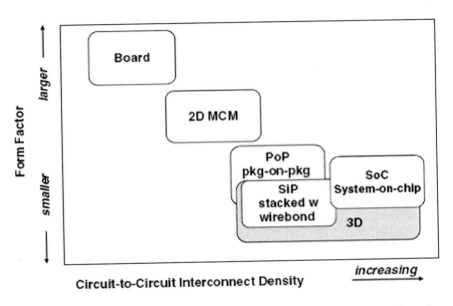

Fig. 1.3 Comparison of various system integration technologies in terms of form factor and circuit-to-circuit interconnect density.[16] © IEEE 2007.

System integration, that is, the integrating together of circuits or intellectual property (IP) blocks, is one of the major applications of 3D integration. As such, 3D integration must compete against a number of established technologies. Figure 1.8 compares the relative capability of several system integration methods (board, 2D multi-chip module – 2D-MCM, package-on-package - PoP, system-in-package – SiP, and 2D system-on-chip – 2D-SoC) in terms of form factor and interconnects density between circuit blocks.[16] 3D integration offers more compact form factor and higher chip-to-chip interconnects density. Comparing with 2D-SoC, 3D integration shortens time-to-market and lowers the system cost. By using larger number of smaller and shorter through silicon via (TSV) as compared to wire bonding in SiP, performance is enhancement via 3D integration due to smaller latency and higher bandwidth, as well as smaller power consumption.

Classification

There are a number of technology options to arrange integrated circuits in a vertical stack. It is possible to stack ICs in a vertical fashion at various stages of processing: (1) post-singulation 3D packaging (*e.g.* chip-to-chip), and (2) pre-singulation wafer level 3-D integration (*e.g.* chip-to-wafer, wafer-to-wafer, and monolithic approaches). Active layers can be vertically interconnected using physical contact such as bond wire or interlayer vertical via (including TSV). It is also possible to establish chip to chip connection via non-contact (or wireless) links such as capacitive and inductive couplings [17]. Capacitive coupling utilizes a pair of electrodes that are formed using conventional IC fabrication. The inductive-coupling I/O is formed by placing two planar coils (planar inductors) above each other and is also made using conventional IC fabrication. The advantages of these approaches are fewer processing steps hence lower cost, no requirement for ESD protection, low power, and smaller area I/O cell.

Since there is substantial overlap between various options and lack of standardization in terms of definition, classification of 3D IC technology is often not straight forward. This chapter makes an attempt to classify 3D IC based on the processing stage when stacking takes place.

Monolithic Approaches

In these approaches, devices in each active layer are processed sequentially starting from the bottom-most layer. Devices are built on a substrate wafer by mainstream process technology. After proper isolation, a second device layer is formed and devices are processed by conventional means on the second layer. This sequence of isolation, layer formation, and device processing can be repeated to build a multi-layer structure.

The key technology in this approach is forming a high quality active layer isolated from the bottom substrate. This bottom-up approach has the advantage that precision alignment between layers can be accomplished. However, it suffers from a number of drawbacks. The crystallinity of upper layers is usually low and imperfect. As a result, high performance devices cannot be built in the upper layers. Thermal cycling during upper layer crystallization and device processing can degrade underlying devices and therefore a tight thermal budget must be imposed. Due to the sequential nature of this method, manufacturing throughput is low. A simpler FEOL process flow is feasible if polycrystalline silicon can be used for active devices; however, a major difficulty is to obtain high-quality electrical devices and interconnects. While obtaining single-crystal device layers in a generic IC technology remains in the research stage, polycrystalline devices suitable for non-volatile memory (NVM) have not only been demonstrated but have been commercialized (for example by SanDisk). A key advantage of FEOL-based 3-D integration is that IC BEOL and packaging technologies are unchanged; all the innovation occurs in 3-D stacking of active layers.

A number of FEOL techniques include: laser beam recrystallization,[18, 19] seeding-assisted recrystallization,[20, 21] selective epitaxy and over-growth,[22] and grapho-exitaxy.[23]

Assembly Approaches

This is a parallel integration scheme in which fully processed or partially processed integrated circuits are assembled in a vertical fashion. Stacking can be achieved with one of these methods: (1) chip-to-chip, (2) chip-to-wafer, and (3) wafer-to-wafer. Vertical connection in chip-to-chip stacking can be achieved using wire bond or through silicon via (TSV).

Wafer level 3D integration, such as chip-to-wafer and wafer-to-wafer stacking, use TSV as the vertical interconnect. This integration approach often involves a sequence of wafer thinning and handling, alignment, TSV formation, and bonding. The key differentiators are:

- Bonding medium — metal-to-metal, dielectric-to-dielectric (oxide, adhesive, etc) or hybrid bonding;
- TSV formation — via first, via middle or via last;
- Stacking orientation — face-to-face or back-to-face stacking;
- Singulation level — chip-to-chip, chip-to-wafer or wafer-to-wafer.

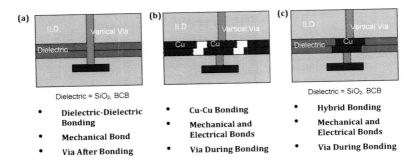

Fig. 1.4 Wafer bonding techniques for wafer-level 3-D integration: (a) dielectric-to-dielectric; (b) metal-to-metal; and (c) dielectric/metal hybrid.

The types of wafer bonding potentially suitable for wafer-level 3D integration are depicted in Fig. 1.4. Dielectric-to-dielectric bonding is most commonly accomplished using silicon oxide or BCB polymer as the bonding medium. These types of bonding provide primary function as a mechanical bond and the inter-wafer via is formed after wafer-to-wafer alignment and bonding (Fig. 1.4(a)). When metallic copper-to-copper bonding is used (Fig. 1.4(b)), the inter-wafer via is completed during the bonding process; note that appropriate interconnect processing within each wafer is required to enable 3D interconnectivity. Besides providing electrical connections between IC layers, dummy pads can also be inserted at the bonding interface at the same time to enhance the overall mechanical bond strength. This bonding scheme inherently leaves behind isolation gap between Cu pads and this could be a source of concern for moisture corrosion and compromise the structural integrity especially when IC layers above the substrate is thinned down further. Figure 1.4(c) shows a bonding scheme utilizing a hybrid medium of dielectric and Cu. This scheme in principle provides a seamless bonding interface consists of dielectric bond (primarily a mechanical bond) and Cu bond (primarily an electrical bond). However, very stringent requirements with regards to surface planarity (dielectric and Cu) and Cu contamination in the dielectric layer due to misalignment are needed.

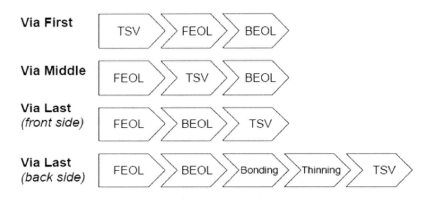

Fig. 1.5 TSV can be formed at various stages of IC processing.

The selection of the optimum technology platform is subject to ongoing development and applications. Cu-to-Cu bonding has significant advantages for highest inter-wafer interconnectivity. As a result, this approach is desirable for microprocessors and digitally-based system-on-a-chip (SoC) technologies. Polymer-to-polymer bonding is attractive when heterogeneous integration of diverse technologies is the driver and the inter-wafer interconnect density is more relaxed; benzocyclobute (BCB) is the polymer most widely investigated. Taking advantage of the viscosity of the polymer, this method is more forgiving in terms of surface planarity and particle contamination. Oxide-to-oxide bonding of fully processed IC wafers requires atomic-scale smoothness of the oxide surface. In addition, wafer distortions introduced by FEOL and BEOL processing introduces sufficient wafer bowing and warping that prevents sufficient contact area to achieve the required bonding strength. While oxide-to-oxide bonding after FEOL and local interconnect processing has been shown to be promising (particularly with SOI wafers that allows for extreme thinning down to the buried oxide layer) the increased wafer distortion and oxide roughness after multilevel interconnect processing require extra attention during processing.

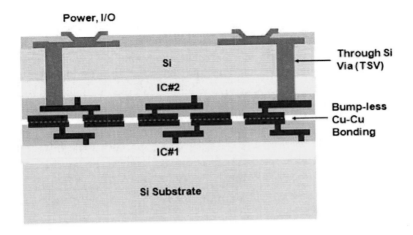

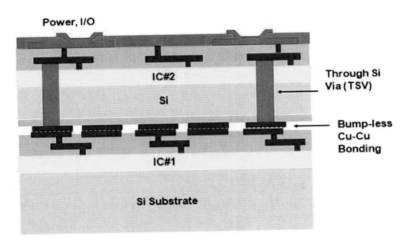

Fig. 1.6 (Top) Face-to-face or face-down stacking; (Bottom) Back-to-face or face-up stacking.

TSV can be formed at various stages during the 3D IC process as shown in Fig. 1.5. When TSV is formed before any CMOS processes, the process sequence is known as "via first". It is also possible to form the TSV when the front end processes are completed. In this "via middle" process, back end processes will continue after the TSV process is completed. When TSV is formed after the CMOS processes are completed, it is known as "via last" process. TSV can be formed from the front side or the back side of the wafer. The above schemes have different requirements in terms of process parameters and materials selection. The choice depends on final application requirements and infrastructures in the supply chain.

Another key differentiator in 3D IC integration is related to the stacking orientation. One option is to perform face-to-face (F2F) alignment and bonding with all required I/Os brought to the thinned backside of the top wafer (which becomes the face of the two-wafer stack). Another approach is to temporarily bond the top wafer to a handling wafer, after which the device wafer is thinned from the back side and permanently bonded to the full-thickness bottom wafer; after this permanent bonding the handling wafer is removed. This is also called a back-to-face (B2F) stacking. These two stacking orientations are shown in Fig. 1.6.

Table 1.1 Comparison between wafer-to-wafer and chip-to-wafer stacking

	Wafer-to-Wafer	**Chip-to-Wafer**
Wafer/die Size	Wafer/die of common size in order to avoid silicon area wastage	Dissimilar wafer/die size is acceptable
Throughput	Wafer scale	Die scale
Yield	Lower than lowest yield wafer, therefore high yield wafer must be used	Known good die can be used if pre-stacking testing is available
Alignment accuracy	<2 μm global alignment	~10 μm for >1000 dph <2 μm for <100 dph

F2F stacking allows a high density layer to layer interconnection which is limited by the alignment accuracy. Handle wafer is not required in F2F stacking and this imposes more stringent requirement on the mechanical strength of the bonding interface in order to sustain shear force during wafer thinning which is often achieved by mechanical grinding or polishing. Since one of the IC layer is facing down in the final assembly, F2F stacking also complicates the layout design as opposed to more conventional layout design whereby IC layers are facing up. Another potential disadvantage of F2F stacking relates to the thickening of the ILD layer at the bonding interface which presents higher barrier for effective heat dissipation. B2F stacking requires the use of a temporary handle and the layer to layer interconnection density is limited by the TSV pitch. Since the device layer is bonded to a temporary handle, the final permanent bond does not sustain damage resulting from wafer thinning. It requires the use of a temporary bonding medium that can provide sufficient strength during wafer handling and can be readily released after successful device layer permanent transfer on the substrate.

In wafer level 3D integration, permanent bonding can be done either in chip-to-wafer (C2W) or wafer-to-wafer (W2W) stacking. A comparison of these two methods is summarized in Table 1.1. As shown in Fig. 1.7, the option of C2W or W2W depends on two key requirements on chip size and alignment accuracy. When high precision alignment is desired in order to achieve high density layer to layer interconnections, W2W is a preferred choice to maintain acceptable throughput by performing a wafer level alignment. W2W is also preferred when chip size gets smaller.

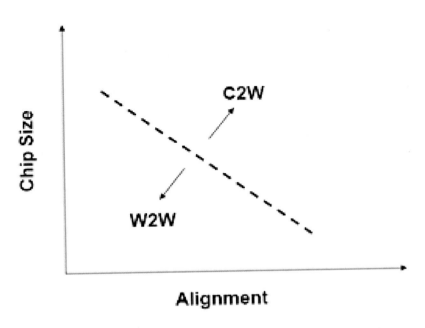

Fig. 1.7 The choice between C2W and W2W depends on the chip size and the required alignment accuracy.

3D Interconnect Technology Definitions by ITRS

Since 3D technology is actively pursued by almost all payers (such as IC foundry, semiconductor assembly and test, printed circuit board, and assembly) in the electronic manufacturing supply chain, a broad variety of technology is being proposed. As a result, the traditional interfaces between all these players are blurring. In order to come to a clear vision on roadmaps for 3D technologies, it is important to come to a clear definition of what is understood by 3D interconnect technology. The International Technology Roadmap for Semiconductor (ITRS), in the 2009 report on Interconnect, has proposed a classification of the wide variety of 3D technologies[24] that capture the functional requirements of 3D technology at the different hierarchical levels of the system and correspond to the supply chain manufacturing capabilities. The following is summary of 3D definitions and naming conventions proposed by ITRS.

3D Interconnect Technology - technology which allows for the vertical stacking of layers of "basic electronic components" that are connected using a 2D-interconnect fabric are listed below. "Basic electronic components" are elementary circuit devices such as transistors, diodes, resistors, capacitors and inductors. A special case of 3D interconnect technology is the Si interposer structures that may only contain interconnect layers, although in many cases other basic electronic components (in particular decoupling capacitors) may be embedded.

3D Bonding - operation that joins two die or wafer surfaces together.

3D Stacking - operation that also realizes electrical interconnects between the two device levels.

3D-Packaging (3D-P) - 3D integration using "traditional" packaging technologies, such as wire bonding, package-on-package stacking or embedding in printed circuit boards.

3D-Wafer-Level-Packaging (3D-WLP) - 3D integration using wafer level packaging technologies, performed after wafer fabrication, such as flip-chip redistribution, redistribution interconnect, fan-in chip-size packaging, and fan-out reconstructed wafer chip-scale packaging.

3D-System-on-chip (3D-SOC) - Circuit designed as a system-on-chip, SOC, but realized using multiple stacked die. 3D-interconnects directly connect circuit tiles in different die levels. These interconnects are at the level of global on-chip interconnects. This allows for extensive use/reuse of IP-blocks.

3D-Stacked-Integrated-Circuit (3D-SIC) - 3D approach using direct interconnects between circuit blocks in different layers of the 3D die stack. Interconnects are on the global or intermediate on-chip interconnect levels. The 3D stack is characterized by a sequence of alternating front-end (devices) and back-end (interconnect) layers.

3D-Integrated-Circuit (3D-IC) - 3D approach using direct stacking of active devices. Interconnects are on the local on-chip interconnect levels. The 3D stack is characterized by a stack of front-end devices, combined with a common back-end interconnect stack.

Table 1.2 (reproduced from ITRS) presents a structured definition of 3D interconnect technologies based on the interconnect hierarchy. This structure also refers to the industrial semiconductor supply chain and allows definition of meaningful roadmaps and targets for each layer of the interconnect hierarchy.

Table 1.2 3D Interconnect Technologies Based on Interconnect Hierarchy.[24]

Level	Name	Supply Chain	Key Features
Package	3D-Packaging (3D-P)	OSAT Assembly PCB	• Traditional packaging of interconnect technologies, e.g., wire-bonded die stacks, package-on-package stacks. • Also includes die in PCB integration. • No through-Si-vias (TSVs).
Bond-pad	3D-Wafer-level Package (3D-WLP)	Wafer-level Packaging	• WLP infrastructure, such as redistribution layer (RDL) and bumping. • 3D interconnects are processed after the IC fabrication, "post IC-passivation" (via last process). Connections on bond-pad level. • TSV density requirements follow bond-pad density roadmaps.
Global	3D-Stacked Integrated Circuit / 3D-System-on-Chip (3D-SIC / 3D-SoC)	Wafer Fab	• Stacking of large circuit blocks (tiles, IP-blocks, memory –banks), similar to an SoC approach but having circuits physically on different layers. • Un-buffered I/O drivers (Low C, little or no ESD protection on TSVs). • TSV density requirement significantly higher than 3D-WLP: Pitch requirement down to 4-16μm.
Intermediate	3D-SIC	Wafer Fab	• Stacking of smaller circuit blocks, parts of IP-blocks stacked in vertical dimensions. • Mainly wafer-to-wafer stacking. • TSV density requirements very high: Pitch requirement down to 1-4 μm
Local	3D-Integrated Circuit (3D-IC)	Wafer Fab	• Stacking of transistor layers. • Common BEOL interconnect stack on multiple layers of FEOL. • Requires 3D connections at the density level of local interconnects.

1.4 TECHNOLOGY PLATFORMS AND STRATEGIES

A number of new enabling technologies must be developed and introduced into the existing fabrication process flow to make 3D integration a reality. Depending on the level of granularity, new capabilities include wafer bonding (permanent or temporary), through silicon/strata via (TSV), wafer thinning and handling, precision alignment, and other related technologies. There are a number of references on technology platforms available in the literature and the references therein.[25, 26] A brief introduction to TSV process flow is given below and detailed coverage can be found in subsequent chapters on deep via etching and Cu filling. This section primarily discusses low temperature Cu-Cu permanent bonding which is the author's core research expertise.

Through Silicon Via

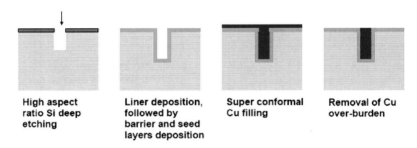

High aspect ratio Si deep etching

Liner deposition, followed by barrier and seed layers deposition

Super conformal Cu filling

Removal of Cu over-burden

Fig. 1.8 A generic process flow of Cu-filled TSV fabrication.

Figure 1.15 is a generic process flow of TSV fabrication flow using Cu as the core metal. It begins with high aspect ratio deep etching of Si. Dielectric liner layer is then deposited on the via sidewall followed by barrier and Cu seed layers deposition. Liner layer, which is made of dielectric layer such as silicon dioxide, provides electrical isolation between Cu core and Si substrate. The liner thickness must be chosen appropriately to control leakage current and capacitance between Cu core and Si substrate. Cu super conformal filling is then achieved with electro-plating process. Super conformal filling is required to prevent void formation in the Cu TSV. Finally, Cu over-burden is removed by chemical mechanical polishing. More information on TSV fabrication can be found in Chapters 4-5.

Cu-Cu Permanent Bonding

3D integration of integrated circuits by means of bump-less Cu-Cu direct bonding is an attractive choice as one accomplishes both electrical and mechanical bonds simultaneously. Cu-Cu direct bonding is desired compared to solder-based connections because: (1) Cu-Cu bond is more scalable and ultra-fine pitch can be achieved; (2) Cu has better electrical and thermal conductivities; and (3) Cu has much better electro-migration resistance and can withstand higher current density in future nodes. Cu-Cu bond has better properties due to the absence of inter-metallic compound (IMC) found in solder based joint.

Direct Cu-Cu bonding has been demonstrated using thermo-compression bonding (also known as diffusion bonding). As the name implies, thermo-compression bonding involves simultaneous mechanical pressing (~ 200 kPa) and heating of the wafers (~ 300-400°C). Two wafers can be held together when the Cu thin films bond together to form a uniform bonded layer. In order for this technique to be applicable to wafers that carry device and interconnect layers, an upper bound of temperature step is set at 400°C to prevent undesired damages particularly to the interconnects. The main objective of this Cu thermo-compression bonding study is to explore its suitability for utilization as a permanent bond that holds active device layers together in a multi-layer ICs stack. Cu is a metal of choice for 3-D ICs application because it is a mainstream CMOS material, and it has better electrical and thermal conductivities compared to Al-based interconnect. Most importantly, Cu bonds to itself under conditions compatible with CMOS back-end processes as initially demonstrated by Fan *et al.*[27]

Wafer Preparation and Bonding Procedures

In this section, wafer bonding by Cu thermo-compression is demonstrated and characterized on blank Si wafers. All wafers used in this experiment were *p*-type 150 mm Si-(100) wafers of 10-20 Ω-cm resistivity. Thermal oxide (5000 Å) was grown on the wafers. All wafers received a 10 min piranha (H_2O_2:H_2SO_4 = 1:3, by volume) solution clean followed by deionized water rinse and spin-dry prior to metallization. The next step was the deposition of tantalum (50 nm) and copper (300 nm) in an e-beam deposition system (a sputtering system can also be used). Ta was used to prevent Cu out-diffusion into the oxide layer. Chamber pressure during metal deposition was 1×10^{-6} Torr. The rms roughness of the $Cu/Ta/SiO_2/Si$ wafers is estimated to be around 1.99 nm from AFM scan.

A pair of wafers was aligned face-to-face in wafer aligner and clamped together on a bonding chuck. Three separation metal flaps were inserted between the wafers at the edges and loaded into bonder. Three cycles of N_2 purge ware done, and the chamber was evacuated to ~ 1×10^{-3} Torr. At this point, a down force was applied on the wafer pair while the flaps were being pulled out. The temperatures of the chuck and top electrode were ramped up to and maintained at 300°C. The contact force was 4000 N when the wafer pair was in full contact at 300°C, and the bonding step lasted for 1 hour. After bonding, the bonded wafers were annealed in atmospheric N_2 ambient for 1 hour at 400°C.

Bonding Mechanism

In order to understand the microstructures of the bonded Cu layer, transmission electron microscopy (TEM) analysis was performed on this sample. Note that the two Cu bonding layers merge and a homogeneous bonded layer is obtained as shown in the TEM image in Fig. 1.9. As can be seen from this image, large Cu grains, which often extend beyond the original bonding interface, are obtained after bonding and annealing. Dislocation lines are also found in the Cu grains.

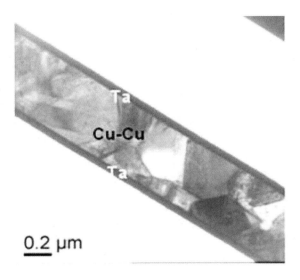

Fig. 1.9 TEM image of bonded Cu layer. Note that the bonding Cu layers merge and a homogeneous Cu layer is obtained after bonding and anneal. Grain structures that extend across the original bonding interface are observed. Dislocation lines are clearly seen in the grains.

A possible bonding mechanism that gives rise to the above grain structures will be proposed in this section. From the TEM image, it is evident that there is substantial grain growth during bonding and annealing. The jagged Cu-Cu interface suggests that inter-diffusion between two Cu layers has taken place. During bonding and subsequent annealing, Cu layers are in intimate contact under the applied pressure. At the bonding temperature, Cu atoms acquire sufficient energy to diffuse rapidly, and Cu grains begin to grow. At the bonding interface, diffusion can happen across the bonding interface and grains growth can progress across the interface. After sufficiently long duration, large Cu grains on the order of 300-500 nm are obtained, and a homogeneous bonded Cu layer is formed. Electron Dispersion Spectroscopy (EDS) analysis of the bonded Cu layer shows that apart from Cu, no appreciable foreign contaminant is found in the bonded layer within the detection limit of EDS. Since e-beam deposition is not used in typical manufacturing environments for metallization, the above demonstration of Cu thermo-compression is repeated using Cu deposited by electro-chemical means. The above experiment is repeated using electroplated Cu and similar observation of the bonding characteristic is made.

Surface Oxide

Since the mechanism for Cu thermo-compression bonding is based on Cu inter-diffusion and grain growth, surface contaminants such as oxide are detrimental to successful bonding especially at very low temperature. However, there is more often than not a time lag between Cu deposition and bonding, and therefore the formation of surface oxide is inevitable. Excessive oxygen incorporation into the bonded Cu layer might also increase the resistivity of the Cu layers and hence degrade the electrical performance of Cu interconnects. Techniques that can be used to reduce the surface oxide prior to bonding include the use of a chemical clean such as HCl[27] and glacial acetic acid followed by a forming gas purge in the bonding chamber[28] prior to bonding. Forming gas anneal can also be performed on Cu wafers prior to bonding and experimental evidence of the reduction of oxygen content in the bonded Cu layer is reported in.[29]

Process Parameters

There are a number of important process parameters that directly determine the quality of the final bond during Cu thermo-compression bonding. Three important bonding parameters, *i.e.,* temperature, duration, and contact pressure, are frequently considered. In the bonding procedures described in above, thermo-compression bonding of Cu is accomplished in two steps, *i.e.,* an initial bonding step to establish bond between pairing wafers and a post-bonding anneal to enhance the bond. Since the bonding step is a single wafer pair step, long bonding duration will decrease through-put in a manufacturing environment. On the other hand, annealing can be accomplished in an atmospheric furnace and it is possible to process batches of wafers during annealing. Therefore, a better way to achieve high through-put Cu wafer bonding is to initiate a preliminary bond with a short bonding step and to enhance the bonding strength with a post-bonding anneal. A number of references that discuss process parameters during thermo-compression bonding of Cu can be found in Refs. 30 and 31.

Non-Blanket Cu Bonding

Since Cu is a conductive medium, a continuous Cu bonding layer between active layers is of no practical application. In an actual multi-layer 3-D ICs implementation having Cu as the bonding medium, Cu bonding should be done in the form of pad-to-pad or line-to-line bonding with proper electrical isolation. Figure 1.10 shows a cross section of Cu lines (2-9 µm) that are successfully bonded.[32] The spacing between bonded lines is 5.3 µm and it is filled with air. Interfacial voids are observed in the bonded lines and they can lead to serious reliability concern. The bonding process should be optimized to minimize the formation of void. Another reliability concern is the empty space between the bonded lines that might reduce mechanical support between the active layers. Moisture in the empty space can also potentially corrode the bonded Cu lines.

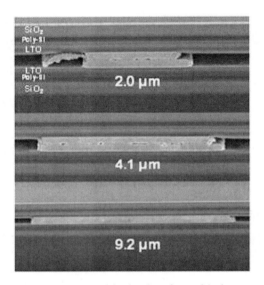

Figure 1.10 Non-blanket bonding of Cu lines.

One solution is to form damascene Cu lines and to perform hybrid bonding of Cu and dielectric. A few examples are:

(i) Jourdain *et al.*[33] at IMEC have successfully demonstrated the 3-D stacking of an extremely thinned IC chip onto a Cu/oxide landing substrate using simultaneous Cu-Cu thermo-compression and compliant glue-layer (BCB) bonding. The goal of this intermediate BCB glue layer between the 2 dies is to reinforce the mechanical and thermal stability of the bonded stack and to enable separation of die pick-and-place operations from a collective bonding step;

(ii) Gutmann *et al.*[34] at RPI have demonstrated another scheme of hybrid bonding using face-to-face bonding of Cu/BCB redistribution layers. The first step is to prepare the single-level damascene-patterned structures (Cu and BCB) by CMP in the two Si wafers to be bonded. The second step is to align the two wafers and bond the two aligned wafers;

(iii) Researchers at Ziptronix have developed a Cu/oxide hybrid bonding technology known as Direct Bond Interconnect (DBI™).[35] Vertical interconnections in direct oxide bond DBI™ are achieved by preparing a heterogeneous surface of non-conductive oxide and conductive Cu. The surfaces are aligned and placed together to effect a bond. The high bond energies possible with the direct oxide bond between the heterogeneous surfaces result in vertical DBI™ electrical interconnections.

Low Temperature Cu Bonding

Thermo-compression bonding of Cu layers is typically performed at a temperature of 300°C or higher. There is strong motivation to move the bonding temperature to even lower range primarily from the point of view of thermal stress induced due to CTE mismatch of dissimilar materials in a multi-layer stack and temperature swing. A number of approaches have been explored:

(i) Surface Activated Bonding[36] — In this method, a low energy Ar ion beam is used to activate the Cu surface prior to bonding. Contacting two surface-activated wafers enables successful Cu–Cu direct bonding. The bonding process is carried out under an ultrahigh vacuum (UHV) condition. No thermal annealing is required to increase the bonding strength. Tensile test results show that high bonding strength equivalent to bulk material is achieved at room temperature. In Ref. 37, the adhesion of Cu-Cu bonded at room temperature in UHV condition was measured to be about $\sim 3 \, J/m^2$ using AFM tip pull-off method;

(ii) Cu Nanorod[38] — Recent investigation on surface melting characteristics of copper nanorod arrays shows that the threshold of the morphological changes of the nanorod arrays occurs at a temperature significantly below the copper bulk melting point. With this unique property of the copper nanorod arrays, wafer bonding using copper nanorod arrays as a bonding intermediate layer is investigated at low temperatures (400°C and lower). Silicon wafers, each with a copper nanorod array layer, are bonded at 200-400°C. The FIB/SEM results show that the copper nanorod arrays fuse together accompanying by a grain growth at a bonding temperature of as low as 200°C;

(iii) Solid-Liquid Inter-diffusion Bonding (SLID)[39] —This method involves the use of a second solder metal with low melting temperature such as Tin (Sn) in between two sheets of Cu with high melting temperature. Typically a short reflow step is followed by a longer curing step. The required temperature is often slightly higher than Sn melting temperature (232°C). The advantages of SLID is that the inter-metallic phase is stable up to 600°C and the requirement of contact force is not critical;

(iv) In the DBI™ technology described in Ref. 35, a moderate post oxide bonding anneal may be used to effect the desired bonding between Cu. Due to the difference in coefficient of expansion between the oxide and Cu and the constraint of the Cu by the oxide, Cu compresses each other during heating and metallic bond can be formed.

(v) Direct Cu-Cu bonding at atmospheric pressure is investigated by researchers at LETI.[40] By means of CMP, the roughness and hydrophily (measure by contact angle) of Cu film are improved from 15 nm to 0.4 nm and from 50° to 12°. Blanket wafers were successfully bonded at room temperature with an impressive bond strength of 2.8 J/m². With a post-bonding annealing at 100°C for 30 min, the bonding strength was improved to 3.2 J/m2;

(vi) A novel Cu-Cu bonding process has been developed and characterized to create all-copper chip-to-substrate input/output (I/O) connections.[41] Electroless copper plating followed by low temperature annealing in a nitrogen environment was used to create an all-copper bond between copper pillars. The bond strength for the all-copper structure exceeded 165 MPa after annealing at 180°C. While this technique is demonstrated as a packaging solution, it is an attractive low temperature process for Cu-Cu bonding;

(vii) In the author's research group, a method of Cu surface passivation using self-assembled monolayer (SAM) of alkane-thiol has been developed. This method has been shown to be effective to protect the Cu surface from particle contamination and to retard surface oxidation. The SAM layer can be thermally desorbed *in-situ* in the bonding chamber rather effectively hence providing clean Cu surface for successful low temperature bonding. Cu wafers bonded at 250°C present significant reduction in micro-void and substantial Cu grain growth at the bonding interface.[42-45]

1.5 ORGANIZATION OF THE BOOK

This book is an edited volume that comprises of chapters contribution from key researchers in the emerging field of 3D integration covering motivations and drivers (mostly in Chapter 1 and 2), technology platforms and strategies (Chapters 3-10), examples of 3D processes and applications (Chapters 11, 12), design issues and opportunities (Chapters 13-15), as well as status and outlook (Chapter 16). The primary focus of this book is 3D technology based on "assembly approaches" utilizing TSV and bonding as described above.

Chapter 2 is a collection of opinion by researchers at IBM led by Emma *et al.* and it serves as a prelude to the rest of the book. Its focus is systems applications and implications: why you may want to make things in 3D, what you might and will not get out of doing so, and the numerous considerations and requirements for making 3D systems in varying degrees of complexity. On the face of it, making a 3D system sounds like a panacea. It *solves* the evolutionary problem of continuing to scale the number of transistors in a chip, known as *Moore's Law*. It seems to be a *no brainer*. The goal of this chapter is to bring some perspectives to this initial reaction.

In Chapter 3, a review is provided by Kim *et al.* from EVG on various bonding techniques suitable for manufacturing 3D-ICs with TSV interconnects, including bond alignment, wafer bonding, advanced chip-to-wafer bonding, and temporary bonding and de-bonding for thin wafer handling and processing.

Through Silicon Via (TSV) Etching is a key process fabrication module employed in 3D IC technologies. Chapter 4 reviews the work already performed in TSV etching by the Compound Semiconductor industry, then examines deep reactive ion etching (DRIE) processes as developed for MEMS device fabrication. Following this review, the chapter presents an analysis of the specific requirements of TSV etching for silicon device 3D technology, and then concludes with a look forward to TSV etching requirements visible on the near-term horizon. This chapter is contributed by Werbaneth *et al.* of Tegal.

The formation of TSV is not complete until it is filled with a conductor. Copper, tungsten, polysilicon, and nickel are candidate materials TSV conductor filling that are under development. Copper filled TSV is emerging as the most widely adopted structure therefore Cu filling by electrodeposition process is the primary subject in Chapter 5 authored by Keigler *et al.* from NEXX Systems.

In Chapter 6, researchers from Brewer Science will explore temporary bonding and TSV creation using several bonding and release (de-bonding) technologies currently in production on thinned wafers. The handling and alteration of a thinned wafer on a carrier must remain intact through high-temperature, high-energy-flow, and highly corrosive environments during backside processing. This is a chapter written by Privett *et al.*

Another important aspect of 3D technology platform is wafer thinning, stress relief, and thin wafer handling. The subsequent chapter (Chapter 7) covers this aspect of 3D technology and it is a contribution from Disco written by Sullivan *et al.*

Unlike conventional alignment method based on optical means, Fukushima *et al.* of Tohoku University discusses self-assembly method for alignment in Chapter 8. In this chapter, an overview of the self-assembly technology for the advanced die-to-wafer 3D integration using the self-assembly technology is introduced and the potential applications of the self-assembly-based 3D integration are described.

Chapter 9 provides a review of advanced direct bond technology and evaluates its suitability as a bond technology platform for 3D integration. It is shown that advanced direct bond technology is an ideal solution for the requirements of a bond technology platform for 3D integration. Enquist *et al.* from Ziptronix will discuss advanced direct bond technology in this chapter.

It is important to decrease the process temperature to around room temperature for application-oriented discrete system packaging, which requires three-dimensional interconnections between diverse substrates. In Chapter 10, Shigetou *et al.* from National Institute of Materials Science (Japan) will concentrate on surface modification methods induced by the beam irradiation process. The technical background, principles, important bondability factors, applications, and future tasks are described.

Over the last few years it was becoming very clear that camera module packaging technology will have to introduce breakthrough solutions to meet the customer expectations in term of physical dimensions. Through silicon via has been the answer to this demand and it is now implemented with some variants by most of the players in this field of applications. Gagnard *et al.* (ST Microelectronics) provide a thorough review on the application of TSV for CMOS image sensor in Chapter 11.

In Chapter 12, researchers from IBM led by Liu *et al.* present a 300-mm wafer-level three-dimensional integration (3DI) process using tungsten (W) through-silicon vias (TSVs) and hybrid Cu/adhesive wafer bonding. A hybrid Cu/adhesive bonding approach, also called transfer-join (TJ) method, is used to interconnect the TSVs to a Cu BEOL in a bottom wafer.

Gutmann *et al.* of RPI address the problem of power delivery in microprocessors, application-specific ICs (ASICs) and system-on-a-chip (SoC) implementations, based upon a wafer-level 3D technology platform with arrays of monolithic DC-DC converter cells in one stratum providing power locally to the signal electronics strata. This forms Chapter 13.

Although emerging 3D technology offers several benefits over 2D, the stacking of multiple active layers in 3D design leads to higher power densities than its 2D counterpart, exacerbating the thermal issue. The next chapter presents an overview of thermal modeling for 3D IC and outlines solution schemes to overcome the thermal challenges at Electrical Design Automation (EDA) and architectural levels. Chapter 14 is a contribution from Xie *et al.* of Penn State University.

Since there are large numbers of devices densely packed in a number of device layers, it brings a significant burden to the supply voltage in 3D ICs. Chapter 15 discusses 3D IC design automation considering both dynamic power and thermal integrity. This chapter is presented by Yu *et al.* of NTU.

Being an emerging technology, 3D integration is a field that is growing rather rapidly. It is therefore worthwhile to provide it present status and make projection on its future outlook. Yang *et al.* from IEK/ITRI have summarized their findings and thoughts in Chapter 16 to provide an update on the development and a critical assessment on the future outlook of 3D technology.

1.6 SUMMARY

In conclusion, the primary drivers for 3D technology development in the high-performance computational space have been reviewed, along with the associated technological approaches. 3D technology is likely to play a variety of roles in improving high-end system performance in the future, and also offers a multitude of exciting opportunities to further enhance system level performance and even create new applications well into the 21st century.

References

1. G. Moore, "Cramming more components onto integrated circuits," *Electronics Magazine*, 1965.

2. R. H. Dennard, F. H. Gaensslen, H. N. Yu, V. L. Rideout, E. Bassous, and A. R. LeBlanc, "Design of ion-implanted MOSFETs with very small physical dimensions," *IEEE J. Solid-State Circuits*, Vol. 9, pp. 256-268 (1974).

3. Litho limits.

4. H.-S.-P Wong, D. J. Frank, P. M. Solomon, C. H. J. Wann, and J. J. Welser, "Nanoscale CMOS," *Proc. IEEE*, Vol. 87, pp. 537-570, 1999.

5. D. J. Frank, "Power constrained CMOS scaling limits," IBM J. Res. & Dev., vol. 46, pp. 235-244, 2002.

6. S. Vangal, J. Howard, G. Ruhl, *Proc. IEEE*, Vol. 87, No. 4, April 1999 537, S. Dighe, H. Wilson, J. Tschanz, D. Finan, P. Iyer, A. Singh, T. Jacob, S. Jain, S. Venkataraman, Y. Hoskote, N. Borkar, "An 80-Tile 1.28TFLOPS network-on-chip in 65 nm CMOS," IEEE International Solid State Circuits Conference (ISSCC), San Francisco, CA, Feb. 11-15, 2007.

7. P. Emma, "How is bandwidth used in computers? Why bandwidth is the next major hurdle in computer systems evolution and what technologies will emerge to address the bandwidth problem," in *High-Performance Energy-Efficient Microprocessor Design*, Springer, US, 2006, pp. 235-287.

8. P. Emma and E. Kersen, "Is 3D chip technology the next growth engine for performance improvement?" *IBM J. Res. & Dev.*, Vol. 52, pp. 541-552 (2008).

9. M. Shapiro, "3D technology: applications and requirements," *3-D Architectures for Semiconductor Integration and Packaging: Examining Routes to Success*, Burlingame, CA, Nov. 17-19, 2008.

10. J. Sun, J.-Q. Lu, D. Giuliano, T. P. Chow, and R. J. Gutmann, "3D power delivery for microprocessors and high-performance ASICs," Twenty Second Annual IEEE Applied Power Electronics Conference (APEC 2007), Anaheim, CA, Feb. 25-Mar. 1, 2007, pp. 127-133.

11. http://www.tezzaron.com/memory/Overview_3D_DRAM.htm

12. K. Bergman, L. Carloni, J. Kash, and Y. Vlasov, "On-chip photonic communications for high performance multi-core processors," 11th Annual Workshop on High Performance Embedded Computing (HPEC 2007), Lexington, MA, Sep. 18-20, 2007.

13. R. Jones, H. Park, A. W. Fang, J. E. Bowers, O. Cohen, O. Raday, and M. J. Paniccia, "Hybrid silicon integration," *J. Mater. Sci.: Mater. Electron.*, Vol. 20, pp. S3-S9 (2009).

14. N. M. Jokerst, *et al.*, "The heterogeneous integration of optical interconnections into integrated microsystems," *IEEE J. Sel. Top. Quant. Electron.*, Vol. 9, pp. 350-360 (2003).

15. M. J. Rosker, "The DARPA Compound Semiconductors on Silicon (COSMOS) Program," CS MANTECH Conference, Chicago, IL, Apr. 14-17, 2008.

16. R.E. Jones et al., "Technology and application of 3D interconnect," In: Proc *IEEE Intern Conf Integrated Circuit Design and Tech*, p. 176, 2007.

17. N. Miura et al., "Capacitive and Inductive-Coupling I/Os for 3D chips," In: *Integrated Interconnect Technologies for 3D Nanoelectronic Systems* (Edited by M. S. Bakir and J. D. Meindl), Artech House, p. 449, 2009.

18. S. Kawamura, N. Sasaki, T. Iwai, M. Nakano, and M. Takagi, "Three-dimensional CMOS ICs fabricated by using beal recrystallization," *IEEE Electron Device Letters*, 4(10), p. 366, 1983.

19. T. Kunio, K. Oyama, Y. Hayashi, and M. Morimoto, "Three Dimensional ICs, Having Four Stacked Active Device Layers," In: *IEDM Technical Digest*, pp. 837, 1989.

20. V. Subramanian, M. Toita, N.R. Ibrahim, S. J. Souri, K.C. Saraswat, "Low-leakage germanium-seeded laterally-crystallized single-grain 100-nm TFTs for vertical integration applications," *IEEE Electron Device Letters*, 20(7), p. 341, 1999.

21. V.W.C. Chan, P.C.H. Chan, and M. Chan, "Three-dimensional CMOS SOI integrated circuit using high-temperature metal-induced lateral crystallization," *IEEE Transaction on Electron Devices*, 48(7): p. 1394, 2001.

22. S. Pae, T. Su, J.P. Denton, and G.W. Neudeck, "Multiple layers of silicon-on-insulator islands fabrication by selective epitaxial growth," *IEEE Electron Device Letters*, 20(5), p. 194, 1999.

23. B. Rajendran, R.S. Shenoy, D.J. Witte, N.S. Chokshi, R.L. DeLeon, G.S. Tompa, R.F.W. Pease, "Low temperature budget processing for sequential 3-D IC fabrication," *IEEE Trans. Electron Devices*, 54(4), p. 707, 2007.

24. International Technology Roadmap for Semiconductor, Interconnect, 2009. (http://www.itrs.net/)

25. C.S. Tan, R.J. Gutmann, and R. Reif, *Wafer Level 3-D ICs Process Technology, Springer*, ISBN 978-0-387-76532-7, 2008.

26. P. Garrou, C. Bower, and P. Ramm, *Handbook of 3D Integrations: Technology and Applications of 3D Integrated Circuits*, Wiley-VCH, ISBN 978-3-527-32034-9, 2008.

27. A. Fan, A. Rahman, and R. Reif, "Copper Wafer Bonding," *Electrochemical and Solid-State Letters* 2(10), pp. 534-536, 1999.

28. R. Tadepalli and Carl V. Thompson, "Quantitative characterization and process optimization of low-temperature bonded copper interconnects for 3-D integrated circuits", Proc. IEEE 2003, International Interconnect Technology Conference, p. 36-38, 2003.

29. C. S. Tan, K. N. Chen, A. Fan and R. Reif, "The effect of forming gas anneal on the oxygen content in bonded Cu layer," *Journal of Electronic Materials*, 34(12), pp 1598-1602, 2005.

30. K. N. Chen, A. Fan, C. S. Tan, and R. Reif, "Temperature and duration effect on microstructure evolution during copper wafer bonding," *Journal of Electronic Materials*, 32(12), pp. 1371-1374, 2003.

31. K. N. Chen, C. S. Tan, A. Fan, and R. Reif, "Morphology and bond strength of copper wafer bonding," *Electrochemical and Solid-State Letters*, 7(1), pp G14-G16, 2004.

32. C. S. Tan, K. N. Chen, A. Fan, R. Reif, and A. Chandrakasan, "Silicon layer stacking enabled by wafer bonding," *MRS Symposium Proceedings*, Vol. 970, pp. 193-204, 2007.

33. A. Jourdai, S. Stoukatch, P. De Moor, W. Ruythooren, S. Pargfrieder, B. Swinnen, and E. Beyne, "Simultaneous Cu-Cu and compliant dielectric bonding for 3D stacking of ICs," Proc. IEEE, International Interconnect Technology Conference, pp. 207-209, 2007.

34. R. J. Gutmann, J. J. McMahon, and J.-Q Lu, "Damascene-patterned metal-adhesive (Cu-BCB) redistribution layers," *Materials Research Society Symposium Proceedings*, Vol. 970, pp. 205-214, 2007.

35. P. Enquist, "High Density Bond Interconnect (DBI) technology for three dimensional integrated circuit applications," *Materials Research Society Symposium Proceedings*, Vol. 970, pp. 13-24, 2007.

36. T. H. Kim, M. M. R. Howlader, T. Itoh, and T. Suga, "Room temperature Cu-Cu direct bonding using surface activated bonding method," *Journal of Vacuum Science and Technology A: Vacuum, Surfaces and Films* 21(2), pp. 449-453, 2003.

37. R. Tadepalli and C. V. Thompson, "Formation of Cu–Cu interfaces with ideal adhesive strengths via room temperature pressure bonding in ultrahigh vacuum," *Appl. Phys. Lett.*, 90, 151919 (2007).

38. P.-I. Wang, T. Karabacak, J. Yu, H.-F. Li, G. G. Pethuraja, S. H. Lee, M. Z. Liu, and T.-M. Lu, "Low temperature copper-nanorod bonding for 3D integration," *Materials Research Society Symposium Proceedings*, 970, pp. 225-230, 2007.

39. P. Benkart, A. Kaiser, A. Munding, M. Bschorr, H.-J. Pfleiderer, E. Kohn, A. Heittmann, and U. Ramacher, "3D chip stack technology using through-chip interconnects," *IEEE Design & Test of Computers*, 22(6), pp. 512-518, 2005.

40. P. Gueguen, L. Di Cioccio, M. Rivoire, D. Scevola, M. Zussy, A. M. Charvet, L. Bally, and L. Clavelier, "Copper direct bonding for 3D integration," IEEE International Interconnect Technology Conference, pp. 61-63, 2008.

41. T. Osborn, A. He, H. Lightsey, and P. Kohl, "All-copper chip-to-substrate interconnects," *Proc. IEEE*, Electronic Components and Technology Conference, pp. 67-74, 2008.

42. D. F. Lim, S. G. Singh, X. F. Ang, J. Wei, C. M. Ng, and C. S. Tan, "Achieving low temperature Cu to Cu diffusion bonding with self assembly monolayer (SAM) passivation," IEEE International Conference on 3D System Integration, art. no. 5306545, 2009.

43. D. F. Lim, S. G. Singh, X.F. Ang, J. Wei, C. M. Ng, and C. S. Tan, "Application of self assembly monolayer (SAM) in Cu-Cu bonding enhancement at low temperature for 3-D integration," Advanced Metallization Conference, Baltimore, October 13-15, 2009. In: D. C. Edelstein, and S. E. Schulz (Eds.), AMC 2009, pp. 259-266, Materials Research Society.

44. C. S. Tan, D. F. Lim, S. G. Singh, S. K. Goulet, and M. Bergkvist, "Cu-Cu diffusion bonding enhancement at low temperature by surface passivation using self-assembled monolayer of alkane-thiol," *Applied Physics Letters*, 95(19), pp 192108, 2009.

45. D. F. Lim, J. Wei, C.M. Ng, and C. S. Tan, "Low temperature bump-less Cu-Cu bonding enhancement with self assembled monolayer (SAM) passivation for 3-D integration," IEEE Electronic Components and Technology Conference (ECTC), Las Vegas, June 1-4, pp. 1364-1369, 2010.

Chapter 2

A SYSTEMS PERSPECTIVE ON 3D DESIGN: WHAT IS 3D? AND WHAT IS 3D GOOD FOR?

Philip Emma and Eren Kursun

IBM Research

This chapter is a prelude to the rest of the book. Its focus is systems applications and implications: why you may want to make things in 3D, what you might and will not get out of doing so, and the numerous considerations and requirements for making 3D systems in varying degrees of complexity. On the face of it, making a 3D system sounds like a panacea. It *solves* the evolutionary problem of continuing to scale the number of transistors in a chip, known as *Moore's law*. It seems to be a *no brainer*. The goal of this chapter is to bring some perspectives to this initial reaction.

Building a system in 3D does enable a few specific opportunities. It also requires solving some completely new sets of problems. As will become clearer in this chapter, you should first decide what problems need to be solved by building a system in 3D before figuring out how to go about building it. Without first identifying what it is you think you are improving by using a 3D solution, it is not possible to tailor a 3D solution to your needs. Without first matching the various processes to the application, you will likely work much harder than needed or derive very little benefit from 3D. 3D is a wide range of technology alternatives that span a spectrum of processes and complexities.[1-3] And that span is not a panacea.

In essence, this chapter is cautionary: Know why you are using 3D before deciding to do so. Identify what you will gain by using 3D. And then tailor the processes to the needs. If you can't identify what, specifically, you are gaining with 3D, don't use it.

Finally, this chapter gives a view as to where we've been and where we are, and a glimpse as to where we might go in systems design. At the time of

3D Integration for VLSI Systems
Edited by Chuan Seng Tan, Kuan-Neng Chen and Steven J. Koester
Copyright © 2012 by Pan Stanford Publishing Pte. Ltd.
www.panstanford.com

this writing, 3D has only been used in limited ways, and our view toward the future is undoubtedly hampered by some of our implicitly held assumptions as to what "must be" when it comes to systems heretofore built in 2D. Since we are well aware of our innate myopia, we will make a conscious effort to look past those blinders when looking into the future.

2.1 SYSTEMS 101: THE BASICS OF COMPUTING SYSTEMS

The term *system* originates from the Greek *systema*, which refers to "a whole, compounded of several parts." A system is an aggregation of subsystems of (perhaps) different functionalities and capabilities, wherein each subsystem can also be a system. On the surface, this seems to be a useless tautology. Why define it thus? Very simply, because this aligns with the way in which we think in abstractions.

This recursive and tautological deconstruction makes even the most complex systems tractable. A system can be abstracted into subsystems. Each subsystem can be abstracted into subsystems, and so on. Eventually out of this aggregation we boil subsystems that are simple enough to specify, understand, design, verify, test, and so on. And then we put them back together to achieve a complex working system.

At the level of thought and designability, this deconstruction requires that we have tools to adequately describe each subsystem in various forms: functional, structural, electrical, thermal, and physical. It also requires that as we compose the aggregate system from its subsystems, we have tools to verify that each aggregation is "correct" in the sense that its pieces cooperate in the manners intended and anticipated.

While such deconstruction and reconstruction ostensibly seem like mere intellectual constructs, a natural artifact of that deconstruction (in fact, an inherent trait of systems that leads to the most natural and logical deconstructions) is that the "connectedness" of the constituent subsystems tends to be simple and narrow relative to any random cut-set that would partition a system arbitrarily.

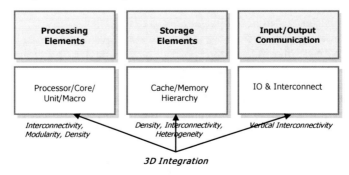

Fig. 2.1 Analyzing the potential effects of 3D in each component/subsystem of a simple computing system.

Therefore, while systems comprise subsystems, there is a method to the madness: those subsystems tend to be relatively self-contained by construction. A typical computer system contains processing elements for data transformations, storage elements, input/output devices for moving data into and out of the computer's memory, and numerous other subsystems.

In turn, for example, what we have called "memory" contains a directory to keep track of the memory's contents, input and output multiplexors to match speeds and feeds, the bulk storage itself, error-checking circuits, drivers and receivers, and so on. And the "bulk storage" contains storage cells that are arranged in a particular way, networks of wires and busses for accessing those cells, sense amplifiers and drivers for reading and storing state, and so on. We can deconstruct all of the named subsystems similarly.

Such a hierarchical deconstruction allows a complicated machine to be tractable and designable. In such a hierarchy, each of the subsystems is much simpler than its aggregate. The subsystems tend to be interconnected in a way that makes sense from the standpoint of how the subfunctions interact. Further, that connectivity is relatively well organized—comprising data busses and control signals. The former are used to move data through the system, and the latter are logical controls to effect the movements and transformations of data.

A key point in identifying the advantages of 3D is to decompose its leverages to different subsystem components and then to try to assess the system-level advantages from the points of view of these individual subsystems. In this context, the leverages of 3D in the storage arrays, for example, can be very different than the advantages found (or not) in other parts of the system, such as the processor itself.

Furthermore, computing systems span a wide range of complexities and costs, from the embedded/mobile domain (such as automobile microcontrollers) and game consoles, to cell phones, to personal computers, database computers, network processors, and supercomputers.[4-6] As a result of this wide variety of applications, each type of system has different requirements in terms of its desirable cost, area, power, temperature, reliability, speed, throughput, compliance to standards, and many other things. Hence, each system is optimized differently. Is 3D good? *The potential advantages of 3D integration are quite different across this wide range of applications.*

How can one use 3D effectively, and what is it good for? Just as there are many kinds of systems and many different criteria being optimized, so there are many different ways to use 3D and a variety of 3D options. 3D technology captures a range of different manufacturing techniques each of which evolves over time -- with increased levels of interconnectivity in general. As a result, it can mean different things to different people. Its value proposition is entirely dependent on what you are doing with it and why. This takes many different forms.

2.2 WHAT 3D CAN BUY YOU: THE POTENTIAL ADVANTAGES OF 3D SYSTEMS

Abstractly and generically, 3D has been used to describe applications in which two or more chips are stacked—put together and connected in the third (vertical) dimension instead of being placed adjacently on a planar package. Depending on the amount of interconnectivity among the layers, 3D can be used as anything from a mere and simple technology that results in fewer components that need final assembly to a disruptive technology that requires pervasive changes in the design flow and even in the microarchitecture. For the current and earlier generations of 3D that had relatively coarse via pitches, the (nondisruptive) packaging arguments still hold. But as a disruptive technology, significant changes will be needed in the way we design systems as 3D evolves.

In the context of a "top-down" approach (meaning one in which 2D systems are conjoined simply and with little forethought), the main benefit of 3D lies in the physical proximity of the chips. Almost by definition, few changes are needed to the chips or to the designs themselves.

Wire bonding is considered an early form of this top-down packaging approach, wherein various dies of different sizes can be interconnected at their peripheries. For the past decade, wire-bonded packages have been available at relatively low costs. They enable stacking multiple chips and wiring them together within that stack by interconnecting (only) peripheral pads on the constituent chips, as shown in Fig. 2.1.

While many people may call this a form of 3D, there is nothing essential about the dimensionality of the system comprising these chips; they are simply interconnected by their peripheral pads. The fact that they are stacked is arbitrary with respect to how they function and cooperate. Each of the chips in the stack can be (and typically is) designed independently. In this simplest form of 3D, what is leveraged is the overall "footprint" of the system. This is valuable in systems having serious volumetric constraints, such as in DIMMs, and embedded mobile applications.

In this simplest form of 3D stacking, if there is any forethought that goes into designing the chips so that they are stackable, it has mostly to do with assigning peripheral pins to signals so as to make the job of wire bonding as simple as possible. The specific planar locations and footprints of the subfunctions on each chip make no difference as to the way that the chips will work when they are stacked; the chips could just as easily be packaged adjacently on a planar package.

Again, the principle advantage of this type of integration is that the final system has a smaller footprint because the chips have been stacked up in 3D instead of wired together in the plane. This also simplifies the circuit board, because it doesn't need the wires to interconnect these chips. In addition to making the final system smaller, the final assembly of the product may be simpler and cheaper. This can provide lots of value to a high-volume product.

Note that to stack the constituent chips (simply) in this way, the number of signals connecting the layers must be small and the total power of the system must be low, or power delivery and heating will cause problems. This is great for many low-cost, low-power consumer applications. Note that with this kind of packaging, all of the wiring (including power delivery) can be peripheral. In this case, there is no need to put vias (metalized holes) in the chips.

The next degree of complexity for what is called "3D stacking" are applications in which the stacking of chips is "essential" in the sense that there is an intrinsic system-level performance benefit derived by stacking the chips. By definition (literally, "by construction"), this can only be true if vias are made in the chips, and if this manner of interconnecting the chips (with vias) provides a direct advantage. Note that if the art of making vias becomes economical enough, this advantage can be purely logistic in simplifying how a system is constructed, although it requires more forethought than we use today.

To do 3D stacking that directly benefits the function of the end product requires that layers be co-designed so as to interconnect in a sensible and natural way. This requires new 3D design tools that facilitate the co-design of the multiple layers that will be stacked and interconnected using signals that are not on the periphery.

Through-silicon vias (TSVs) provide two main advantages. First, if subsystems on the constituent chips are spatially co-designed, then the wiring between them can be shorter if we allow signals to go through the chips. Second (and more fundamentally), power delivery becomes an important limiter above a threshold (which is of more importance for server applications as opposed to embedded systems).

For 3D systems of this type (in which chips are co-designed and connected with through-vias) there is a wide range of required interconnectedness and power, and hence a correspondingly wide range of techniques that are best suited to making them. In applications having a strong functional separation between the layers, mere thousands of vias may be sufficient, while for other more complex applications, millions of vias may be needed. While both sets of applications would be "3D," the design and processing steps used for each set of applications would be very different.

Finally, the highest "tier" of 3D complexity is applications in which the system is not conceived as a set of interconnected 2D planes (although it might be implemented as such); rather, the system is conceptually an almost uniformly interconnected mesh of devices that are placed in—and wired in—"3D space" so as to optimize it. Suffice it to say that we are not at that point yet, and our current understanding of design disciplines makes this scenario hard to fathom (practically) today. But we cannot dismiss it either; it may well become feasible in the future.

2.3 THE 3D PROCESSING SPECTRUM

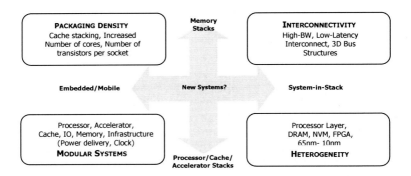

Fig. 2.2 System-level benefits of 3D for different market segments.

So far we focused on the different benefits of 3D in a wide range of technologies and applications, a common set of advantages have been the main drivers: (i) increased packaging density and footprint/volumetric miniaturization, (ii) improved interconnectivity and bandwidth among layers, (iii) enhanced modularity in putting together systems from subsystems, and (iv) the heterogeneous integration of disparate technologies.

The capability of packing more transistors onto the same die/footprint is the driving force for the famous Moore's law. Yet, the costs and complexities associated with technology scaling have been presenting greater challenges in each successive generation. 3D is now considered to be an alternative and potentially more cost-effective solution for increasing transistor density per socket. Similarly, the effective density improvement provides enhanced interconnectivity among the layers, especially in fine-pitch TSV technologies. With increased interconnectivity, on-chip wirelengths can be reduced considerably while enabling high-bandwidth and low-latency communication among layers. This has significant performance advantages.

For servers, modularity is a major advantage along with the density improvement. In vertical integration, a balanced separation of functionality in different device layers can be combined with the ease of putting together existing IP blocks of disparate technologies or layers in a modular way. In this sense, 3D provides an improved SoC concept.[7] *This, in turn, improves the efficiency of designing custom systems from standard components in order to meet the needs of different systems effectively.* Such modularity can provide cost and complexity advantages in addition to time-to-market benefits. Moreover, the device layers can be manufactured in different technology generations and even disparate technologies, which can also enable countless new applications. However, it is clear that such modularity comes with additional design stages as well as with significant changes to the existing design stages to incorporate the infrastructure capabilities that enable the above flow. In the upcoming sections we will focus on corresponding challenges in this new paradigm.

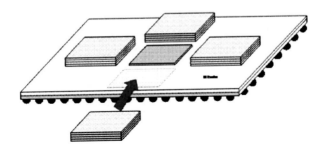

Fig. 2.3 Stacked 3D processor/memory stacks.

On one end of the computing spectrum—low-end mobile/embedded applications—miniaturization is a major advantage. The vertical stacking of thinned layers can provide much denser computation in the same die thickness and footprint than the original 2D case. Given the same footprint, a standard 700–800μm 2D layer can be replaced with up to seven to eight layers of 100 μm thickness in a 3D stack. (Further thinning could even provide denser stacking options as in 50 μm thick layers.) Nonetheless, the same density improvement is the cause of power density and cooling challenges for the high-power applications.

Note that cost sensitivity in the lower end of the processing spectrum is of primary importance. 3D here enables a "miniaturization," which acts as a strong driver in mobile/embedded applications that were traditionally interested in low-cost wire bonding, or in package-on-package systems. Following that trend, there is increasing interest in adapting 3D in flash memories, which can help the process technology to mature at a level required for more complex applications. The potential success of 3D in the cost-sensitive low end will likely have a significant impact on the overall semiconductor market.

For the higher end of the spectrum such as server architecturesthe processing spectrum, a slightly different set of benefits emerge as potential drivers. These drivers include improved bandwidth and an enhanced storage hierarchy with larger and potentially lower latency caches. In addition the system can be put together in a mode modular way, where the different design layers can be optimized in different technologies. The increased costs and complexities associated with this segment of the market causes a slower adaptation of vertical integration, cost and complexity. Furthermore, the infrastructural needs in design flow and the needed tool changes take longer compared to embedded/low-end applications. For high-power and high-power-density cases, power, thermal, and reliability challenges are of great concern, where specialized techniques needed to facilitate them.

2.4 THE ARCHITECTURAL LEVERAGES OF 3D

Over the years, microarchitecture trends have been shaped (principally) by density scaling and device scaling.[8] With each technology generation, more transistors were available for various functions on the chip, and more functions were added to the chip, most of which ended up being used to enhance single-thread performance by adding more parallelism. With each technology generation, the goal became incorporating more cores, and more powerful cores, i.e. higher performance. As a result, *feeding* the increased number of (faster) cores has become more of a challenge; the pressure on the memory hierarchy, and the bandwidths used to access it, is elevated.

Exploiting 3D integration appears to be a new and promising way to alleviate the corresponding memory hierarchy pressure. Bringing larger and faster caches closer to the processor cores can improve the off-chip bandwidth requirements, while improving the power and performance characteristics of the system.[9] For existing microprocessor architectures, the off-chip bandwidth limitations have become quite prominent.

Caches: One of the most promising options investigated is the integration of cache layers with processor layers.[10] In systems like this, one can even leverage heterogeneous technology as well (e.g., logic with DRAM). Heterogeneous integration of cache layers can enable denser, higher-performance, or more power-efficient levels in the memory hierarchy. Emerging nonvolatile memory stacking and DRAM and embedded DRAM stacks are under investigation.

Stacked DRAM is another alternative that brings optimization trade-offs such as density and interconnectivity.[5] Through-silicon vias or via-farm areas are likely to compromise the inherent density advantages of existing DRAM organizations. Furthermore, the TSV process limits the integration of vendor DRAM with microprocessor systems. Cost is another critical factor for 3D DRAM integration that is likely to limit the practicality of some alternatives. And increasing DRAM temperature by stacking it in a processor system could also exacerbate DRAM leakage and retention.

Recent research studies have explored ways of utilizing the inherent wire-length reduction in 3D to improve interconnect-limited performance within the processor core and other functional units.[4] Since the cost associated with finer granularities of integration is higher, this finer grain partitioning of functionality onto multiple layers incurs higher costs and risks. Though it might be possible to design around the inter-layer variation with proper design tools and margins, the performance advantages are also likely to diminish considerably.

Testing fine-grain functional partitioning becomes a major challenge, as pieces of functionality are scattered around the stack. Similarly, the costs and complexities associated with such fine-grain integration may be quite high for top-down 3D approaches. The cost benefits and trade-offs of such systems is still being studied. The jury is still out.

Beyond the core and unit/macro levels, the interconnectivity advantages of 3D span the buses at the chip level as well. Current microprocessor architectures are under pressure to add more metal layers to satisfy higher levels of interconnectivity.[11] The interconnect-limited nature of processing can be clearly observed especially in the wire and port-limited blocks like register files in the processing cores. Many systems will see considerable benefits due to the addition of interconnectivity in the z direction as well as reduced wire lengths in the x and y directions.

It is important to note that improved interconnectivity in 3D isn't a default benefit but requiresspecial floorplanning and routing optimizations to take advantage of the additional dimension in the 3D design space.[12] Interconnectivity among cores is another point of optimization for multicore architectures. 3D-optimized bus structures can provide much faster and more energy-efficient communication between the cores and other processing units on different layers (e.g. two device layers can share a bus infrastructure in face-to-face integration) .

Using 3D layers as an infrastructure aid is also an approach with clear benefits. The area overhead in 3D voltage regulator circuits header/footer devices and deep-trench capacitors can be minimized by placing them in a separate layer on the path of power delivery from the board. Such low-risk, high-value 3D integration can be more promising from almost any point of view providing higher energy efficiency.

2.5 SYSTEM-LEVEL CHALLENGES: THE FUTURE OF POWER-CONSTRAINED PROCESSORS IN 3D

It recent years 3D manufacturing technology has made significant progress, especially in low-end embedded applications that have already been using this technology. But there are numerous issues and challenges to be addressed before the acceptance of 3D becomes widespread—especially in the higher-end and lower-volume market segments, such as server systems.

One limiting issue is the availability of design tools and methodologies required for 3D system design. Although the need for 3D design tools and techniques span the entire spectrum the corresponding overhead is dependent on the initial complexity of the system; the higher design complexity translates to a higher cost and overhead with more challenging dependencies. In many cases, 3D integration requires extensive changes in the design flow: 3D timing, placement, routing, tools to support clocking, and even architecture-level items like performance modeling. Furthermore, it requires more frequent feedback loops than in the traditional design flow, because of the increased complexity in this new z axis.

Beyond tool and design flow readiness, many other things are needed to build a system in 3D. As the technology matures, the technology characterization data are available at the system level, where basic decisions regarding performance, power, thermal, and reliability characteristics need to be handled. The process of designing an efficient system in a potentially disruptive technology using characterization data is a nontrivial task. The success in designing high-performance, energy-efficient, and reliable systems with a given set of technology constraints will also be a major factor in deciding the future of the emerging 3D systems. While existing studies provide reasonable outlook on the power, the thermal, and the reliability characteristics of 3D systems, there are many unknowns at the time.

When it comes to reliability, 3D-specific structures such as micro-C4s and TSVs need to be characterized to assess the resulting stack characteristics. Current studies indicate that the natural reduction in interconnect length and capacitance is likely to reduce simultaneous switching noise: both TSVs and micro-C4s exhibit favorable noise characteristics. As 3D technology evolves detailed noise and cross-talk characterization for signal farm structures, power supplies, ground noise within the shared-stack infrastructure, and the yield and electro-migration characteristics of TSVs are among a few items from a much longer list of future 3D analysis requirements.

In integrating individual dies to wafers (as in die-to-wafer integration) and wafers to wafers, 3D integration inevitably creates more exposure to the challenges of increasing process variation. The variation among different dies and wafers is much higher than within the 2D chip itself. As a result, the functionality and efficiency of the 3D system gets affected. Even though functional partitioning can alleviate this problem (at least partially), by separating the layers of logic and DRAM or eDRAM, the finer-grain partitioning of logic stacking is constrained by its variation characteristics.

The power and thermal characteristics of 3D microprocessor stacks have been of particular interest.[13] Whether 3D integration will provide better power efficiency or not is still heavily debated, while the answer is likely to have many parameters. Many factors contribute to the power–efficiency breakeven of a system, such as the increased leakage power due to the higher temperatures, the power cost of on-chip versus off-chip communications, and the infrastructural benefits of 3D, such as voltage regulators and/or voltage converter layers.

On-chip heating also becomes a challenge in 3D. It is clear that the temperature profiles degrade in stacked architectures because of the lower thermal conductivity of the materials used within the interlayer boundaries of a stack. Temperature increases can vary considerably depending on those specific implementation details. Recent studies have shown that a cache-on-processor stacking option is likely to see tolerable increases in the peak temperatures, while a processor-on-processor option is likely to be thermally limited.[14]

As we move into the implementation details of 3D, performance, power, temperature and reliability characteristics show more inter-dependency. The performance improving steps in a 3D design are likely to worsen the power/thermal/reliability characteristics and visa versa. For instance the higher power density regions (such as logic macros or cores at large scale) can create significant thermal gradients and cause reliability issues in thinned device layers (due to CTE mismatch in the materials in the 3D stack).

In another example, the performance driven ordering of device layers (where the processor layer is placed close to the package/off-chip communication) results in thermal problems due to the increased distances/vertical impedances to the heat sink. The reverse ordering could potentially increase the IR drop across the stack (over the power delivery TSVs) as the power hungry processor layer is placed at the top of the stack.

This introduces the need for sophisticated optimization techniques and significant changes in the design flow.

Recent studies have highlighted this need to carefully examine and plan for the intricate dependencies and optimization challenges in the 3D spectrum.

2.6   EMERGING TOPICS: SECURITY

The inherent inter-connectivity/containment of structures in the relatively smaller footprint area can be leveraged for security applications as well. Specialized 3D layers can be used to protect critical chips from side-channel attacks.

In current systems, side-channel attacks (Simple or differential power/EM analysis), thermal imaging can reveal important information about the characteristics of the chip and the applications.

Specialized 3D structures can protect the critical device layers from thermal imaging differential power/electromagnetic analysis by intelligently adjusting the amount of information revealed from the full stack as the additional device layers effectively scramble the power/EM profile of the critical layer to be protected. The 3D layers can effectively serve as shield structures for next generation server architectures.

2.7   3D SYSTEMS: HOW DO WE GET THERE? DESIGN AND TEST FOR 3D

In general, the 3D integration of dies known to be good can potentially improve the yield of a system compared with its planar counterpart. But it is important to note that such an improvement is constrained to the lab environment for now; the time, cost, and complexity associated with achieving this yield improvement in production are questionable in any high-volume product. The challenges and opportunities in testing vary significantly in the 3D manufacturing spectrum from wafer-to-wafer, wafer-to-die, and die-to-die stacking.[16-19] In all cases, the combination of wafer and module-level testing is required, with a special emphasis on novel challenges in 3D.

Beyond the traditional wafer and module testing stages, 3D integration is likely to introduce new stages in the test flow, as well as significant changes to the existing tests. One such stage is the concept of single-stratum testing, where a single device layer is tested before the integration of the entire stack.

Though there are major limitations to this stage, especially in fine-grain integration cases, single-stratum testing is essential to improving the overall stack yield. Stack testing, which follows the single-stratum test stage, is more likely to resemble the traditional test flow, although the "flow" for the test might be artificial.

In the top-down 3D approach partitioning the functionality onto multiple strata in 3D introduces new test challenges such as accessing the signals that cross stratum boundaries.[20] Cross-strata signals in 3D stacking can connect various logic circuits within any latch-to-latch path across the layers (in principle). By using latches on either side of a via (interlayer interconnect), or by using the landing pads, one can access and probe these signals. Yet it is not possible to AC-test the logic-to-logic signals (i.e., signals not on a latch boundary) without introducing new latches that are only there for testing.

The test probe option implies an electrostatic discharge diode (ESD) with a large capacitance and a landing pad to be placed for safe access with the test probe, which is going to cause the signal to slow down. Because of the overhead associated with such scheme, the number of signals tested through probing is quite limited. In either case, cross-stratum signal testing requires some significant innovations before it becomes practicable.

A 3D-optimized design environment is necessary in such cases to enable better partitioning of functionality across strata boundaries, both from performance and from DFT points of view. Placing latches on either side of the interstratum interface is ideal from a test point of view. For the cases that are not latch to latch (also that have sufficient timing slack), boundary scan latches can be used to capture the interlayer signals.

While AC tests with boundary-scan latches on a single stratum are essential, they still fail to account for the impedance of the through-silicon via and interstratum interconnect in the final stack. A natural way of partitioning for test requires that signals traversing the stack through interstratum boundaries be latch-to-latch to be testable. That being said, one can differentiate between the signals that are accessible only through scan rings and those that a tester can probe. Most signals in a large system must be of the former type only, because there are too many of them to probe.

Another concern regarding test probes in 3D structures is the amount of force exerted on the chip being tested. The existing 3D manufacturing flows differ significantly in the way and/or order in which various processing stages are executed. In the manufacturing flows where individual device layers are thinned first to expose the connections, this may cause mechanical problems such as bowing of the thinned layers. While layers with thicknesses of hundreds of micrometers are not affected by this, thinning device layers to tens of micrometers makes stress a great concern. It is also the case that test probes cause damage to the surface of the landing pad, which probably will require further planarization prior to stacking.

The testable state in the finished stack can be large. Therefore, partitioning the scan chains effectively through the stack is important in

3D systems. In trying to access this potentially massive amount of testable data, scan chains are commonly used in coordination with built-in self-test structures. The number and length of the scan chains is of primary importance in determining the challenge. Scan chains can be leveraged for both single-stratum testing and post-bonding stack tests. Reconfiguring the scan chains to target both single-stratum test and full stack is highly desired to maximize the efficiency, though this requires careful planning.

A straightforward approach is to connect the scan chains from each of the 2D layers and to preserve the existing scan order for each layer. On the other hand, this serial approach results in long scan chains, which may not be practical. An alternative way is to enable scanning across the layer boundaries, with 3D-specific infrastructures that enable accessing the existing 2D chains hierarchically. Although the latter makes the parts of the finished product more directly accessible for testing, there must be reasonable certainty that the vertical interconnection infrastructure (vias) will work. Adding redundant vias and interlayer interconnects can enable this approach without a high overhead.

2.7 CONCLUSIONS

The term "3D" has been used by many people to mean many very different things. In this chapter, we have outlined some of those things, articulated some broad categories of what is called 3D, and adumbrated the additional complexities that will be required to design general 3D systems.

Much of what is called 3D has already been in practice for some time now. These are applications in which low-power chips having only peripheral connections are stacked up to make a cost-effective product. The fact that these systems are "built in 3D" is pure convenience: the power density doesn't make it problematic, and the "connectedness" of the chips doesn't leverage the 3D "closeness" in any particular way.

The era that is emerging is one in which the "3D-ness" of the system is designed into the system so as to leverage the third dimension for some aspect of performance. These applications generally require more connections between the chips—connections that are not merely peripheral—and power vias that deliver power through the stack. Because the 3D-ness is specifically done to leverage some aspect of performance, the chips in the stack must necessarily be co-designed so as to create spatial correspondences between their subsystems. Again, having a spatial correspondence necessarily implies that vias within the chip (i.e., not merely on the periphery) interconnect the chips.

When done correctly, 3D designs (i.e. with careful design planning and trade off analysis), while necessarily having more tools, models, and tests required to make them work, should produce systems that simplify the logistics and economics of dealing with large 2D chips. Or they should produce systems that are not reasonable to produce in 2D.

Very large 2D chips are expensive to produce for numerous reasons. But they are complete systems that are "ready to go" (principally) because they are single components. And generally speaking, the circuitry on them is specifically designed so that it interacts only with other proximate (to that circuitry) circuitry. The large area of a 2D chip is made (principally) to contain "enough stuff" in a single component so as to simplify a larger system. A 2D chip cannot readily leverage the "on-chipness" of componentry that is not proximate, although that componentry can be connected on chip without adding wiring complexity to the package.

Therefore, the leverage of a 2D chip is principally one of "containment." It simplifies the packaging infrastructure. Some, but certainly not all, of the wiring is made "simpler" (smaller wires with simpler connections) in 2D, especially if it connects things that are proximate in 2D. And most of our systems are conceived in 2D: we draw their schematics this way.

Literally, 3D offers a new dimension. Intellectually, our current generation of designers doesn't conceive systems in this new way. When we did our first designs in school, and when we did our last design this year, we sketched them on a plane. Ironically, using the third dimension well requires that we add two new dimensions to our thinking. The first is opportunity (new function and structure), and the second is limitation (power and heat).

Of course, you may think that some 3D applications are obvious. Many will point to stacks of DRAM as an example. Really? In all concepts involving things like this being marketed today, the only aspect of 3D being exploited is density. The opportunity here (not being discussed) is that some applications (that don't exist yet) may want a memory system that can directly fetch structured data using different dimensions of association. Sure. Today we write programs to figure things like this out. But today we certainly don't access memory in any manner other than linear.

Some new thinking is needed.

And finally, we alluded to "future 3D systems" that are not necessarily built as stacks of 2D systems that are (crudely, relative to on-chip wiring) wired together by putting holes through them. If we conceived real 3D systems that were not constrained to be made in accordance with our rich legacy in 2D concepts, designs, and manufacturing flows, what would we come up with? Is there an optimal design point for interconnectivity/computation in the computing systems? Maybe a brain?

Again, some new thinking is needed.

References

1. Banerjee, K., Souri, S., Kapur, P., and Saraswat, K. (2001) 3-D ICs: a novel chip design for improving deep sub-micrometer interconnect performance and systems-on-chip integration, *Proc. IEEE*, 89(5), 602-633.

2. Burns, J. A., Chen, C. K., Knect, J. M., and Wyatt, P. W. (2006) A wafer-scale 3-D circuit integration technology, *IEEE Trans. Electron Devices*, **53**(10), 2507-2516.

3. Guarini, K. W., Topol, A. W., Ieong, M., Yu, R., Shi, L., Newport, M. R., Frank, D. J., Singh, D. V., Cohen, G. M., Nitta, S. V., *et al.* (2002) Electrical integrity of state-of-the-art 0.13μm SOI CMOS devices and circuits transferred for three-dimensional integrated circuit fabrication, in *Technical Digest of the International Electron Devices Meeting*, pp. 943-945.

4. Black, B., Annavaram, M., Brekelbaum, N., Devale, J., Jiang, L., Loh, G., Mccauley, D., Morrow, P., Nelson, D. W., Pantuso, D., *et al.* (2006) Die stacking microarchitecture, in *Proceedings of the 39th Annual IEEE/ACM International Symposium on Microarchitecture*, Association for Computing Machinery, New York, pp. 469-479.

5. U. Kang, H.-J. Chung, S.Heo, S.-H. Ahn, H. Lee, S.-H. Cha, J. Ahn, D. Kwon, J.H. Jim, J. W. Lee, *et al.* (2009) 8Gb 3D DDR3 DRAM using through-silicon-via technology, in *IEEE International Solid-State Circuits Conference: Digest of Technical Papers, 2009*, pp. 130-131, 131a.

6. Suntharalingam, V., Berger, R., Clark, S., Knect, J., Messier, A., Newcomb, K., Rathman, D., Slattery, R., Soaers, A., Stevenson, C., *et al.* (2009) A 4-side tileable back illuminated 3D-integrated MPixel CMOS image sensor, in *IEEE International Solid-State Circuits Conference: Digest of Technical Papers, 2009*, pp. 38-39, 39a.

7. Patel, C., Tsang, C., Schuster, C., Doany, F. E., Nyikal, H., Baks, C. W., Budd, R., Buchwalter, L. P., Andry, P. S., Canaperi, D. F., *et al.* (2005) Silicon carrier with deep through-vias, fine pitch wiring and through cavity for parallel optical transceiver, in *Proceedings of the 55th IEEE Electronic Components and Technology Conference*, pp. 1318-1324.

8. Emma, P., and Kursun, E. (2008) Is 3D integration the next growth engine after Moore's Law, or is it different? *IBM J. Res. Dev.*, **52**(6), 541-552.

9. Saito, H., Nakajima, M., Okamoto, T., Yamada, Y., Ohuchi, A., Iguchi, N., Sakamoto, T., Yamaguchi, K., and Mizuno, M. (2010) A chip-stacked memory for on-chip SRAM-rich SoCs and processors, *IEEE J. Solid-State Circuits*, **45**(1), 15–22.

10. Tsai, Y. F., Xie, Y., Vijaykrishnan, N., and Irwin, M. J. (2005) Three-dimensional cache design using 3D Cacti, in *Proceedings of IEEE International Conference on Computer Design: VLSI in Computers and Processors*, pp. 519-524.

11. Bernstein, K., Andry, P., Cann, J., Emma, P., Greenberg, D., Haensch, W., Ignatowski, M., Koester, S., Magerlein, J., Puri, R., *et al.* (2007) Interconnects in the third dimension: design challenges for 3D ICs, in *Proceedings of the 44th Annual Design Automation Conference*, Association for Computing Machinery, New York, pp. 562-567.

12. Xie, Y., Loh, G. H., Black, B., and Bernstein, K. (2006) Design space exploration for 3D architecture, *ACM J. Emerging Technol. Comput. Syst.*, **2**(2), 65-103.

13. Cong, J., Wei, J., and Zhang, Y. (2004) A thermal-driven floorplanning algorithm for 3D ICs, in *IEEE/ACM International Conference on Computer Aided Design*, pp.306-313.

14. Zhu, C., Gu, Z., Shang, L., Dick, R. P., and Joseph, R. (2008) Three-dimensional chip-multiprocessor run-time thermal management, *IEEE Trans. Comput. Aided Des. Integr. Circuits Syst.*, **27**(8), 1479-1492.

15. Knickerbocker, J. U., Patel, C. S., Andry, P. S., Tsang, C. K., Buchwalter, L. P., Sprogis, D., Gan, H., Horton, R. R., Polastre, R., Wright, S. L., *et al.* (18-21 September 2005) Three dimensional silicon integration using fine pitch interconnection, silicon processing and silicon carrier packaging technology, in Proceedings of the IEEE Custom Integrated Circuits Conference, pp. 659-662.

16. Colgan, E. G., Furman, B., Gaynes, M., Graham, W. S., La Biance, N. C., Magerlein, J. H., Polastre, R. J., Rothwell, M. B., Bezama, R. J., Choudhary, *et al.* (2007) A practical implementation of silicon microchannel coolers for high power chips, in *IEEE Trans. Compon. Packaging Technol.*, **30**(2), 218-225.

17. Reif, R., Fan, A., Kuan-Neng, C., and Das, S. (2002) Fabrication technologies for three-dimensional integrated circuits, in *Proceedings of the IEEE International Symposium on Quality Electronic Design*, pp. 33-37.

18. Swinnen, N., Ruythooren, W., De Moor, P., Bogaerts, L., Varbonell, L., De Munck, K., Eyckens, B., Stoukatch, S., Sabuncuoglu Tezcan, D., Tokei, Z., *et al.* (2006) 3D Integration by Cu-Cu thermo-compression bonding of extremely thinned bulk-Si die containing 10 μm pitch through-Si-vias, in *IEEE International Electron Devices Meeting*, issue 11, pp.1-4.

19. Topol, A. W., La Tulipe, D. C., Jr., Shi, L., Frank, D. J., Bernstein, K., Steen, S. E., Kumar, A., Singco, G. U., Young, A. M., Guarini, K. W., *et al.* (2006) Three-dimensional integrated circuits, *IBM J. Res. Dev.*, **50**(4/5), 491-506.

20. Ma, Y., Liu, Y., Kursun, E., Reinman, G., and Cong, J. (2008) Investigating the effects of fine-grain three-dimensional integration on microarchitecture design, *ACM J. Emerging Technol. Comput. Syst.*, **4**(4),1-30.

Chapter 3

WAFER BONDING TECHNIQUES

Bioh Kim, Thorsten Matthias, Viorel Dragoi, Markus Wimplinger and Paul Lindner

EV Group

3D stacked IC (3D-IC) has various potential advantages over conventional system-on-chip (SoC) or system-in-package (SiP) approaches when it comes to device performance, form factor, interconnect density, heterogeneous integration, and even manufacturing costs. The continuation of Moore's law by conventional complementary metal oxide semiconductor (CMOS) scaling is becoming more and more challenging, requiring huge capital investments. 3D-IC with through-silicon via (TSV) interconnects provides another path towards "More Moore" and "More than Moore" with relatively smaller capital investments.

The 3D-IC approach using TSV interconnects, which provides electrical pathways through the thickness of a Si die, calls for a combination of via processes, extreme wafer thinning, and various wafer-to-wafer and chip-to-wafer bonding techniques with thin wafer handling mechanisms. There are various factors related to device integration, which should be considered before adopting TSV technology design. Those include (a) TSV process flow type: via-first, via-middle or via-last, (b) bonding process type: chip-to-chip, chip-to-wafer or wafer-to-wafer, (c) substrates stacking approach: face-to-back or face-to-face, (d) using or not using interposer, (e) substrate type: bulk silicon wafer or silicon-on-insulator (SOI) wafer, and so on. For the majority of 3D-IC integration applications, full wafers of a specific function are produced separately and then stacked vertically with accurate alignment schemes in order to create multi-functional devices.

The increased complexity of new generations of 3D-ICs imposes challenging requirements for wafer bonding and handling. Among those are

3D Integration for VLSI Systems
Edited by Chuan Seng Tan, Kuan-Neng Chen and Steven J. Koester
Copyright © 2012 by Pan Stanford Publishing Pte. Ltd.
www.panstanford.com

low process temperature (typically, < 400°C), precise alignment of substrates, ability to bond a large variety of substrates, and the possibility to bond with defined intermediate layers. In addition, making reliable 3D-ICs usually requires a significant reduction in wafer thickness, which, when combined with a large wafer diameter, necessitates new wafer handling mechanisms. The use of thinner wafers enhances IC performance, allows TSV processing optimization (via definition) and enables innovative packages, but handling thin wafers or even thin dies raises new challenges to semiconductor equipment and processes. Therefore, both temporary bonding of device wafers on carrier substrates for handling through specific steps in the process flow and debonding of the device wafer from the carrier wafer are indispensible processes to 3D-IC manufacturing.

In this chapter are reviewed various bonding techniques suitable for manufacturing 3D-ICs with TSV interconnects, including bond alignment, wafer bonding, advanced chip-to-wafer bonding, and temporary bonding and debonding for thin wafer handling and processing.

3.1 INTRODUCTION

The development of IC technology is driven by the needs to increase performance and functionality while reducing size, weight, power consumption and manufacturing cost.[1,2] 3D-IC or wafer-level 3D integration is an emerging architecture and technology developed to overcome the limitations of current SoC and SiP approaches. Key enabling technologies for TSV integration include (a) TSV formation for electrically isolated interconnects through wafers with deep reactive ion etch (DRIE), insulator and barrier deposition, via filling and chemical mechanical polishing (CMP), (b) extreme wafer thinning and edge trimming for obtaining damage-free thin wafers, and (c) aligned wafer bonding with thin wafer handling mechanisms.[3-4]

Wafer bonding was first developed for the micro-electro-mechanical systems (MEMS) market, primarily as a wafer-level capping technique.[5] The manufacturer would produce a MEMS wafer with some fragile structures (e.g., membranes and cantilever beams) and would then use another wafer to cap off and protect the MEMS structures, sealing off the cavities around them to ensure the functional environment and to prevent damage and contamination. Today wafer bonding is being used not only for capping MEMS wafers, but also for essentially stacking wafers with different functionalities for homogeneous and heterogeneous 3D integration.[6-7] The industry can take various wafers (such as MEMS wafers, memory wafers, and CMOS processor wafers), stack them on each other, and then connect them by using TSV interconnects. Currently, a wide variety of wafer bonding methods are being used for various device packaging and engineered substrate manufacturing, which include molecular bonding, anodic bonding, metal bonding, glass frit bonding and adhesive bonding depending upon whether an intermediate layer is used or not, as summarized in Figure 3.1.[8]

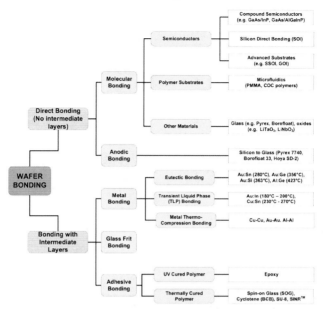

Fig. 3.1 Overview of different wafer bonding methods.

However, many established bonding methods are not compatible with CMOS devices due to several issues. For instance, anodic, glass frit, Au-Au thermo-compression and some eutectic bonding methods are not considered due to the risk of metal ion contamination. The bonding temperature has to be compatible with the thermal budget of the devices, e.g., less than 400°C for CMOS devices. The TSV diameter is limited due to space constraints, which requires the intermediate layer to be very thin in order to minimize the aspect ratio of the TSV. The bond layer needs very good thickness uniformity. Due to those issues, the bonding processes compatible with TSV-interconnected CMOS wafers are limited to such bonding methods as direct oxide bonding (SiO_2), metal bonding (Cu-Cu or Cu-solder-Cu), adhesive bonding, and several hybrids of those methods as illustrated in Figure 3.2.

3D-IC stacking can be performed through chip-to-chip, chip-to-wafer or wafer-to-wafer approaches, each method having its own benefits and disadvantages. Chip-to-chip stacking is being used very broadly today, but is regarded as a very costly process for TSV applications. Wafer-to-wafer stacking is most practical for high yield devices. Chip-to-wafer stacking is believed to be best suited for low yield devices or heterogeneous stacking although pick-and-place, align, and bond assembly time is relatively costly when done chip-by-chip since there are no economics at the wafer level. If any subsequent post-bond processing is required, surface planarity issues can create added processing complexity and cost compared to wafer-to-wafer alignment and bonding.

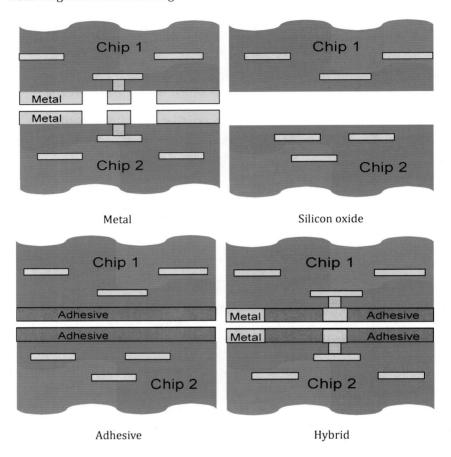

Metal   Silicon oxide

Adhesive   Hybrid

Fig. 3.2 Bonding methods suitable for TSV integration (courtesy of Fraunhofer IZM).

Another important aspect of wafer bonding that has become pivotal to ensuring a successful bonded structure is the ability to maintain the accuracy of alignment between the two wafers being bonded. Especially, 3D stacking applications, which help improve device packaging density and signal delays without negatively impacting yield, are particularly dependent upon the alignment accuracy of wafers. The electrical performance of devices can be greatly affected by bond misalignment.

The use of a thinned substrate for TSV integration enhances IC performance and enables various innovative packages. Recently, an increasing number of applications started to take advantage of reduced bulk silicon thickness. Wafer thinning removes a part of the bulk silicon and leaves an active circuit layer with a thinned bulk silicon layer. Typical benefits of using thinned wafers include package miniaturization, better heat and power dissipation, improved device reliability (due to more flexibility of dies) and thinner vertical stack.[9] However, using a thinned wafer raises additional challenges to wafer handling and processing due mainly to less mechanical stability causing wafer bowing and warping.

As wafer thickness is getting thinner and thinner, adopting dedicated chucks, robot end-effecters or special cassettes is not sufficient to cope with a majority of application types and wafer thicknesses. Therefore, the most promising and most widely investigated handling solution for thin or ultra-thin wafers is to use temporary bonding and debonding techniques utilizing a carrier wafer to provide sufficient mechanical support for the subsequent processes.[10–11]

3.2 ALIGNED PERMANENT BONDING PROCESSES

Wafer bonding is the only semiconductor process that requires a high level of uniformity of temperature and pressure. Due to the specific features of wafers used for 3D-IC fabrication any non-uniformity in temperature may result in building high stresses in the bonded wafers. In some cases the stress can be so high that it may result in high bow of the bonded pair (with the impact on handling during next processes) or can even produce wafer breakage during post-bonding processes such as grinding. Such high uniformity is achieved through specialized wafer chucks with double-sided heating to control the temperature to the wafers and the gradient across the stack. Wafer alignment and bonding are well-established technologies originating from MEMS manufacturing as mentioned previously. Aligned wafer bonding enables heterogeneous functional stacks like CMOS logic with memory, mixed signal or bipolar even with heterogeneous substrate material combinations e.g.,-silicon with compound semiconductor.[12] Aligned wafer bonding consists of three process steps; the first step is the preparation of the wafer surface in terms of cleanliness, surface chemistry and intermediate layer properties, the second step is to align two wafers within the alignment specification, and the last step is the permanent bonding. In the case when direct oxide bonding is chosen, additional batch annealing step is required to increase the bond strength.

Bonding voids may be generated by particles, surface protrusions, or trapped air. Those voids can be observed after surface contact is made and they are not changed during annealing. The resulting void is usually two to three orders of magnitudes larger than the particle itself. For manufacturing purposes it is necessary to control surface properties tightly. With high bond strength, neither debonding nor delamination would occur during the following process steps or during the complete product life cycle. The key equipment parameters for high bond strength are pressure and temperature uniformities as mentioned above. The bond quality can be analyzed by several methods. Typically, infrared (IR) microscopy and scanning acoustic microscopy (SAM) are being used as part of manufacturing steps.

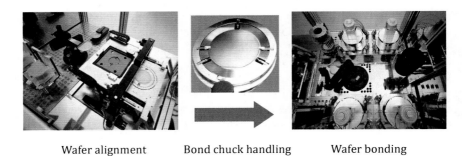

Wafer alignment   Bond chuck handling   Wafer bonding

Fig. 3.3 Process separation principle for aligned wafer bonding (courtesy of EVG).

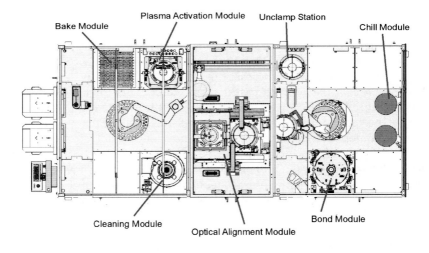

Fig. 3.4 Schematic of 300 mm EVG Gemini® wafer bonding system with up to four pre-processing modules, a wafer-to-wafer alignment station and up to four bond modules (courtesy of EVG).

For production purposes the individual processes of wafer alignment and bonding are separated into different process modules. The technical requirements of rapid and uniform heating, high and uniform pressure, and full flexibility regarding process gases or vacuum are not compatible with high precision alignment stages. The economical reasoning is that one alignment station with cycle times of 3 – 6 min can easily support several bond chambers with cycle times of 10 – 60 min. The process separation principle is illustrated in Figure 3.3. With a proper bond chuck design, the inherent risk that the wafer alignment can be deteriorated during heating up the wafer stack can be minimized. The results in high volume manufacturing over the past 15 years have shown that high performance wafer bonding systems can maintain the alignment accuracy safely. An example of the schematic of the production bonder platform is shown in Figure 3.4, which has three types of process modules including pre-treatment, alignment and bonding, where thermo-compression and *in situ* direct oxide bonding can be performed in a single machine.

3.2.1 Wafer alignment

Wafer alignment should be performed precisely to realize a high level of device integration. Since wafer bonding process is not limited to bonding blanket wafers, most of the applications require the alignment and bonding of patterned wafers instead. Various alignment methods are available, but most suffer from one or more issues that prevent them from being utilized as a generic solution. Two main types of wafer-to-wafer alignment methods, mechanical alignment and optical alignment, are currently used. In mechanical alignment, the main flats or notches of the two wafers are aligned to each other mechanically. Typically, a setup of special pins is used to accomplish this task. For SEMI standard wafers, the guaranteed alignment accuracy is typically within ±50 μm shift and 0.1° rotation for 300 mm wafers. Mechanical alignment is used for a cap wafer being bonded to a device wafer in some MEMS applications or for non-patterned wafers bonding (e.g., SOI substrate fabrication).

Optical alignment methods are essential for 3D-IC stacking due to the high alignment accuracy required. For 3D-IC stacking, TSVs come into contact with the corresponding bond pads after bonding and following processes. The alignment tolerance is determined by the diameter of the vias and the size of the contact pads. In order to minimize the area of the chip consumed for 3D interconnects, applications with high via density require very small via sizes e.g., below one micrometer, whereas for devices with moderate via density larger via diameters are acceptable. Optical wafer-to-wafer alignment is based on the alignment of two specific points or alignment keys on both wafers. The alignment keys can be either on the front side of the wafers within the bond interface or on the backside. In the case when the keys of both wafers are within the bond interface and at least one of the wafers is transparent for visible light or infrared light, a direct alignment of those two wafers can be performed. An example is illustrated in Figure 3.5, where direct alignment keys were used for silicon oxide bonding.

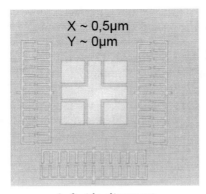
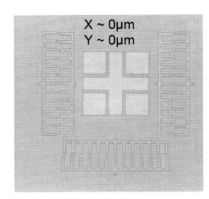

Left side alignment   Right side alignment

Fig. 3.5 Direct alignment of silicon oxide bonds (courtesy of CEA-Leti).

As illustrated in Figure 3.6, direct alignment means that a live image of both corresponding alignment keys is used for alignment. The feedback based on the live image enables a closed control loop and results in the highest alignment accuracy. Typically, the substrates used for 3D stacking are not transparent for visible light. However, silicon wafers with thickness in the range of few hundreds micrometers are transparent to IR radiation. The IR transparency of silicon depends on the band gap; thus, for constant wavelength the transparency decreases consistently with increasing the doping of silicon (corresponding to decreasing the band gap). The best application for IR alignment is a 2-layer wafers stack. For multiple layer stacks the image quality degrades due to multiple diffraction, reflection and interference.

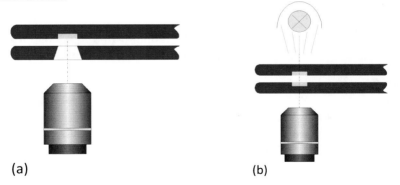

Fig. 3.6 Direct alignment methods (courtesy of EVG); (a) backside alignment with transparent wafer and (b) IR alignment.

In the case when live alignment is not possible due to the use of wafers which are optically non-transparent, an indirect alignment method can be applied. Figure 3.7 shows two types of indirect alignment methods. A backside alignment with the digitized image is illustrated in Figure 3.7(a). In this alignment approach one wafer has to have defined alignment keys on the backside. Since no live image is available in this alignment, each wafer has to be aligned individually to two external reference points. Usually the positions of the microscopes, specifically the center of the respective field of views, are taken as the reference points. Sequentially each wafer is aligned to the reference points and finally the aligned wafers are brought into contact. Then those are clamped on the bond chuck and are ready for transfer to the bond chamber.

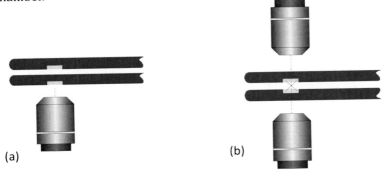

(a)   (b)

Fig. 3.7 Indirect alignment methods (courtesy of EVG); (a) backside alignment with digitized image and (b) SmartView® alignment.

The SmartView® alignment method uses dual microscopes with a common focal plane as shown in Figure 3.7(b). This technique enables the use of any alignment keys on the front-side of non-IR transparent wafers for face-to-face alignment, where the front-side or face means the surface or side with the alignment keys. The operation principle is illustrated in Figure 3.8. In the alignment position both wafer interfaces are in the same focal plane, which is guaranteed by a wedge compensation mechanism. The positions of microscopes are the reference points for the alignment. First, the bottom wafer is moved into a pre-defined alignment position and the microscopes are positioned according to the alignment keys. The position of microscopes is locked and the bottom wafer chuck is moved away. Then the top wafer is moved into alignment position and aligned to the microscopes. The top wafer stage is moved precisely in $x, y,$ and θ. Due to the vector-based pattern recognition, the alignment procedure is very robust, which can handle contrast and even slight shape variations from wafer to wafer. The main advantage of SmartView® alignment is its flexibility and versatility. It is suited for all types of wafers independent of the surface, bulk properties or thickness and it can be repeated multiple times for multi-layer stacking without any loss of quality. Additionally this face-to-face alignment does not require the definition for backside alignment keys, which is one of the first sources of misalignment in the backside alignment method. The pre-bond alignment accuracy data with SmartView® aligner is shown in Figure 3.9, where all measured data with 400 repeats showed an alignment accuracy (dx and dy) of less than 120 nm (3σ) for both left side and right side alignments.

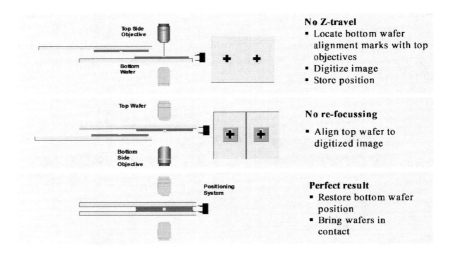

Fig. 3.8 Operating principle of SmartView® alignment (courtesy of EVG).

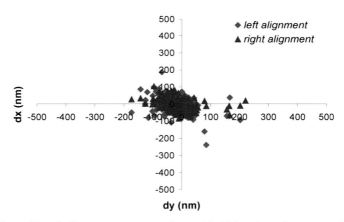

Fig. 3.9 SmartView® alignment accuracy data with 400 repeats (courtesy of EVG).

3.2.2 Wafer bonding

All envisioned process flows for 3D integration on wafer level contain wafer bonding as a key step. Over the last two decades, various bonding methods have been developed and implemented for volume production. As mentioned earlier, due mainly to the thermal budget issue of CMOS devices, the bonding processes compatible with CMOS processing are limited to metal bonding, low-temperature direct oxide bonding, polymer adhesive bonding and several hybrids of those methods.

One important topic in aligned wafer bonding is alignment accuracy. Due to its specific features, wafer bonding process step has an influence on alignment accuracy. If it is somehow obvious that in a bonding approach using a compressive bonding layer the alignment accuracy provided by optical alignment equipment is slightly deteriorated, the same effect is present also when directly bonding two rigid surfaces due to the different thermal expansion of the patterned layers at the surface of the substrates. Table 3.1 compares three major bonding processes for TSV integration in terms of post-bond alignment accuracy, process time and throughput.

Table 3.1 Comparison of three major bonding processes (courtesy of EVG).

	Cu-Cu	SiO_2	Polymer
Post-bond alignment accuracy (μm, 3σ)	Today : 1.0 Roadmap : 0.5	Today : 0.5 Roadmap : 0.25	Today : 1.0–1.5* Roadmap : 0.5–1.0*
Process time (min)	60 – 120	2.5 – 4	30 – 60
Throughput (bonds/min)	2 – 4	15 – 24	4 – 8

* *In the case of polymer bonding, alignment accuracy varies with polymer adhesive thickness. Thinner adhesive layers would allow better alignment accuracy.*

3.2.2.1 Metal bonding

Patterned metal thermo-compression bonding is one of the most explored bonding methods for TSV integration because it can facilitate fine-pitch, high-density stacking of various devices leading to lower electrical resistance and higher mechanical strength. Various metal-metal bonding approaches based on different principles have been demonstrated. Metal bonding based on the formation of eutectic alloys (*i.e.*, eutectic bonding) or intermetallic compounds (IMCs, *i.e.*, transient liquid phase bonding) as bonding layers were developed mainly for MEMS applications requiring vacuum encapsulation in device cavities. Apart from these two processes, metal thermo-compression bonding was used in compound semiconductors and MEMS applications and more recently in wafer stacking for 3D-IC fabrication. In this case two surfaces are brought into close contact by applying high contact force and heat. The applied force and heat causes the fusion of the opposing surfaces. For TSV integration, both Cu-Cu direct bond and Cu-solder-Cu bond are used for various applications. Figure 3.10 compares the cross-section of two types of bonds.

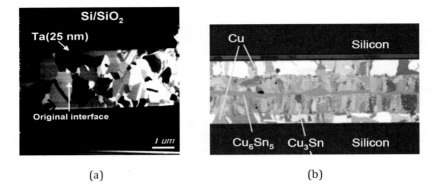

(a)   (b)

Fig. 3.10 Two types of metal bonds; (a) Cu-Cu direct bond (courtesy of EVG and Andong National University) and (b) Cu-Sn-Cu bond (courtesy of Fraunhofer IZM).

Solder bonding for TSV integration uses a layer of Sn or Sn alloy solder (such as SnAg) to bond two wafers. The wafers are brought into close contact and are heated above the melting temperature of the solder material. The solder reflows and wets both wafer surfaces, which causes intimate contact and bonding of the surfaces. The advantages of solder bonding are low bonding temperatures and the ability to join various wafer materials with a hermetic bond. Sn-based solder bonding has been widely used to create electrical contacts in flip-chip bonding. However, solder-based bonds have weak joint interfaces and low electrical performances due to the formation of brittle IMCs and Kirkendall voids.[13]

Another variation of solder bonding developed recently is using the IMCs as direct joint materials. IMCs result from the reaction between Cu and Sn, where a thin layer of Sn with several micrometers of thickness is used as a reaction layer between Cu pads. Two types of IMCs, Cu_3Sn and Cu_6Sn_5, can be formed, where the ratio between two types of IMCs changes with reaction time and temperature. This process is typically called SOLID (Solid Liquid InterDiffusion) from Infineon and ICV-SLID (Inter Chip Via - Solid Liquid InterDiffusion) from Fraunhofer IZM. The stronger bonds are enriched in Cu_3Sn and its resistivity is much lower than that of conventional solder (8.9 $\mu\Omega$-cm as opposed to approximately 16.5 $\mu\Omega$-cm for a SnPb solder).[14] In this bond, Sn (or Sn alloy) is applied between two chips, heat and pressure is applied to those chips to make reaction between Cu pads and Sn solder, and then IMCs are formed between Sn and Cu. Those IMCs are working as electrical and mechanical joints between two chips. Figure 3.11 shows an ICV-SLID bond between the two chips.

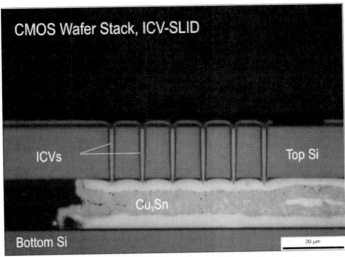

Fig. 3.11 ICV-SLID bond of two chips (courtesy of Franuhofer IZM).

Direct Cu-Cu bonding has been demonstrated using thermo-compression bonding via parallel applications of heat and pressure. Among all CMOS-compatible bonding processes, Cu-Cu bonding looks most favored by the industry for TSV integration because Cu-Cu joints can provide much lower electrical resistivity, higher density and higher electromigration (EM) resistance than any other joints.[15-16] The two key process parameters for Cu-Cu wafer bonding are temperature and pressure uniformities where the maximum temperature is limited by the thermal budget of the devices. The precise pre-bond alignment of the wafers has to be maintained during heating. The temperature non-uniformities in the wafer surface would create local distortions and induce stress into the bond layer. With better temperature uniformity, faster heating ramp is achievable. Good pressure uniformity is necessary in order to compensate for wafer bow and warp as well as for thickness variations. Figure 3.12 illustrates one of the examples of the post-bond alignment accuracy of TSV patterned wafers, where SmartView® alignment was used for face-to-face alignment and Cu-Cu thermo-compression bonding was performed at 400°C. Through the proper control of pre-bond alignment, temperature and pressure uniformity, and heating and cooling rates of both sides, the post-bond alignment accuracy of better than 1 µm (3σ) was demonstrated.[17]

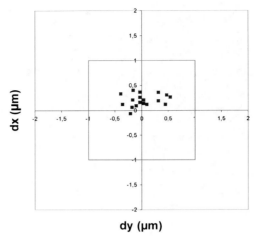

dy (μm)

Fig. 3.12 Post-bond alignment accuracy data of Cu-Cu thermo-compression bonding (courtesy of EVG).

The reliable Cu-Cu bonding for most industrial applications comes from high temperature, high pressure and long process time mainly because of its tendency to form a native oxide which strongly impacts device reliability. Studies by Kim, *et al.*[18] showed that with increasing bonding temperature at a given bonding condition, the interfacial adhesion energy (Figure 3.13(a)), measured by the 4-point bending test,[19] increases and the original bond interface (Figure 3.13(b)) tends to disappear due to an activated in terdiffusion through two layers. That is the main reason why high temperature is applied for Cu-Cu bonding.

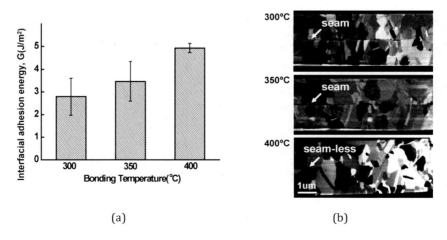

(a)   (b)

Fig. 3.13 Effect of bonding temperature on bond properties (courtesy of EVG and Andong National University); (a) interfacial adhesion energy and (b) microstructures.

Today the significant drawback of Cu-Cu thermo-compression bonding is a very low throughput compared to other processes as shown in Table 3.1. This bonding process involves heating the bonded wafers up to 400°C for 30 – 60 min under pressure, requiring that the pre-bonded wafer pair must spend considerable time at one bonding module. To address this bottleneck, equipment suppliers have developed various platforms with multiple bonding modules as illustrated in Figure 3.4. Much research is also underway to reduce Cu-Cu bonding temperature and thus improve throughput, device reliability and yield. Thermal annealing, *i.e.*, any heat treatment in which the microstructure and therefore the properties of a material are altered, seems to be a good way to improve bond quality. As presented in Refs. 18,20–22, when thermo-compression bonding was tried at lower temperatures and process times (e.g., 300°C for 30 min in that experiment), the post-bond annealing drastically improves the interfacial adhesion energy (to 8.9– 12.2 J/m²) and reduces the interfacial seam voids as illustrated in Figure 3.14. The feasibility of direct hydrophilic Cu-Cu bonding at room temperature and atmospheric pressure without applying any external stress are also being investigated as presented in Refs. 23–24. This bonding process does not need any additional processing steps, but seems to need stringent planarization and surface treatment techniques to ensure smooth hydrophilic surfaces for good bonding conditions.

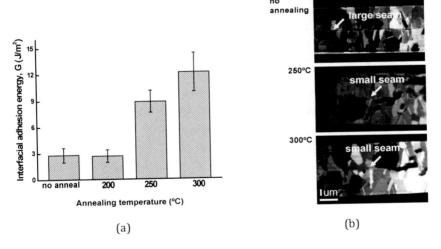

(a)   (b)

Fig. 3.14 Effect of post-bond annealing temperature on bond properties at a given bonding condition (courtesy of EVG and Andong National University); (a) interfacial adhesion energy and (b) microstructures.

3.2.2.2 Direct oxide wafer bonding

Direct oxide bonding depends on molecular forces that naturally attract both surfaces together when they are very smooth and flat. In direct or fusion bonding the two wafers come into contact without any assistance of significant pressure, electrical field, or intermediate layers. Direct wafer bonding typically involves wafer surface preparation and cleaning, room temperature contacting of the wafers, and an annealing step to increase the bond strength. In order to achieve reliable and high-yield bonds, typical requirements for direct oxide bonding include microscopic smoothness (typically, RMS < 0.5 nm), macroscopic flatness, surface cleanliness and surface chemistry. When such SiO_2 surfaces are placed in contact at room temperature, they initially form van der Waals bonds which are too weak to allow further processing of bonded wafers. Subsequent heating to high temperatures (typically, >1000°C) is necessary to achieve a high bond strength through the formation of covalent bonds replacing the hydrogen bonds formed at room temperature.

One of the major advantages of this bonding method is that pre-bonding (*i.e.*, wafer contact and hydrogen bonds formation) occurs at room temperature, so a run-out error in alignment can not be induced during bonding and thus the achievable post-bond alignment accuracy is better than that with any high temperature bonding methods, as shown in Table 3.1. As the thermal annealing can be done as a batch process, the process time is much shorter than that for wafer bonding methods which have to go through the complete bond cycle of heating, bonding and cooling. The classical methods require annealing schemes with peak temperatures of up to 1100°C, which is unacceptable for CMOS processes. For such applications there was a need for developing low temperature processes (conventionally "low temperature wafer bonding" refers to processes performed at maximum 400°C) based on additional surface preparation.

The surface preparation methods developed for low temperature processes are based on surface activation, resulting in different types of bonds at room temperature and requiring less energy during the thermal annealing step. Modifying the surface chemistry by plasma activation (LowTemp® plasma activation) allows the formation of chemical bonds at significantly lower annealing temperatures at 200 – 400°C.[25] Figure 3.15 illustrates the variation of bond strength (expressed as surface energy) with thermal annealing temperature and time. In a standard fusion bonding process using high temperature annealing the full bond strength (in the range of substrate bulk fracture strength, ~2.5 J/m^2) is reached after few hours of annealing at temperatures higher than 900°C (typically, 1100 – 1200°C). In Figure 3.15 can be observed that for plasma activated wafers annealing temperature of 100°C results in surface energies of ~1.5 J/m^2 after few hours annealing (a surface energy of ~1.1 J/m^2 is needed for bonded wafers to survive grinding) while annealing processes of only 1 – 2 hours at temperatures in the range of 200°C to 400°C result in maximum bond strength (in the range of bulk fracture strength of silicon). The possibility to decrease the temperature below 400°C brings not only major technical advantages e.g., high precision alignment and device reliability, but also major economic advantages e.g., the use of commercially available wafer bonders and low temperature annealing equipment with high throughput (by enabling low temperature batch annealing processes).

The IR transmission images of oxide bonds after plasma activation of the surfaces are shown in Figure 3.16 for 200 mm blanket and patterned wafers, respectively, where oxide layers were deposited by plasma-enhanced chemical vapor deposition (PECVD). The bond energy measured was ~ 2.5 J/m^2.[26] Neither voids nor delamination was observed. As summarized in Table 3.1, as long as throughput is concerned, direct oxide bonding is much better than Cu-Cu bonding. However, this bonding technique requires additional steps for electrical connections after bonding is performed.

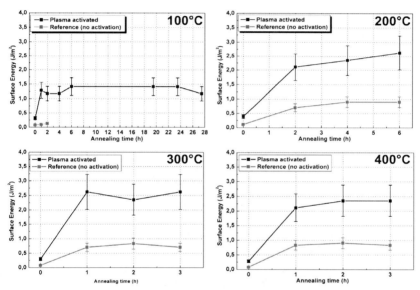

Fig. 3.15 Surface energy as a function of thermal annealing time at various annealing temperatures (100°C to 400°C) (courtesy of EVG).

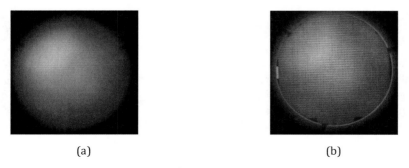

(a)   (b)

Fig. 3.16 IR transmission images of PECVD oxide bonds (courtesy of EVG); (a) blanket oxide wafer and (b) patterned oxide wafer.

3.2.2.3 Adhesive wafer bonding

In this bonding process, an adhesive layer, which is ultraviolet (UV)-curable or thermally-curable, is used as a bonding layer between two surfaces. Typically, a polymer layer is applied to one or both of the wafer surfaces to be bonded. After bringing in contact the two wafer surfaces coated with the polymer material, the two wafers are heated or UV-exposed according to the polymer type. The polymer adhesive is thus converted from a liquid or viscoelastic state into a solid state. The main advantages of adhesive bonding include the low bonding temperature compared to metal bonding (200 – 300°C depending upon polymer materials), the tolerance to the topography or conditions of wafer surfaces (even minor particles contamination can be acceptable if particles diameter is lower than polymer thickness), the compatibility with standard CMOS wafers, and the ability to join any wafer materials. Benzocyclobutene (BCB) is one of the most common polymer materials used for polymer bonding pertinent to 3D integration. A schematic of 3D chip stacking with BCB is shown in Figure 3.17. In this figure, two processed wafers are bonded with a face-to-face integration approach using an adhesive layer. The top wafer is thinned and then high aspect ratio TSVs are etched through the backside of the thinned wafer to provide vertical electrical connections between the two wafers.[27]

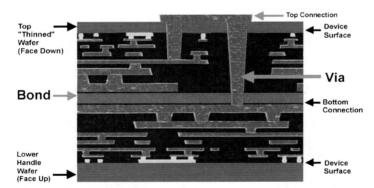

Fig. 3.17 Schematic view of 3D chip stacking with adhesive bonding (courtesy of J.-Q. Lu at RPI and Ref. 28).

As BCB reflows during heating, the particles on the wafer surface are embedded in BCB and do not pose any significant problems for bond qualities. However, the material reflow imposes some challenges for maintaining the alignment accuracy.[29] Depending upon the degree of pre-bond cross-linking, the material goes through a liquid or Sol/Gel rubber type phase during heating. A liquid layer between the two wafers is always a problem as any shear force on one wafer would immediately result in a shift of one wafer, *i.e.*, translational misalignment. However, studies by Niklaus, *et al.*, showed that by tailoring the pre-bond cross linking status, the alignment accuracy can be maintained during bonding. As BCB has a low thermal conductivity, another important aspect is the simultaneous heating of both wafers. Any temperature mismatch between top and bottom heaters would result in asynchronous thermal expansion of the wafers, which may deteriorate alignment accuracy. Therefore simultaneous heating of top and bottom wafer is of highest importance for BCB aligned wafer bonding. A scanning acoustic microscopic image of 200 mm BCB bonded wafers is shown in Figure 3.18.

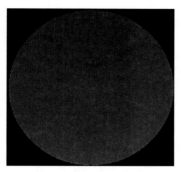

Fig. 3.18 Scanning acoustic microscopy (Sonix AW Vision 3000) of defect-free 200 mm wafers bonded with BCB layers (courtesy of EVG).

3.2.2.4 Hybrid wafer bonding

Several hybrids of those three methods can be proposed. One example is the simultaneous thermo-compression reaction of Cu-BCB as shown in Figure 3.19. In this approach, the original bonding process, consisting of pure Cu-Cu thermo-compression bonding, has been slightly modified by the inclusion of an additional patterned and compliant glue layer between stacked dies.[30] The polymer layer mechanically stabilizes the extremely thin top dies after bonding and carrier release, and also mechanically supports the stacked dies in areas where there are little or no electrically functional interconnections between the different chip levels. Hence, the inclusion of the dielectric layer enables stacking and electrical interconnection of dies with a highly non-uniform distribution of TSVs across the stack.

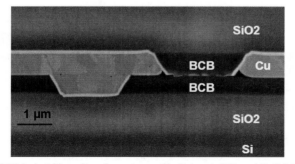

Fig. 3.19 Cu-BCB hybrid bonding (courtesy of J.-Q. Lu at RPI and Ref. 31).

3.3 ADVANCED CHIP-TO-WAFER BONDING

Due to its inherent advantages in terms of enabling high alignment accuracy and subsequently bonding of higher density of TSV interconnects, wafer-to-wafer bonding was outperforming the chip-to-wafer approach. However, wafer-to-wafer stacking schemes are inherently limited to stacking the dies with equal sizes. Moreover, when 2D device yield is limited, it leads to an important system yield loss.

The main alternative to wafer-to-wafer integration is chip-to-wafer integration. Chip-to-wafer stacking may be of more interest for the fabrication of heterogeneously integrated systems as it does not impose the requirement of equal die size. The method is also compatible with the selection of known good dies (KGDs) prior to stacking and, therefore, is of great interest in the cases where one of the components in the stacked system is a product with limited yield. However, the cost of chip-to-wafer stacking for most bonding methods is limited by the process throughput, especially when heat needs to be applied to achieve the reliable bond.

Chip-to-wafer bonding through metal-metal thermo-compression mechanisms such as Cu-polymer hybrid bonding (Figure 3.19) or Cu-Sn IMC bonding (Figure 3.11) is expected to be used for future heterogeneous chip stacking. In those bonding methods, as the diffusion rate is proportional to temperature and pressure, it is not economically feasible to perform the bonding process at a single-die level. Accordingly, advanced chip-to-wafer (AC2W) bonding is preferred for the applications with a lengthy process time. The basic idea behind AC2W is to split the bonding process into two sub-steps as shown in Figure 3.20. The temporary pre-bonding with alignment is performed on a typical pick-and-place machine, whereas the permanent bonding of the dies is performed as a batch process in a dedicated chip-to-wafer bond chamber.

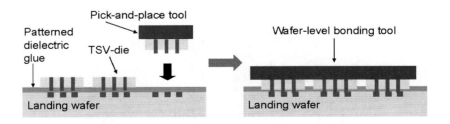

Fig. 3.20 Principle of advanced chip-to-wafer bonding (courtesy of EVG).

Today the chip alignment accuracy on the wafer depends highly upon the pick-and-place tool. Depending upon alignment type and tool supplier, it varies from 0.5 to 10 μm (3σ). The permanent bonding process, which is the second step, requires a controlled, homogeneous, perpendicular force to be applied on every single chip due to non-uniform distribution of KGDs on the base device wafer. A non-uniform KGD distribution can result in a possible shift in the center of gravity away from the actual center of the bottom substrate. This occurrence requires a controlled shift in the center of applied force during the bonding process in a chamber. In addition to the variable center of force, different process technologies require a different absolute force and a variable coverage of the bottom wafer with top chips. In any cases, the force application has to be controlled in such a way that only perpendicular forces are applied on each die. The bonding system also needs to be tolerant to the variation of chip thicknesses. Even though the chips might have different thicknesses, the force application still needs to be purely perpendicular and uniform. This is achieved by the use of a flexible compliant layer. Therefore, a true KGD stacking can be achieved through the control of the center position, the absolute value, and the direction of the applied force.

3.4 TEMPORARY BONDING AND DEBONDING

Another big issue in TSV integration is final device or wafer thickness. As semiconductor manufacturers continue to reduce the thickness of devices and wafers, innovative methods have to be developed to overcome the manufacturing challenges associated with new products and processes. Making reliable TSV interconnects for 3D-IC manufacturing requires a reduction in wafer thickness, which, when combined with a large wafer diameter, necessitates new wafer handling mechanisms.

The most promising and most widely investigated handling solution for thin or ultra-thin wafers is by using temporary bonding and debonding techniques utilizing a carrier wafer which provides sufficient mechanical support. Upon the device wafer being temporarily bonded to the carrier wafer, it is ready for backside processing including thinning, via processing and metallization. After completing the backside processing steps, the device wafer can be released from the carrier wafer and proceed to final packaging processes.

The carrier wafer, either glass or silicon, gives the device wafer mechanical stability and protects the wafer edge. In addition, the carrier wafer prevents the thin device wafer from bending or warping. A glass carrier has a longer history in the IC industry than a silicon carrier, but 3D-IC manufacturers are switching from the glass carrier to the silicon carrier due mainly to thermal issues. The silicon carrier has much better thermal conductivity than the glass carrier and better matching of the coefficient of thermal expansion (CTE) to the device wafer. The geometry of the wafer stack can be tailored in such a way that the stack mimics a standard wafer. All these effects enable device manufacturers to use standard fab equipment for further processing. The cost for the thin wafer handling with a carrier wafer is usually lower than that for dedicated thin wafer chucks, end-effectors and wafer cassettes on each individual machine.[32] After backside processing the adhesive force between the two wafers is released and the thin wafer is either sent to assembly or stacked on another wafer.

As illustrated in Figure 3.21, a typical process flow for temporary bonding first involves fully processing the device wafer on the front side. Subsequently, the carrier wafer and/or the device wafer is coated with an adhesive. Both wafers are then transferred to a bond chamber, where they are carefully centered and vacuum-bonded at an elevated temperature. The selection of the appropriate coating process and technology is influenced by the wafer topography that must be embedded by this material. Following temporary bonding, the wafer stack undergoes TSV backside processes. Finally, the thinned device wafer is debonded from the carrier wafer.

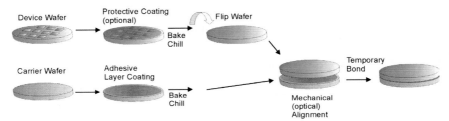

Fig. 3.21 Typical temporary bonding process flow (courtesy of EVG).

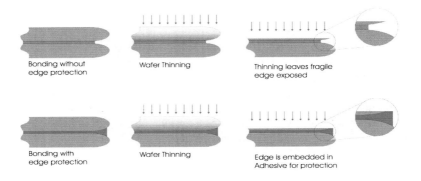

Fig. 3.22 Edge profile of wafers stacked with two different adhesives (courtesy of EVG).

There are a variety of different intermediate materials for temporary bonding. The two main classes are spin-on adhesives and laminated tapes. The choice of the correct intermediate material depends mainly on (a) chemical resistance of adhesive materials, (b) maximum operating temperature and thermal budget during following processes, and (c) protection of sharp wafer edge and coverage of surface topography. For TSV integration, the process based on a spin-on adhesive is becoming more preferred over that with a lamination tape due mainly to better edge protection (Figure 3.22), compatibility with topographic surfaces (Figure 3.23), and better stability at higher process temperatures.

Fig. 3.23 High topographic surface with bumped structures to be bonded onto a carrier wafer (courtesy of EVG).

The selection of the appropriate coating process and technology is influenced by the wafer topography that must be embedded by the adhesive material. When using BSI's WaferBond™ HT10.10 adhesive for wafer topographies less than e.g., 20 µm, a single spin-coating process is generally sufficient for coating, followed by a baking step to remove the solvent. For topographies greater than 20 µm, spin-coating both device wafer and carrier wafer are used to planarize and cover features. For topographies beyond e.g., 40 µm, such as wafers with bumps or solder balls as well as wafers with chips bonded onto their surfaces, spray coating is used to planarize and cover features.

For commercially available adhesives, there are three classes of debonding mechanisms; chemical release, thermal release and UV release. Chemical release, typically dissolution in solvent, has the main disadvantage that a thin wafer is floating uncontrolled in the solvent bath after debonding. The debonding of this type of adhesive can also be done using a perforated carrier through which solvent can dissolve the adhesive material. In the case of thermal release, the release temperature of thermal release materials can be higher than the maximum operating temperature, which is sometimes not compatible with the thermal budget of devices. UV release materials require a transparent carrier wafer, which increases the cost and has the disadvantage that the thermal expansion properties of device wafer and carrier wafer are different. This may result in the bow or warp of the stack. Additionally the thick carrier wafer can dominate the thermal expansion behavior of the whole stack. Therefore, a proper material and release method should be chosen depending upon applications and process flows. Tape materials are usually debonded by the wedge lift-off process, whereas spin-on materials are de-bonded by the slide lift-off process as depicted in Figure 3.24. A typical debonding process flow, starting with wafer stack debonding, followed by cleaning and unloading the thin wafer into the film frame is illustrated in Figure 3.25.

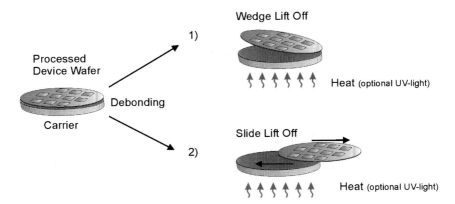

Fig. 3.24 Lift-off methods with heat or UV (courtesy of EVG).

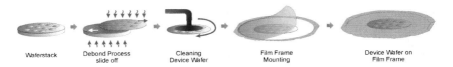

Fig. 3.25 Typical debonding process flow (courtesy of EVG).

In the IC manufacturing and packaging industry, there still exist several challenges with temporarily bonded wafers. For instance, bonded wafer stack creates challenges to robots, older front opening unified pods (FOUPs), wafer ports, notch alignment and other portions of fabrication chains. However, industry's preliminary study showed that temporary bonding and debonding has no significantly detrimental effects on device performances.[10-11] Joint development research among CEA-Leti, BSI and EVG demonstrated successful TSV creation for CMOS image sensor (CIS) applications using temporary bonding and debonding techniques. The main purpose of this research was to access the TSV creation on temporarily bonded complex active wafers. A device wafer was bonded to a carrier wafer using BSI's WaferBond™ HT10.10. The active wafers used were those of an ST Microelectronics CIS with a TSV diameter of 65 μm and the thicknesses of 70 or 120 μm (*i.e.*, an aspect ratio of 1:1 or 2:1, respectively). A wafer stack undergoes a series of backside processes comprising backside grinding, stress relief, via etching after lithography, PECVD insulation (with silane at 150°C for 3 min for 1:1 aspect ratio vias and with tetraethyl orthosilicate (TEOS) at 255°C for 3 min for 2:1 aspect ratio vias, respectively), via filling by electroplating, BCB passivation (at 200°C for ten hours for 1:1 aspect ratio vias and at 250°C for one hour for 2:1 aspect ratio vias, respectively), and solder bumping. The thinned wafers were debonded by slide-off at 180°C and cleaned with appropriate solvent on the automated EVG tools.

The preliminary results from that research showed that (a) temporary bonding and debonding offers time and cost efficiency for 3D packaging integration processes utilizing existing and established equipment and technologies, (b) the new adhesive meets all the requirements for bonding strength, chemical resistance, and thermal stability, (c) it was also confirmed that debonding and cleaning processes do not scratch the wafer or break the ultra-thin bottom of the TSVs, and (d) the electrical tests performed showed that via resistance values were the same for TSVs processed using a temporary bonding layer compared to permanent bonding. However, more studies are needed to verify the full impact of using thin wafer, based on temporary bonding and debonding, on the performances of major ICs at larger manufacturing scales, which is under way in the 3D-IC industry.

3.5 SUMMARY

We reviewed various bonding techniques for 3D-IC manufacturing with TSV interconnects, such as wafer alignment, wafer-to-wafer bonding, advanced chip-to-wafer bonding, and temporary bonding and debonding for thin wafer handling and processing. Due mainly to thermal, contamination and alignment issues, the bonding processes compatible with TSV-interconnected CMOS wafers are limited to such bonding methods as direct oxide bonding, metal bonding (Cu-Cu or Cu-solder-Cu), adhesive bonding, and several hybrids of those methods. Each method has its own benefits and disadvantages in terms of alignment accuracy, process time, throughput and reliability. For production purposes the individual processes of wafer alignment and bonding are separated into different process modules.

There are two types of optical alignment processes for wafer bonding, direct alignment and indirect alignment. In the case when the keys of both wafers are within the bond interface and at least one of the wafers is transparent for visible light or IR light, a direct alignment of those two wafers can be performed with live images. An indirect alignment method has to be used when wafers are not transparent and one wafer is aligned using the digitized image of the second wafer as a reference point. The SmartView® alignment method, the most advanced indirect alignment scheme for a face-to-face integration approach, uses dual microscopes with a common focal plane. This technique enables the use of any unique features on the front-side of non-IR transparent wafers for face-to-face alignment with the pre-bond alignment accuracy of 0.2 µm (3σ).

The main alternative to wafer-to-wafer integration is chip-to-wafer integration. This approach is compatible with the selection of KGDs prior to stacking and, therefore, is of great interest in the cases where one of the components in the stacked system is a product with limited yield. However, the cost of chip-to-wafer stacking for thermo-compression type bonding is limited by the process throughput, especially when heat needs to be applied for a lengthy process time to achieve reliable bonds. Therefore, advanced chip-to-wafer bonding is preferred for the applications with a long process time. The basic idea behind this technique is to split the bonding process into two sub-steps. The temporary pre-bonding with alignment is performed on a typical pick-and-place machine, whereas the permanent bonding of the dies is performed as a batch process in a dedicated chip-to-wafer bond chamber.

Another big issue in TSV integration is final device or wafer thickness. Making reliable TSV interconnects for 3D wafer stacking usually requires a reduction in wafer thickness, which, when combined with a large wafer diameter, necessitates new wafer handling mechanisms. The most promising and most widely investigated handling solution for thin or ultra-thin wafers is by using temporary bonding and debonding techniques utilizing a carrier wafer which provides sufficient mechanical support. Once the device wafer is temporarily bonded to the carrier wafer, it is ready for backside processing including thinning, via processing and metallization. After completing the backside processing steps, the device wafer can be released from the carrier wafer and proceed to final packaging processes. A variety of research is ongoing to define the impact of using thin wafers (temporarily bonded onto carrier wafers) on the overall process flows and device performances.

References

1. P. E. Garrou and E. J. Vardaman, *3D Integration at the Wafer Level*, TechSearch International, Austin, Texas, p. 4–8 (2006).

2. J. C. Eloy, *3D TSV Interconnects : Devices and Systems – 2008 Report*, Yole Development, Lyon, France, p. 5–13 (2008).

3. S. Cheramy, J. Charbonnier, D. Henry, A. Astier, P. Chausse, M. Neyret, C. Brunet-Manquat, S. Verrun, N. Sillon, L. Bonnot, X. Gagnard, and J. Vittu, "3D integration process flow for set-top box application: description of technology and electrical results", ISBN 978-1-4244-4722-0 IEEE, in *Proc. European Microelectronics and Packaging Conference (EMPC)* (2009).

4. T. Jiang and S. Luo, "3D integration – present and future", ISBN 978-1-4244-2118-3 IEEE, in *Proc. 10ᵗʰ Electronics Packaging Technology Conference (EPTC)* (2008).

5. P. Lindner, "The name's bonding, wafer bonding", *EuroAsia Semiconductor*, vol. 30, no. 12, p. 15, December 2008/January 2009.

6. P. Lindner, V. Dragoi, S. Farrens, T. Glinsner, and P. Hangweier, "Advanced techniques for 3D devices in wafer-bonding processes", *Solid State Technology*, June, p. 55 (2004).

7. P. Leduc, M. Assous, L. Di Cioccio, M. Zussy, T. Signamarcheix, A. Roman, M. Rousseau, S. Verrun, L. Bally, D. Bouchu, L. Cadix, A. Farcy, and N. Sillon, "First integration of Cu TSV using die-to-wafer direct bonding and planarization", ISBN 978-1-4244-4511-0 IEEE, in *Proc. IEEE International 3D Systems Integration Conference* (2009).

8. F. Niklaus, G. Stemme, J. Q. Lu, and R. J. Gutman, "Adhesive wafer bonding", *J. Appl. Phys.*, vol. 99, 031101, p. 1–28 (2006).

9. D. Lu and C. P. Wong, *Materials for Advanced Packaging*, p. 219, Springer, New York (2008).

10. A. Jouve, S. Fowler, M. Privett, R. Puligadda, D. Henry, A. Astier, J. Brun, M. Zussy, N. Sillon, J. Burggraf, and S. Pargfrieder, "Facilitating ultrathin wafer handling for TSV processing", ISBN 978-1-4244-2118-3 IEEE, in *Proc. 10th Electronics Packaging Technology Conference (EPTC)* (2008).

11. J. Charbonnier, S. Cheramy, D. Henry, A. Astier, J. Brun, N. Sillon, A. Jouve, S. Fowler, M. Privett, R. Puligadda, J. Burggraf, and S. Pargfrieder, "Integration of a temporary carrier in a TSV process flow", in *Proc. Electronic Components and Technology Conference (ECTC)* (2009).

12. V. Dragoi, P. Lindner, M. Tischler, and C. Schaefer, "New challenges for 300 mm Si technology: 3D interconnects at wafer scale by aligned wafer bonding", in *Proc. Mat. Sci. in Semicond.*, vol. 5 (4–5), p. 425 (2003).

13. J. M. Koo, B. Q. Vu, Y. N. Kim, J. B. Lee, J. W. Kim, D. U. Kim, J. H. Moon, and S. B. Jung, *J. Electron. Mat.*, vol. 37, p. 118 (2007).

14. P. E. Garrou and E. J. Vardaman, *3D Integration at the Wafer Level*, TechSearch International, Austin, Texas, p. 27 (2006).

15. A. Fan, A. Rahman, and R. Reif, *Electrochemical and Solid-State Lett.*, vol. 2, p. 534 (1999).

16. P. Morrow, C. M. Park, S. Ramanathan, M. J. Kobrinsky, and M. Harmes, "Three-dimensional wafer stacking via Cu-Cu bonding integrated with 65-nm strained Si/low-k CMOS technology", *IEEE Electron Device Lett.*, vol. 27, no. 5, p. 335 (2006).

17. W. H. Teh, C. Deeb, J. Burggraf, M. Wimplinger, T. Matthias, R. Young, C. Senowitz, and A. Buxhaum, "Recent advances in submicron alignment 300 mm copper-copper thermocompressive face-to-face wafer-to-wafer bonding and integrated infrared, high-speed FIB metrology", *to be presented and published at 2010 International Interconnect Technology Conference (IITC)* (2010).

18. B. Kim, E. Cakmak, T. Matthias, M. Wimplinger, P. Lindner, E. J. Jang, J. W. Kim, Y. B. Park, S. Hyun, and H. J. Kim, in *Proc. 6th International Wafer Level Packaging Conference (IWLPC)*, p. 122 (2009).

19. Z. Huang, Z. Suo, G. Xu, J. He, J. H. Prévost, and N. Sukumar, "Initiation and arrest of an interfacial crack in a four-point bend test", *Eng. Fracture Mech.*, vol. 72, p. 2584 (2005).

20. E. J. Jang, J. W. Kim, B. Kim, T. Matthias, H. J. Lee, S. Hyun, and Y. B. Park, "Effect of N_2+H_2 forming gas annealing on the interfacial bonding strength of Cu-Cu thermo-compression bonded interfaces", *J. Microelectronics & Packaging Soc.*, vol. 16, no. 3, p. 31 (2009).

21. E. J. Jang, S. Pfeiffer, B. Kim, T. Matthias, S. Hyun, H. J. Lee, and Y. B. Park, "Effect of post-annealing conditions on interfacial adhesion energy of Cu-Cu bonding for 3-D IC integration", *Kor. J. Mater. Res.*, vol. 18, no. 4, p. 204 (2008).

22. E. J. Jang, Y. B. Park, S. Pfeiffer, B. Kim, and T. Matthias, "Effect of post-annealing conditions on Cu-Cu wafer bonding characteristics", *J. Kor. Phy. Soc.*, vol. 54, no. 3, p. 1278 (2009).

23. T. H. Kim, M. M. R. Howlader, T. Itoh, and T. Suga, "Room temperature Cu-Cu direct bonding using surface activated bonding method", *J. Vac. Sci. Technol.*, vol. 21, Issue 2, p. 449 (2003).

24. P. Gueguen, L. Di Cioccio, F. Rieutord, P. Gergaud, J. P. Barnes, M. Rivoire, D. Scevola, M. Zussy, A. M. Charvel, L. Bally, D. Lafond, and L. Clavelier, "Copper direct-bonding characteristics and its interests for 3D integration circuits", *J. Electrochem. Soc.*, vol. 156, Issue 10, p. H772 (2009).

25. V. Dragoi, S. Farrens, and P. Lindner, "Low temperature MEMS manufacturing processes: plasma activated wafer bonding", in *Proc. MRS Series*, vol. 872, p. J7.1.1. (2005).

26. V. Dragoi, G. Mittendorfer, C. Thanner, and P. Lindner, "Wafer-level plasma activated bonding: new technology for MEMS fabrication", *J. Microsystem Tech.*, vol. 14 (4–5), p. 509 (2008).

27. Y. Kwon, A. Jindal, J. J. McMahon, J. -Q. Lu, R. J. Gutman, and T. S. Cale, "Dielectric glue wafer bonding for 3D ICs", in *Proc. MRS Series*, vol. 766, p. E5.8.1 (2003).

28. J.-Q. Lu, A. Kumar, Y. Kwon, E.T. Eisenbraun, R.P. Kraft, J.F. McDonald, R.J. Gutmann, T.S. Cale. P. Belemjain, O. Erdogan, J. Castracane, A.E. Kaloyeros, "3-D Integration Using Wafer Bonding", in Advanced Metallization Conference 2000 (AMC 2000), MRS Proc. Vol. V16, Eds. D. Edelstein, G. Dixit, Y.

29. F. Niklaus, R. J. Kumar, J. J. McMahon, J. Yu, T. Matthias, M. Wimplinger, P. Lindner, J. -Q. Lu, T. S. Cale, and R. J. Gutman, "Effects of bonding process parameters on wafer-to-wafer alignment accuracy in benzocyclobutene (BCB) dielectric wafer bonding", in *Proc. MRS Series*, vol. 863, p. B10.8.1 (2005).

30. J. J. McMahon, J. -Q, Lu, and R. J. Gutman, "Wafer bonding of damascene-patterned metal/adhesive redistribution layers for via-first three-dimensional (3D) interconnect", in *Proc. IEEE Electronic Components and Technology Conference (ECTC)*, p. 331 (2005).

31. J.-Q. Lu, J.J. McMahon and R.J. Gutmann, "Via-First Inter-Wafer Vertical Interconnects utilizing Wafer-Bonding of Damascene-Patterned Metal/Adhesive Redistribution Layers," *Proc. of 3D Packaging Workshop at IMAPS Device Packaging Conference*, paper # WP64, Scottsdale, AZ, March 20–23 (2006).

32. D. Kharas and N. Sooriar, "Cycle time and cost reduction benefits of an automated bonder and debonder system for a high volume 150 mm GaAs HBT back-end process flow", in *Proc. CS MANTECH Conference* (2009).

Chapter 4

TSV ETCHING

Paul Werbaneth

Tegal Corporation

Through Silicon Via (TSV) Etching is a key process fabrication module employed in 3D IC technologies. While there are yet to be significant volumes of commercially available TSV-based packaged devices in silicon IC technologies today, there are scores of TSV development activities currently underway within programs organized by semiconductor capital equipment consortia, university research centers, Integrated Devices Manufacturers, packaging houses, and various national and international government-sponsored efforts. Many 3D technology topics require substantial new progress be made on difficult areas of process technology, for example low-temperature deposition of dielectric films in silicon vias for Via-Last fabrication schemes, TSV etching, however, benefits greatly from the large body of knowledge readily available from two important sources. For deep etching of features and structures in silicon, the plasma etch processes developed for Microelectromechanical (MEMS) device fabrication can be readily ported over to the TSV etch module for 3D technology platforms. And, for a general understanding of High Volume Manufacturing (HVM) requirements and challenges for through-wafer processing technologies, the Compound Semiconductor industry has routinely and successfully employed backside via structures for many years. This chapter reviews the work already performed in TSV etching by the Compound Semiconductor industry, then examines deep reactive ion etching (DRIE) processes as developed for MEMS device fabrication. Following this review, the chapter presents an analysis

3D Integration for VLSI Systems
Edited by Chuan Seng Tan, Kuan-Neng Chen and Steven J. Koester
Copyright © 2012 by Pan Stanford Publishing Pte. Ltd.
www.panstanford.com

of the specific requirements of TSV etching for silicon device 3D technology, and then concludes with a look forward to TSV etching requirements visible on the near-term horizon.

4.1 THROUGH WAFER VIA ETCHING IN COMPOUND SEMICONDUCTOR DEVICE FABRICATION

Here is a familiar sequence of processing steps:

- A wafer is attached backside-up to a carrier substrate using a high temperature thermal-plastic adhesive and a vacuum bonding machine.
- Wafer thickness measurements are collected on the bonded wafer and the wafer then undergoes a mechanical thinning process followed by a chemical polish to remove any grind damage to the backside surface.
- A photo pattern incorporating an optical edge bead is applied to the wafer backside.
- Through-substrate vias are created using an ICP dry etch process, followed by a low temperature photoresist ash.
- Adhesion and seed metal layers are deposited in the through substrate vias.
- A thick film of metal is electroplated in the through-substrate vias.
- The wafer is demounted from the substrate carrier and passes on to die singulation.

Familiar as this process flow seems, these processing steps do not describe a manufacturing sequence performed during 3D silicon device processing in 2009; instead, they describe the operations performed for GaAs Heterojunction Bipolar Transistor (HBT) fabrication in HVM ca. 2001. [1,2]

Through-substrate via technology was developed for GaAs device processing for several different reasons, including the need for low resistance / low impedance grounds for very high frequency (30 GHz to 110 GHz) RF devices made in GaAs, as well as providing thermal dissipation paths for heat generated by operation of these devices.[3]

GaAs through-substrate vias typically have dimensions on the order of 100μm deep (thinned wafer - full thickness) by 60μm to 80μm in diameter. [4,5] For a variety of reasons, including lower chemical costs, ease of automation, and, most importantly, the ability to precisely control the dimensions and profiles of the GaAs vias, which has a direct impact on high frequency circuit performance, plasma etching GaAs vias is the technology of choice in the compound semiconductor industry.

Standard plasma etching processes for GaAs through-substrate vias are based on Inductively Coupled Plasma (ICP) reactors running BCl3 / Cl2 chemistries etching GaAs wafers that have been transported into the ICP plasma tool on sapphire or quartz wafer carriers. Maintaining control over wafer temperature is important in GaAs through-wafer via etching, so the ICP tools are typically equipped with electrostatic clamps and other cooling hardware in order to promote heat transfer from the etch surface through the

GaAs wafer, the adhesive used to bond the wafer to the carrier substrate, and then through the carrier itself.

GaAs etch rates >6μm/min are commonly achieved, with good selectivity to photoresist masks, effective control of both vertical and tapered via profiles, and minimal amounts of etch residue. [6] An SEM micrograph of a representative result for a GaAs via is shown in Fig. 4.1.

Fig. 4.1 SEM micrograph of etched GaAs via with tapered / vertical etch profile.

So, with a robust and extensive manufacturing base of through-wafer via plasma etch experience and knowledge readily available and portable from commercial GaAs manufacturing, what prevents the same ICP etch tools and process chemistries from being used for through-silicon via etching? Answer: the etch rate "ceiling" of silicon etching in chlorine-based plasmas.

4.2 SILICON ETCHING IN CHLORINE PLASMAS

The etching of silicon in a chlorine plasma environment, where the silicon surface is subject to ion bombardment, is determined by the action of three components: a sputter etch component, a thermal or "spontaneous" etch component, and an ion-enhanced etch component. [7] Typically, plasma etching of silicon for applications like Shallow Trench Isolation (STI) formation is carried out at moderate (20-80°C) temperatures, at which the spontaneous etch rate of silicon by atomic chlorine is negligible. [8] Ion-assisted or ion-enhanced etch is the dominant mechanism by which the etch front proceeds into the silicon surface in any plasma reactor in which there is an adequate supply of halogen reactant. Physical sputtering of silicon can also account for the removal of an appreciable volume of silicon at higher ion energies, although care must be taken to avoid the damage that might result from high energy ions impinging on the device wafer.

The volatile products evolved during the etch of silicon in a Cl_2-Ar plasma will be some combination of silicon and the reactive halogen atom, along with atomic silicon sputtered from the mixed layer of adsorbed chlorine and silicon from the substrate. The silicon chloride etch product, $SiCl_x$, will be distributed among several possible forms: $SiCl$, $SiCl_2$, $SiCl_3$, and $SiCl_4$, for example, with $SiCl_4$ being the fully saturated product molecule.

The exact etch product mix will be a function of several factors. Chiefly, for the system considered here (silicon etching in a Cl_2-Ar plasma), the distribution of etch products is a function of incoming ion bombardment energy and the neutral/ion flux ratio at the etch surface. As ion bombardment energy is decreased, $SiCl_4$ production is favored over $SiCl_2$; or, as ion bombardment energy is increased sufficiently, the etch product desorbed from the silicon surface will transition further from $SiCl_2$ to $SiCl$ and even Si. [9,10] ($SiCl_3$ does not seem to be observed much.)

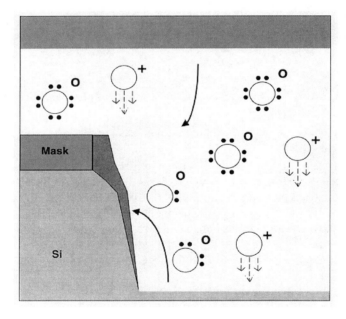

Fig. 4.2 Local environment of trench etched into silicon using chlorine-based etching plasma

The neutral/ion flux ratio at the silicon surface has an effect similar to ion bombardment. Running with a very low neutral/ion flux ratio will favor the path of silicon removal from the etch surface by sputtering. Increasing the ratio will favor in turn the evolution of $SiCl$, $SiCl_2$, and $SiCl_4$ etch product species. [11]

The precise mixture of etch product species can have a profound effect on silicon etch profile evolution. Etch product redeposition along the evolving feature sidewall can result in tapering of the trench as the trench develops into the silicon. [12] Consider the local environment depicted in Fig. 4.2.

As etch product evolves along the silicon etch front and desorbs from the surface under the influence of ion bombardment, there will be a possibility that, depending on the "stickiness" of the product molecule $SiCl_x$ (or of the sputtered Si), there will be redeposition on the wall of the etched feature. Stickiness is represented as the sticking coefficient, S_p, which can range from a value of 0 (no propensity to stick) to 1 (guaranteed to stay in place). At wafer temperatures typical of commercial plasma etch reactors, 20-80°C, the sticking coefficient of a sputtered silicon atom is near 1. SiCl and $SiCl_2$ sticking coefficients range between 0.1 and 0.3 and $SiCl_4$ has a sticking coefficient estimated at ≤0.002. [13,14]

Figure 4.3 is an example of a completed STI etch with a complex profile created by controlling the relative amount of etch product redeposition created during various phases of the chlorine-based plasma etch.

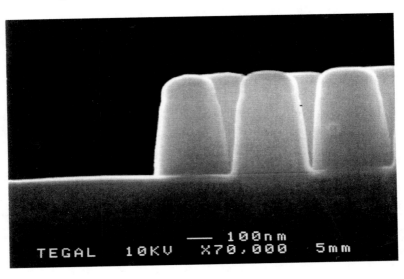

Fig. 4.3 STI features with complex profiles etched into silicon using chlorine-based etching plasma

Why then, with proven supply chain components for through-substrate via processing in compound semiconductor fabrication, using the chlorine-based plasma etch chemistries that are so effective etching silicon trench structures for shallow trench isolation, do we come up short when it comes to porting all this over to through-silicon via etching? The basic answer is the throughput limitation imposed on TSV etching as a result of the chlorine etch rate "ceiling."

In their paper "Comparison of Cl2 and F-based dry etching for high aspect ration Si microstructures etched with an inductively coupled plasma source," Tian, Weigold, and Pang state: [15]

> Another widely used gas to etch Si is Cl2. With Cl2 etching, many processes were developed for producing vertical sidewalls, smooth surface morphology, fine critical dimension control, and high aspect ratio microstructures for MEMS. The main advantage of Cl2 etching is that the etching is anisotropic since it is an ion assisted process rather than a spontaneous etching process. Vertical profile can be easily obtained in Cl2 etching without polymer passivation or extensive water cooling. The disadvantages of this approach are the lower etch rates and lower selectivity to etch mask compared to F-based etching.

> For bulk micromachining of MEMS, where large portions of the substrate must often be etched, the F-based etching can be used to form deep trenches with high etch rate and high selectivity to a mask.

> However, for larger features the lower etch rate and selectivity ([for CL2 etching] compared to F-based etching make it harder to etch very deep trenches in a reasonable time with a practical mask thickness.

4.3 SILICON ETCHING IN FLUORINE PLASMAS

Silicon etches readily in plasma reactors running fluorine-based (e.g. SF_6, CF_4) etch chemistries. The reaction of silicon (a solid) with atomic fluorine (a gas) and with fluorine-containing species (gases) produced by the etching plasma to form silicon fluoride etch products (gases, or high vapor pressure compounds) proceeds as a result of several different etch mechanisms, which include both spontaneous and ion-enhanced mechanisms. [16] Whether fluorine is etching silicon purely spontaneously, or whether there is some ion-enhanced component to the etch, the end result is almost always rapid (>1.0 µm/min) silicon etching with a propensity to etch isotropically, i.e. producing an undercut from the original etch mask dimensions.

There are several different ways to obtain the high etch rates of fluorine-based silicon etching without sacrificing control over the silicon etch profile. In cryogenic etch processes, the wafer temperature is held to the range of -120°C to -80°C during etch. At this very low temperature, fluorine-based plasma chemistries to which oxygen have been added create thin layers of silicon-oxygen-fluorine passivation on the sidewalls of the etched feature, preventing lateral, or isotropic, etching, and allowing etch mask dimensions to be preserved. The advantages of cryogenic silicon etch processes include low sidewall roughness, fast etch rates, and high selectivities to either photoresist or silicon dioxide etch masks. [17]

The second, and by far most widely used, fluorine-based etch process for deep silicon etching is a cyclical (or time-multiplexed) process of alternating etch (etching with SF_6, or other fluorine-rich reactant gases) and deposition steps (deposition with C_4F_8, or other polymer-forming gases) known as the Bosch process: [18]

A method of anisotropic plasma etching of silicon to provide laterally defined recess structures therein through an etching mask employing a plasma, the method including anisotropic plasma etching in an etching step a surface of the silicon by contact with a reactive etching gas to removed material from the surface of the silicon and provide exposed surfaces; polymerizing in a polymerizing step at least one polymer former contained in the plasma onto the surface of the silicon during which the surfaces that were exposed in a preceding etching step are covered by a polymer layer thereby forming a temporary etching stop; and alternatingly repeating the etching step and the polymerizing step. The method provides a high mask selectivity simultaneous with a very high anisotropy of the etched structures.

4.4 BOSCH PROCESS ETCHING FOR MEMS DEVICE FABRICATION

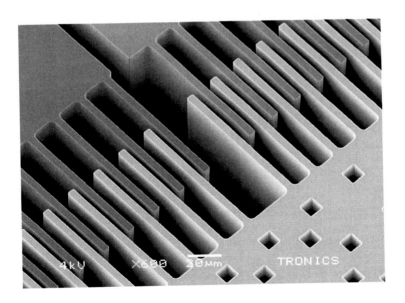

Fig 4.4 MEMS device (silicon gyroscope) etched using the Bosch process.

Here is a partial list of MEMS devices currently in commercial HVM for which Bosch process etching of silicon structures is an integral part of the device fabrication sequence: two- and three-axis accelerometers for consumer and industrial applications; silicon gyroscopes for anti-rollover systems,

GPS navigation, and vehicle stability control in the automotive market; pressure sensors for tire pressure monitoring and for medical instrument applications; inkjet print heads for desktop color printing; image sensors for visible and infrared imaging applications; and electro-acoustic filters for cell phone handsets and other wireless communication applications.

Figure 4.4 is an example of a typical MEMS device silicon structure created using the Bosch process to etch deeply into the silicon.

Added together, these commercial MEMS device shipments total billions of units shipped per year by companies that include Hewlett-Packard, STMicroelectronics, Robert Bosch, Avago Technologies, Analog Devices, Seiko Epson, Denso, and Tronics.

The Bosch process is also used in several semiconductor IC fabrication processes, including deep silicon trench isolation in Power Device fabrication, and for creating three-dimensional capacitors for passive devices like electrostatic discharge (ESD) protection circuits.

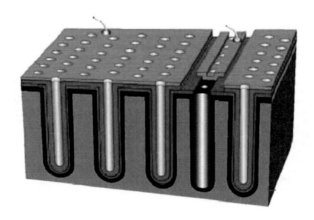

Fig. 4.5 Passive device three-dimensional capacitor created using the Bosch process

Figure 4.5 shows a pictorial representation of a three-dimensional capacitor (Bosch process silicon etching) for a passive device employed in an ESD protection circuit.

Added together, these Bosch-process based commercial MEMS, power, and passive device shipments total billions of units shipped per year.

4.5 DEEP SILICON ETCHING USING THE BOSCH PROCESS: HARDWARE AND PROCESS CONSIDERATIONS

There are several different kinds of plasma reactor technology used in plasma etch processing for commercial microelectronic device fabrication, and there are many different semiconductor capital equipment tool suppliers offering multiple examples within those various reactor technologies. Generally speaking, single wafer plasma etch reactors using capacitively coupled plasma sources were the first commercially important single wafer plasma etching tools employed in commercial semiconductor fabrication, followed by various examples of magnetically-enhanced plasma etching systems. Magnetic enhancement of capacitively coupled plasmas creates plasmas with higher plasma densities, density here being a term that incorporates both the degree to which the electrically neutral non-dissociated etch gases introduced into the plasma reactor become dissociated and ionized.

In a kind of "space race" among plasma etch tool manufacturers, the quest for ever denser plasmas led to several important plasma reactor developments, including the introduction of Electron Cyclotron Resonance (ECR) plasma reactors, Inductively Coupled Plasma (ICP) reactors and their variants, and plasma etch reactors based on helical resonator plasma sources.

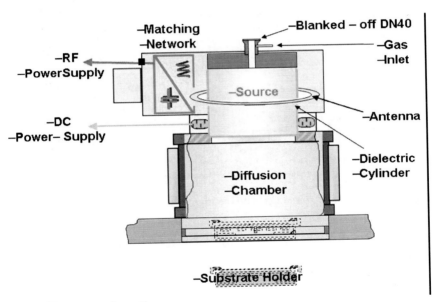

Fig. 4.6 ICP reactor schematic

Some general requirements for plasma sources are their capability to produce high plasma densities at low gas pressures, since high plasma densities are required in order to ensure high ion fluxes are seen at the wafer for producing adequate etch rates, and therefore, adequate system throughput. The low gas pressures are useful in that scattering collisions seen by ions crossing the plasma sheath, from the bulk plasma to the wafer are reduced (mean free paths increases) as operating pressures are reduced. On-axis ions (ions striking perpendicular to the wafer surface) are important to etching high aspect ratio structures. Also, in order to control and reduce plasma damage possibly created by energetic ion bombardment of the wafer surface, plasma sources capable of producing low plasma potential plasmas have great benefits. [19]

Inductively couple plasma sources are well-qualified for all the important plasma reactor considerations described above. [20] A schematic representation of a typical ICP reactor used for Bosch process etching of deep silicon structures is show in Fig. 4.6.

Figure 4.7 shows a cluster configuration plasma etch tool in which one (of a possible three total) ICP process module has been installed.

Fig. 4.7 ICP reactor mounted in cluster configuration plasma etch system

Here, the plasma source generates a high density inductively coupled plasma by coupling 13.56 MHz radio frequency power through an antenna and dielectric isolation cylinder into the source chamber. Efficient and reproducible inductive power coupling is ensured through the use of an rf matching network in-line between the rf generator and the source antenna.

Process gases are introduced into the top of the plasma source chamber, where they are dissociated and ionized as a result of interaction with the inductively coupled rf power. Vacuum pressure in ICP plasma systems is typically maintained by using turbomolecular vacuum pumps, which can achieve the desired conditions of high process gas flows and low (mTorr) operating pressure. In this example, the turbomolecular pump is located below (out of sight) the diffusion chamber and substrate holding, creating a net gas flow from the gas inlet, at the top of the reactor, through the active plasma zone in the source chamber and through and across the etching wafer, located on the substrate holder at the bottom of the diffusion chamber.

Unreacted gas, and etch products, are then removed from the ICP reactor by the turbo pump below the wafer.

Since the Bosch process (or other time-modulated deep reactive etch processes for silicon) is composed of rapidly (on the order of seconds, or tenths of seconds) alternating process steps where SF_6, typically, is used to etch silicon and C_4F_8, typically, is used to passivate the etched silicon feature sidewalls, fast-response mass flow controllers (MFCs) are used in the gas delivery system feeding the ICP reactor. Fast-response MFCs are one hardware component that differentiates Bosch process ICP systems from ICP systems used for other plasma etch applications.

The silicon wafer being etched in an ICP reactor running the Bosch process for deep silicon etching sees a significant heat load. In addition to the energy the wafer receives as a result of momentum transfer from energetic ions bombarding the front surface of the wafer during the etch, there will also be heating contributions from plasma radiation reaching the wafer surface, and from hot surfaces within the reactor, particular for ICP reactors equipped with heated reactor liners.

Beyond the wafer heating contributors described above, it is the etching process itself which is the single largest contributor to the total heat load a wafer sees during Bosch process etching. Overall, silicon etches in fluorine containing plasmas according to:

$$Si(solid) + 4F(gas) = SiF_4(gas) + \Delta H$$

This chemical reaction is vigorously exothermic: ΔH (the heat of formation of SiF_4) is -386 kcal/mole (or -1615 kJ/mole), which means a great deal of energy is liberated for every mole of SiF_4 formed during the etch. For comparison, in the case of etching silicon with a chlorine-containing plasma, the equivalent chemical reaction is:

$$Si(solid) + 4Cl(gas) = SiCl_4(gas) + \Delta H$$

Where ΔH, the heat of formation of $SiCl_4$, is -158kcal/mole. Given that silicon will etch much more rapidly in Bosch process (fluorine) plasmas than in chlorine plasmas, and given that etching silicon with fluorine liberates considerably more energy per product molecule formed than is produced when etching silicon with chlorine, the engineering required to ensure good control of wafer temperatures during long Bosch process etches is not trivial.

The multiple roles played by the substrate holder (wafer electrode) now become much clearer. First, the substrate holder must be capable of repeatable and reliable mechanical wafer transport, receiving the wafer from the transfer robot and presenting the etched wafer back to the transfer robot at the end of the etching process. Second, the substrate holder needs to be able to couple rf or low frequency bias power into the wafer. Wafer bias, either rf or low frequency, is the means by which energetic ions are extracted from the plasma to the wafer surface; depending on how the plasma etching process has been optimized, these ions become an important, if not critical, component of etching deep features into silicon structures.

Fig. 4.8 Mechanically clamped wafer substrate holder for ICP reactor.

Fig. 4.9 Electrostatically clamped wafer substrate holder for ICP reactor

Finally, the substrate holder needs to be able to control wafer temperature uniformly across the wafer to a given wafer temperature setpoint, all the while dealing with the significant amount of heat generated by the (exothermic) Bosch process etch. Typically, the wafer is clamped, either mechanically, or with an electrostatic clamp, to the wafer electrode / substrate holder, while helium flows between the wafer backside and the temperature controlled electrode. Examples of mechanically-clamped and electrostatically clamped (ESC) substrate holders are shown in Figs. 4.8 and 4.9.

An interesting sidenote: since, by definition, through-silicon via etch processes etch mostly through, if not completely through, the silicon wafer, if the wafer edge is exposed during the TSV etching process the net effect will be to either reduce the total wafer diameter during the TSV etch (etch completely through the wafer) or, to create a zone of thin, brittle silicon along the periphery of the wafer, which is undesirable from a yield standpoint. Mechanical clamps typically shield the wafer edge from etching by the plasma (impacting yield as a result of the clamp "shadow"), but ESCs, by design, expose the entire wafer surface to the etching plasma. Since most customers use photoresist edge bead removal processes before the DRIE step, wafers etched with ESC substrate holders may need to be protected along their edge by some kind of deliberate shadowing ring implemented in the mechanical design of the ICP reactor.

There are several general trends about Bosch process DRIE of silicon that can be observed in Fig. 4.10.

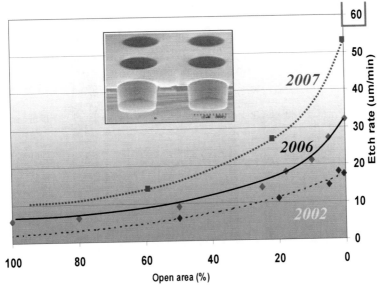

Fig. 4.10 Bosch process silicon DRIE etch rate and open area trends.

First, as is typical of plasma etch processes whose rates are dependent on the amount, or flux, of available reactive species delivered to the wafer, the plasma tends to "load" as the open area, here, the amount of silicon exposed to the plasma, increases. For any of the curves above the silicon etch rates for very small exposed silicon areas, say 5%, are 5× to 10× higher than are the silicon etch rates for large, say 80%, exposed silicon.

Through-silicon via patterns tend toward low exposed areas, on the order of 10% exposed silicon or less, which is a great natural gift resulting, as the low exposed areas do, in high silicon etch rates for the deep etches inherent in creating through-silicon via structures.

The second trend observable in Fig. 1.10 is the great progress plasma etch tool makers have made in increasing Bosch process silicon etch rates for a given exposed area of silicon. In the small exposed area range, ICP tool process and hardware improvements, for example higher power ICP sources, higher gas throughput, and better wafer temperature control, have resulted in a 3× to 4× increase in effective silicon etch rates over the space of five years.

Beyond work on silicon etch rates, Bosch process etching of through-silicon vias must also be optimized for sidewall profile and sidewall smoothness.

Figure 4.11 is an SEM micrograph of an etched silicon feature, in cross-section, showing the sidewall "scallops" characteristic of Bosch process etching.

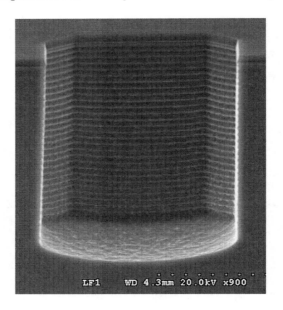

Fig. 4.11 SEM micrograph of etched silicon feature showing characteristic Bosch process scallops

Achieving the right amount of sidewall roughness in deep silicon etching is a matter of balancing the etching and passivation sequences in the time-multiplexed (Bosch process) silicon etch. [21]

Figure 4.12 presents the characteristic relationship between silicon etch rate (controlled by the relative amount of time the time-modulated silicon etching process spends in any single etch sequence) and scallop depth.

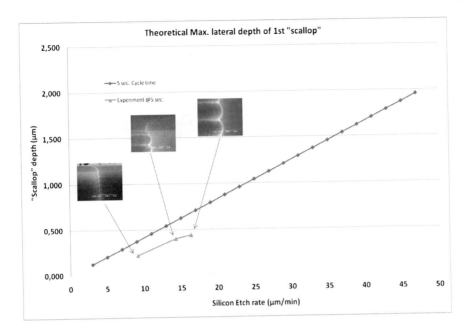

Fig. 4.12 Scallop depth (sidewall roughness) versus silicon etch rate for a time-multiplexed silicon etch process.

4.6 ADVANCED PROFILE CONTROL IN TSV ETCHING

There is a competing set of TSV profile requirements in 3D technology applications. On the one hand, sloped TSV profiles allow for more flexibility in the post-etch via filling process steps, where some combination of dielectric isolation film, seed layer / barrier layer metal, and metal (tungsten or copper) depositions occur. On the other hand, however, when TSV packing density is an item of concern, tapered via profiles work against via density, since the tapered profiles necessarily consume more real estate at the top surface of the via. The real estate issue is depicted in the following two figures, Figs. 4.13 and 4.14, showing ample via-to-via spacing in the case of vertical TSV sidewalls (Fig. 4.13), and relatively tight spacing in the case of tapered vias (Fig. 4.14).

The TSV etch process is equally accommodating for producing either vertical or tapered vias; it is perhaps more up to the process integration engineers to determine exactly where the best balance between TSV profiles and TSV packing density lies.

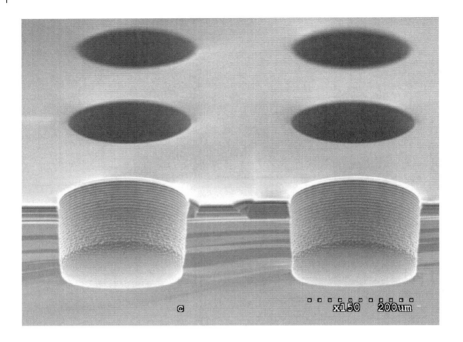

Fig. 4.13 Generic array of TSV structures with vertical via profiles, showing ample via-to-via spacing.

Fig. 4.14 Generic array of TSV structures with sloped via profiles, showing marginal vie-to-via spacing.

4.7 TSV ETCHING FOR 3D TECHNOLOGIES: PROCESS PERFORMANCE ROADMAP

Table 4.1 summarizes the process performance roadmap expected for TSV etching over the next several years, with the calendar year 2010 process results being typical of TSV etching performance being currently demonstrated in production (image sensor) TSV fabrication processes, or being demonstrated now in advanced development / pilot production (stacked memory) manufacturing processes.

Table 4.1 TSV Etching Process Performance Roadmap.

3D Market	Application	Open Area	Process Requirement											
			Etch Rate (um/min)			Uniformity (%)			Roughness (nm)			Tilt (deg)		
			2010	2011	2012	2010	2011	2012	2010	2011	2012	2010	2011	2012
Image Sensor TSV	Tapered Street	10%	12	18	20	5	3	2	1000	1000	1000			
	Tapered Via	10%	10	12	15	3	2	2	1000	1000	1000			
	Vertical Via	10%	12	15	20	3	2	2	150	100	100			
Stacked Memory TSV	Via First Before FEOL	<5%	4	5	8	<1	<1	<1	80	50	50	0.5	0.3	0.2
	Via First Before BEOL	<5%	13	18	25	<3	<2	<2	150	100	100			
	Via Last After BEOL	<5%	20	25	40	<3	<2	<2	300	150	150			

We can expect that innovation will continue to occur with Bosch (or other time-multiplexed, or cryogenic) plasma etch processing, given the many engines of innovation currently at work on TSV etch process optimization for 3D technologies.

4.8 TSV ETCHING FOR 3D TECHNOLOGIES: CONCLUSIONS

Through Silicon Via (TSV) Etching is a key process fabrication module employed in 3D IC technologies that benefits greatly from the large body of knowledge readily available from two important sources. For deep etching of features and structures in silicon, the plasma etch processes developed for Microelectromechanical (MEMS) device fabrication can be readily ported over to the TSV etch module, and, for a general understanding of High Volume Manufacturing (HVM) requirements and challenges for through-wafer processing technologies, the Compound Semiconductor industry has routinely and successfully employed through-substrate via structures for many years. This chapter reviewed the work already performed in TSV etching by the Compound Semiconductor industry, and then examined deep reactive ion etching (DRIE) processes as developed for MEMS device fabrication. Following this review, the chapter presented an analysis of the specific requirements of TSV etching for silicon device 3D technology, including hardware and process considerations for the widely-used Bosch process for deep silicon etching, reviewed some aspects of advanced profile control in TSV etching, and then concluded with a look forward to TSV etching developments visible on the near-term horizon roadmap. Given the rapid progress being made in TSV etching, there are likely to be exciting developments being reported month-by-month as the field of TSV etching for 3D technologies continues to evolve. The basics of TSV plasma etching, though, will likely remain the same, and were, hopefully, usefully covered in this short chapter.

Acknowledgments

The author thanks the many important contributions to this chapter provided by his colleagues past and present, including Michel Puech, Yannick Pilloux, Franck Miguel-Torres, Nicolas Launay, John Almerico, Genevieve Beique, and Robert Ditizio. Also, the author is indebted to the Compound Semiconductor manufacturing community, who demonstrated years ago that cost-effective high volume manufacturing of through-substrate vias is not only technically feasible, but that it can be routinely done as a result of the dedicated cooperation and community effort between device makers themselves and commercial capital equipment and materials supply chain participants.

References

1. Klingbeil, L.S., Kirschenbaum, K.L., Rampley, C.G., Young, D. (2001). 150mm through substrate interconnect conversion one year later, *Digest of Papers 2001 International Conference on Compound Semiconductor Manufacturing Technology*, GaAs MANTECH, pp. 41–44.

2. Hendriks, H., Crites, J., D'Urso, G., Fox, R., Lepowski, T., Patel, B. (2001). Challenges in rapidly scaling up back-side processing of GaAs wafers, *Digest of Papers 2001 International Conference on Compound Semiconductor Manufacturing Technology*, GaAs MANTECH, pp. 181–184.

3. Nam, P., Tsai, R., Davison, D., Allen, B., Barsky, M., Grundbacher, R., Lai, R., Olson, S. (2003). Impact of backside via dimension changes on high frequency GaAs MMIC circuit performance, *Digest of Papers 2003 International Conference on Compound Semiconductor Manufacturing Technology*, GaAs MANTECH, pp. 265–268.

4. Anderson, D., Knoedler, H., Tiku, S. (2003). Cycle time reduction during electroplating of through wafer vias for backside metallization of III-V semiconductor circuits, *Digest of Papers 2003 International Conference on Compound Semiconductor Manufacturing Technology*, GaAs MANTECH, pp. 169–171.

5. Bonneau, D., Borkowski, P., Shelley, R., Fortier, A., Young, M. (2002). Capability and cycle time improvements in Nortel Networks' substrate via fab, *Digest of Papers 2002 International Conference on Compound Semiconductor Manufacturing Technology*, GaAs MANTECH, pp. 113–116.

6. Clayton, F., Westerman, R., Johnson, D. (2002). Characterization of a manufacturable high rate GaAs via etch process, *Digest of Papers 2002 International Conference on Compound Semiconductor Manufacturing Technology*, GaAs MANTECH, pp. 121–124.

7. Chang, J., and Sawin, H. (1997) Kinetic study of low energy ion enhanced polysilicon etching Using Cl, Cl_2, and Cl^+ beam scattering, *J. Vac. Sci. Technol. A*, Vol. 15, No. 3, , May/Jun 1997, p. 610.

8. Chang, J., Arnold, J., Zau, G., Shin, H., Sawin, H. Kinetic study of low energy argon ion-enhanced plasma etching of polysilicon with atomic/molecular chlorine, *J. Vac. Sci. Technol. A*, Vol. 15, No. 4, Jul/Aug 1997, p. 1853.

9. Rossen, R., and Sawin, H. Time-of-flight and surface residence time measurements for ion-enhanced $Si-Cl_2$ reaction products, *J. Vac. Sci. Technol. A*, Vol. 5, No. 4, Jul/Aug 1987, p. 1595.

10. Goodman, R., Materer, N., Leone, S. Ion-enhanced etching of Si(100) with molecular chlorine: reaction mechanisms and product yields, *J. Vac. Sci. Technol. A*, Vol. 17, No. 6, Nov/Dec 1999, p. 3340.

11. Coburn, J. Ion-assisted etching of Si with Cl_2: the effect of flux ratio, *J. Vac. Sci. Technol. B*, Vol. 12, No. 3, May/Jun 1994, p. 1384.

12. Lane, J., Klemens, F., Bogart, K., Malyshev, M., Lee, J. Feature evolution during plasma etching. II. polycrystalline silicon etching, *J. Vac. Sci. Technol. A*, Vol. 18, No. 1, Jan/Feb 2000, p. 188.

13. Chang, J., Mahorowala, A., Sawin, H. Plasma-surface kinetics and feature profile evolution in chlorine etching of polysilicon, *J. Vac. Sci. Technol. A*, Vol. 16, No. 1, Jan/Feb 1998, p. 217.

14. Tuda, M., Ono, K., Nishikawa, K. Effects of etch products and surface oxidation on profile evolution during electron cyclotron resonance plasma etching of Poly-Si, *J. Vac. Sci. Technol. B*, Vol. 14, No. 5, Sep/Oct 1996, p. 3291.

15. Tian, W., Weigold, J., Pang, S. Comparison of Cl2 and F-based dry etching for high aspect ratio Si microstructures etched with an inductively coupled plasma source, *J. Vac. Sci. Technol. B*, Vol. 18, No. 4, Jul/Aug 2000, p. 1890.

16. Winters, H., Coburn, J., Chuang, T. Surface processes in plasma-assisted etching environments *J. Vac. Sci. Technol. B*, Vol. 1 No. 2, Apr/June 1983, p. 469.

17. Pruessner, M., Rabinovich, W., Stievater, T., Park, D., Baldwin, J. Cryogenic etch process development for profile control of high aspect-ration submicron silicon trenches *J. Vac. Sci. Technol. B*, Vol. 25 No. 1 Jan/Feb 2007, p. 21.

18. Laermer, F. and Schilp, A. Method of anisotropically etching silicon, US Patent 5,501,893, 26 March 1996.

19. Stewart. R., Vitello, P., Graves, D. Two-dimensional fluid model of high density inductively coupled plasma sources *J. Vac. Sci. Technol. B*, Vol. 12 No. 1 Jan/Feb 1994, p. 478.

20. Paranjpe, A. Modeling an inductively couple plasma source *J. Vac. Sci. Technol. A*, Vol. 12 No. 4 Jul/Aug 1994, p. 1221.

21. Blauw, M., Zijlstra, T., van der Drift, E. Balancing the etching and passivation in time-multiplexed deep dry etching of silicon *J. Vac. Sci. Technol. B*, Vol. 19 No. 6 Nov/Dec 2001, p. 1.

Chapter 5

TSV FILLING

Arthur Keigler
NEXX Systems

5. 1 INTRODUCTION

Copper, tungsten, polysilicon, and nickel are candidate materials TSV conductor filling that are under development. Copper filled TSV is emerging as the most widely adopted structure therefore Cu filling by electrodeposition process is the primary subject of this chapter.

Electrodepostion as a means of filling vias and trenches has been developed since the mid 1990's as part of the copper damascene process for back end of line (BEOL) wiring so the first section compares in general the TSV and Cu-Damacene filling applications. Electro-filling is a fascinating technology that encompasses diffusion, adsorption, and chemical reaction mechanisms, which have been studied for 15 years as applied to Cu-Damascene applications and are only recently being studied with respect to TSV applications, the next section describes the primary Cu-Damascene filling models. The third section reviews some of the literature on TSV filling and discusses how the Cu-Damascene models may be extended to TSV geometries.

Much of TSV process development as of 2009 remains quite empirical and the final section includes experimental methods and examples of TSV filling processes, future developments and outlooks for the economics of TSV filling.

3D Integration for VLSI Systems
Edited by Chuan Seng Tan, Kuan-Neng Chen and Steven J. Koester
Copyright © 2012 by Pan Stanford Publishing Pte. Ltd.
www.panstanford.com

5.2 COPPER FILLING: TSV COMPARED TO DAMASCENE

Several first order estimates show why TSV filling is different from the well known Cu-Damascene gap-fill process and points out what more is needed to extrapolate Cu-Damascene gap-fill mechanisms to fully explain TSV filling. TSV structures are typically cylindrical, either a solid cylinder or a tubular cylinder, and from 10 to 200 microns in depth, where the depth is set by the required thickness for downstream handling processes during stacking. The aspect ratio (AR = depth / width) is generally set by the economics of fabricating the liner/barrier/seed or the filling process, both of which become considerably more expensve as AR>5. There are myriad TSV geometries in development that differ due to a wide variety of constraints, but they may usefully be grouped into three general types of TSV geometry as shown in Table 5.1.

Table 5.1 General types of TSV.

Application	Plating	Depth	Diameter	Aspect Ratio
Image sensor	Conformal	50 to 100	30 to 50	1 to 3
Interposer	Full-fill	50 to 150	20 to 30	4 to 8
Device	Full-fill	20 to 60	2 to 10	5 to 15

Copper dual damasene structures differ from TSVs in two ways: (1) the feature size width is on the order of 0.1 micron and depth is on the order of 1 micron, and (2) shape consists of not only interlayer vias but also lines, and wiring lines that may vary considerably in density on the same wafer layout. As will be later discussed, the filling behavior is a function not only of individual TSV geometry but also the TSV density, and so it is advantageous to restrict device layouts to a minimum number of regions with different pattern densities.

Figure 5.1 shows how the significant difference in geometric size between TSV and Cu-Damascene influences the process time scale as determined by the associated diffusion times. For features associated with Cu-Damascene processing the diffusion time constant is on the order of milli-seconds, several orders of magnatude less than for the TSV filling process. Diffusion delivers reactants to the growing surface where adsorbtion and reaction kinetics cause electrodeposition, the ratio between these diffusion and kinetic characteristic times is much higher for TSV then for copper damascene processes.

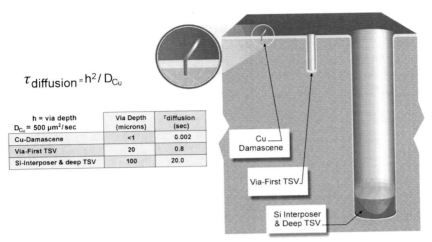

$$\tau_{\text{diffusion}} = h^2 / D_{Cu}$$

h = via depth D_{Cu} = 500 μm^2/sec	Via Depth (microns)	$\tau_{\text{diffusion}}$ (sec)
Cu-Damascene	<1	0.002
Via-First TSV	20	0.8
Si-Interposer & deep TSV	100	20.0

Fig. 5.1 Contrasting scale of geometry of TSV structures with the geometry of Copper Damascene structures indicates why optimal TSV process is more then simply an extension of Cu-D process, this also gives an indication of why TSV filling times are >10 minutes whereas Cu-D < 1 minute.

Conformal copper plating for low aspect ratio partially filled TSVs for low density I/O applications uses plating chemistry and process conditions similar to those used for copper pattern plating in resist masks. Copper is plated in a layer of between 5 and 10 microns thick into a resist mask structure that both follows the topography of the tapered via hole to connect through the silicon as well as providing a wiring pattern on top of the silicon. These processes are similar to those used for redistribution layers (RDL) or wafer level chip scale package (WLCSP) wiring and pads, with some optimization made to achieve a satisfactory ratio of bottom to top thickness; a target ratio of 50% to 80% is typical, with a higher ratio being more expensive due to longer plating times (i.e. slower deposition rate.) Since these pattern plating applications are relatively well understood and supplied by commercial chemistry[1] and tool[2] vendors they will not be further discussed here.

5.2.1 Bottoms-Up Filling

Full filling of the TSV requires a deposition growth front that is faster in the bottom then near the top, this has been termed "bottoms-up" or "super-conformal" growth. For bottoms-up fill to occur, the vertical moving deposition from the bottom must be significantly faster then the radial deposition on the sidewall lest the sidewalls grow inward and touch, or "pinch-off", and prohibit further deposition deeper in the via. A simple approximation shows that the bottoms-up rate must be much faster then the radial rate near the top of the via—remarkably, for a 10:1 AR TSV this indicates a 40-fold difference in growth rates.

The local deposition rate at any point on the growing surface is determined by the local concentrations of different types of surface sites on the copper and reactants near the copper, which will be discussed in more detail below, however to make an estimate of the required growth rate ratio assume that growth near the top is characterized by a radial deposition rate is R_{radial} and near the bottom by a bottoms-up deposition rate is $R_{bottoms-up}$, then

$$t_{pinch-off} = \frac{(d/2)}{R_{radial}}$$

$$t_{bottoms-up} = \frac{h}{R_{bottoms-up}} = \frac{AR * \theta}{R_{bottoms-up}}$$

where AR=aspect ratio, d=diameter, and h=TSV height or depth, and the times are for pinch-off at the via top ($t_{pinch-off}$) and straight bottoms-up growth ($t_{bottoms-up}$) respectively.

Ideal bottoms-up filling requires that

$$t_{bottoms-up} \lhd t_{pinch-off}$$

for example, assume a factor of two process operating margin, then

$$2 * t_{bottoms-up} = t_{pinch-off}$$

therefore

$$\frac{R_{bottoms-up}}{R_{radial}} = 4 * AR$$

This shows why higher aspect ratio TSVs require a larger ratio between bottom and top region growth rate to achieve reliable bottoms-up filling; for example an aspect ratio of 10 this means deposition at the via mouth must be suppressed to $1/40^{th}$ of that at the via bottom. Engineering of the TSV filling process depends on controlling the applied voltage profile and the mixture of molecular structures to cause this substantial difference in growth rates as the TSV is filled.

5.2.2 Conformal and Bottoms-Up Growth

TSV filling occurs as a combination of conformal deposition and bottoms up deposition as shown in Figures 2(a) and 2(b). Conformal growth fills the volume faster but at the risk of creating a "seam-void". Bottoms-up filling is more robust against voids but the filling time is longer: purely bottoms-up is 2*AR the time of purely conformal. In practice a combination of the two modes is used, with the ideal growth front having a flat bottom, or U shape.

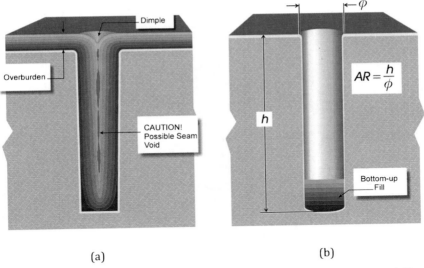

Fig. 5.2 TSV filling copper growth front shapes: (a) Shows primarily conformal filling, (b) shows bottoms-up filling. Also shown are features that must be controlled during filling: the overburden thickness, the dimple depth, and avoidance of seam void.

A lower bound for filling time in purely bottoms-up mode is to consider the filling time that would result from purely bottoms-up filling if deposition rate were limited by only by the availability of copper diffusing into the via. The diffusion limited current density on a planar surface is determined by the fluid boundary layer thickness as follows:

$$ i_{\text{diffusion}} = \frac{K_{Cu} * Z * D_{Cu}}{l_{b.l.}} $$

where K_{Cu} is mols/liter of copper, Z is coulombs/mol of Cu^{+2}, D_{Cu} is diffusivity of copper ion, and l_{bl} is the effective boundary layer thickness. During the filling step the fluid boundary layer is the sum of the surface boundary layer and the distance to the growth front inside the via, as the TSV fills up l_{bl} gets smaller. Figure 5.3 shows the filling time for purely bottoms-up deposition as a function of via depth assuming deposition rate at half the limiting current (K_{Cu} = 1 mols/liter, Z = 1.9e5 coulombs/mol, D_{Cu} = 4e-6 cm²/sec) for fluid boundary layer thicknesses outside the TSV of 10 and 50 microns, this shows the importance of using process cell configurations with active boundary layer thinning to reduce the filling time.

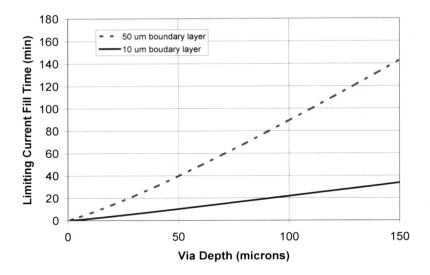

Fig. 5.3 Fill time versus TSV depth for deposition at half the average limiting current deposition rate for 1 mol/liter copper concentration.

The foregoing calculation significantly underestimates the filling time because in the TSV filling process the deposition rate must be strongly reduced, or "suppressed", below the diffusion limit in order to achieve the difference between rates at the via top and bottom required to cause bottoms-up filling. How super-conformal growth is achieved is the subject of the next section.

5.3 SUPERCONFORMAL ELECTRODEPOSITION

Superfilling was developed by researchers at IBM[3] during the 1990's as part of an effort to replace aluminum in IC on-chip wiring with copper to achieve lower resistance and better electromigration performance. Lack of a production worthy means of precision etching copper prohibited use of the blanket deposition and dry etch patterning methods used for aluminum so the damascene process was developed for copper. Bottoms-up electrodeposition and chemical mechanical polishing were two new wet processes which have now become primary unit processes in BEOL fab lines.

While production linewidths for Cu-damascene have shrunk from 0.5 to 0.03 microns over the last 15 years the scientific understanding of superfilling has become more precise and theoretical models predict much of the observed results. This remains an area of active research both as applied to sub-micron copper-damascene features and recently as applied to TSV scale features. While electroplating scientists have more work to do to provide a fully predictive model for copper-damascene superfilling, and have only begun developing a predictive model for TSV filling, the various theories emphasize different aspects of the superfilling mechanism. All of these shed light on the process engineering required for the TSV production development. These are reviewed below.

5.3.1 Chemistry of copper plating baths for TSV

5.3.1.1 Acids and salts

Sulphuric acid (H_2SO_4) and methylsulphonic acid (MSA) are the two primary systems used for TSV filling. Copper salt is dissolved into the solution in a range from 10-70 grams/liter for a sulphric acid system and from 50 to 120 grams/liter in an MSA system. Chloride is essential, typically added as hydrochloric acid (HCl), in the range of 50-100 parts per million; the chloride ion specially absorbs on copper surfaces ($Cl^-_{aq} \rightarrow Cl^-_{adsorbed}$) and catalyzes many adsorption and reaction paths on the cathode, it is also important at the anode where a $CuCl_2$ layer is formed to protect the organic molecules from oxidation at the anode.[44]

5.3.1.2 Organic Molecules Determine Growth

Small (ppm) additions of organic molecules have been used for many years or more to influence the morphology of electroplated metal films, the patent literature goes back at least to Hull's 1939 patent[5] for an apparatus to study "addition agents". In the 1940's copper baths used additives such as molasses[6] and wetting agents like sodium-laurel-sulpahte that form the basis of hair shampoo; although research required another 50 years to unearth the catalytic mechanism of trace chloride ion concentrations, these too were known empirically. Developing bright copper underlayers for automobile applications in the 1950's, Fellows *et.al.*[7] identified the advantage of combining "brightener" and "leveler" additives to provide a broad process operating window for fine grained copper electrodeposition and some of these molecules, like Janus Green B, are used today to study TSV filling.[8] By the early 1990's most of the additives which would later be used for copper-damascene and TSV filling were being used for both decorative bright copper finishes as well as for printed circuit board and through hole applications.[9,10] Early research into copper-damascene[3] showed that by manipulating the different diffusion and reaction characteristics of these additives superfilling can be generated.

The required top to bottom difference in deposition rates is caused by differences in transport to, and adsorbtion of, organic molecules onto the copper surface: these molecuules change the deposition rate from that which would occur on a pure copper surface at the same applied potential. A number of different types of organic additives are used to control copper deposition, and the descriptive names of these species can be confusing because their functional behavior can be shifted by the precsence of other species in the chemistry, or names vary by historical use of the additives. Important attributes of the organic molecules are diffusivity, adsorbtion energy on the copper surface, and catalytic influence on both copper deposition itself as well as displacement of other additives. In general, the organics are chosen to avoid co-deposition into the copper because of unwanted stress and impurity effects, however at higher rate deposition, or for reasons of influencing the grain structure, a small fraction may be purposefully co-deposited.

TSV filling chemical systems typically use three organic additives: a slow diffusing and rapidly adsorbing suppressor, a fast diffusing accelerator that can slowly displace the adsorbed suppressor, and a slow diffusing leveler that can de-activate the accelerator. Table 5.2 summarizes key aspects of the organic additives which will be discussed in more detail below. Commercial additive systems use proprietary versions of these type of molecules, wherein the molecular weight distributions are optimized for diffusion effects and the polarity and functional group properties are optimized for adsorption behavior.

Table 5.2 Key features of organic additives for TSV filling.

Functional Name	Main Function	Mechanism	Example Molecule	Mol Wt	Diffusion Rate (cm²/sec)	Adsorption Rate
Suppressor Inhibitor Carrier	Block Cu deposition	Physio absorption	Polyether PEG: Poly-etheylene-glycol	3,000 - 10,000	Slow 5e-7	Fast
Accelerator Brightener	Catalyze Cu reduction	Chemical absorption	Sulphur containing SPS		Fast 1e-5	Slow Displaces PEG slowly
Leveler	Compete with accelerator or deactivate accelerator	Chemical absorption	Nitrogen containing PVP: polyvinylpyrrolidone Polyvinyl-imadazolium Janus-Green B		Slow	Medium

Suppressors

Supressor or inhibitor molecules are long chain polyether, polypropyl or polyacetal polymers with molecular weight of between 1000 and 100,000. There is not a strict demarcation between suppressor and levelers, both are long chain polymers of many types, but in general suppressors are considered to have little charge polarization and levelers have a charged head group. Polyethylene glycol (PEG) has been the most extensively studied.[5, 15, 24, 36] The large chain molecules physio-absorb on the Cu surface and inhibit charge transfer. Numerous studies have shown the importance of 50-100 ppm of chloride ion to faciliate PEG absorption.

Using a microfluidic apparatus, Wiley and West[15] shows that the molecule absorbs in a kinked, or relaxed, configuration on the Cu surface rather than standing up like brush bristles. Operating in potentiastic mode at -0.2 V with C_{PEG}=300 ppm, Mw_{PEG}=3350 g/mol the current density drops from 90 ma/cm^2 to 10 ma/cm^2 in 0.5 sec; using larger molecules with Mw_{PEG}=10^5 g/mol the blocking reaction time constant slowed to 1.0 sec. The longer chain molecules require more time to arrange themselves on the Cu surface such that their "sticking points" block the Cu-reduction. Similarly the desorption behavior was characterized indicating that the longer chain molecules take longer to untangle and desorb. Wiley and West propose a multi-step process that includes a step when molecules act independently on the surface, and a series of steps where the molecules reconfigure.

Accelerators
Accelerators are comparitively small fast diffusing sulphur containing molecules, the two most commonly used being bis(3-sulfopropyl) disulphide (SPS) or mercaptopropane sulfonic acid (MPSA).[4] Both SPS, MPSA can dissociate to form an active complex of Cu(I)thiolate which catalyzes the copper reduction reaction on the copper surface, the absorbtion and activation time constants depend on the amount of other additives in the near surface layer.[18]

Levelers
Leveler molecules, like suppressors, are also high molecular weight polymers, with the primary functional difference being stronger absorption. Leveller molecules typically contain a nitrogen tertiary or quatenary compound which has a positive charge and is therefore attracted to regions of high local potential, the large, 100-10,000 mol-wt, polymer blocks further copper growth in that region, thereby "leveling" the deposit. Examples are they dye molecule Janus Green B,[8] polyethyleneimine (PEI), dodecyltrimethyl ammonium chloride (DTAC), and polyvinylpyrrolidone (PVP).

In copper damascene applications levellers provide an important role in quenching overshoot growth of accelerator rich regions[24,25] by displacing the adsorbed accelerator complex. Levelers play more diverse roles in TSV filling and are important for inhibiting deposition in the upper section of the TSV[8,27] where strong inhibition is required to avoid pinch-off. Also the engineering of the relative polariazation of the functional groups of levelers in conjunction with polymer chain length and flexibility play an important role in the TSV filling bath as it influences the longer time scale, 100-1000 seconds, interactions of competitive adsorption with the accelerator, as discussed below this particularly influences the final phase of filling that is important for eliminating the dimple.

Selection of leveler molecules is a very active area of commercial development for TSV filling baths.

5.3.2 Transport and Adsorption

Ideally the wafer surface and top regions of TSV are coated with suppressor molecules to inhibit growth and the bottom surfaces of TSV are coated with accelerator molecules that counteract suppression or accelerate growth, this is depicted in Figure 4. Setting up, and maintaining, these surface gradients will be discussed in this section.

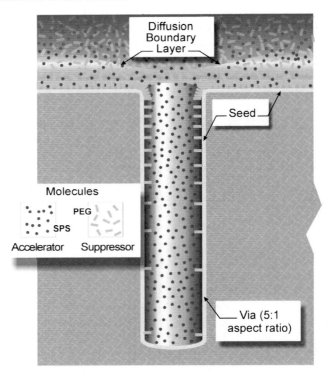

Fig. 5.4 TSV cross section showing idealized distribution of suppressor and accelerator adsorbed species.

5.3.2.1 Diffusion

Transport of reactants to the growing copper surface occurs in the bulk solution by fluid mixing and near the surface by diffusion. Fluid mixing transports species to the static boundary layer which may be from 10 microns to 60 microns thick on the planar surface depending on the degree of active boundary layer thinning accomplished by the plating tool. Diffusion to the top of the TSV will be quicker then to the bottom, as given by

$$\tau_{\text{Diffusion},s,z} = \frac{(z + l_{bl})^2}{D_s}$$

where s is reactant species (Cu^{2+}_{aq}, SPS, PEG, etc.), z is depth position in the via ($z = h$ at via bottom), l_{bl} is the diffusion boundary layer thickness outside the via, and D_k is the species diffusivity which ranges from approximately 1e-5 cm^2/sec for Cu^{2+}_{aq} and SPS to 5e-7 cm^2/sec for PEG. This shows for example that for a 100 micron deep TSV with a 10 micron fluid boundary layer the diffusion time constant at the via bottom is approximately 12 seconds for Cu^{2+}_{aq} or SPS and 240 seconds for PEG. An important relationship is the ratio of $\tau_{Diff,s,Bot}$ to $\tau_{Diff,s,Top}$ for the organic additives that mediate the local depositon rate and how these time constants compare with adsorption time constants to influence the adsorbed species gradient necessary for bottoms-up growth.

Most of the published literature on superfilling is focused on copper damascene applications, so it is instructive to compare the characteristic diffusion times of various TSV structures with a 1 micron deep damascene structure, this is shown in Table 5.3.

Table 5.3 Diffusion time comparison for various TSV depths. Typical fill times are shown a broad average for a 10:1 aspect ratio TSV.

Diffusion time estimates

Parameter	Units	D cm^2/sec	Via depth:	1	25	50	100
SPS, Cu	cm^2/sec	1.0E-05	Tdiff SPS, Cu top (sec)	0.10	0.1	0.1	0.1
PEG	cm^2/sec	5.0E-07	Tdiff SPS, Cu bottom (sec)	0.12	1	4	12
SPS to PEG ratio		20	Tdiff PEG top (sec)	2	2	2	2
Boundary Layer	ums	10	Tdiff PEG bottom (sec)	2	25	72	242
			Bot to Top PEG diffusion time ratio	1.2	12	36	121
			Typical filling time (min)	1.0	20	60	200
			Ratio of fill time to Tb/Tt	50	98	100	99

In practice, typical TSV filling times are about double that which would be predicted by the a simple ratio to the PEG diffusion time constant for a 1 micron copper-damascene filling time. This is indication of the importance of longer time scale reactions involved in the TSV fill that don't occur for shorter times, in particular some of the slower aspects of competitive absorption among organic additives.

5.3.2.2 Adsorption and Reaction

Many researchers are involved in different aspects of the reactions which occur on or near the copper surface. Vereecken *et.al*,[4] discuss 27 different reactions that play roles in copper damascene superfilling and review the literature available in 2004. Alkire's research group is proceeding now with a comprehensive numerical model[37-39] of the copper surface evolution during trench filling that incorporates 17 species and reactant intermediaries participating in a network of 15 surface reactions and 3 bulk solution reactions; their modeling extends to detailed growth morphology,[40,41] which is also reviewed by Barkey.[27, 28, 49] At the risk of oversimplifying this work, the main reaction paths are presented here to provide a sense of the importance of interphase reactions.

$$CuCl_{ads} + PEG \rightarrow CuClPEG_{ads} \quad (1)$$

$$SPS_{aq} + 2e^- \rightarrow 2 \text{ thiolate}^-_{aq} \quad (2)$$

$$Cu^+_{aq} + \text{thiolate}^-_{aq} \rightarrow Cu(I)\text{thiolate}_{ads} + H^+_{aq} \quad (2b)$$

$$Cu^+_{aq} + Cu(I)\text{thiolate}_{ads} + e^- \rightarrow Cu(I)\text{thiolate}_{ads} + Cu_{solid} \quad (3)$$

$$Cu(I)\text{thiolate}_{ads} + HIT_{aq} \rightarrow Cu(I)HIT_{ads} + MPS_{aq} \quad (4)$$

Reaction (1) shows PEG combining with an adsorbed CuCl complex to form an adsorbed site-blocking PEG. Reaction (2a) shows the SPS splitting into two thiolates, preliminary to adsorbing and becoming activated as the adsorbed comples by reaction (2b). Reaction (3) shows the activated and adsorbed portion of the accelerator molecule catalyzing the reduction of a acquasous copper ion into solid copper. Reaction (3) shows the adsorbed accelerator being displaced by the leveler molecule HIT.

5.3.2.3 Curvature Enhanced Accelerator Coverage

Using only kinetic parameters taken from planar surface electrochemical measurements Moffat *et.al.* have developed a well accepted model for superfilling based on curvature enhanced accelerator coverage (CEAC) as the growth interface changes geometry as the feature is filled.[22-26] They have shown this model is also predictive for trench filling in other systems, gold and silver electroplating as well as copper CVD.[26] Experimental verfication using a two step process wherein the surface is first "derivatized", or coated with accelerator, and then filled in an electrodepostion bath without accelerator shows that superfilling does not require an ongoing supply of accelerator, the shrinking surface area of the metal growth front along with surface diffusion of the accelerator causes the via bottom to grow faster then the top.

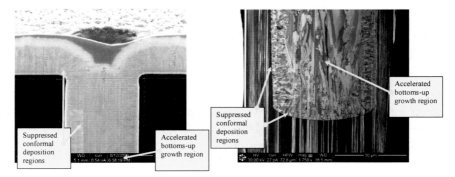

Fig. 5.5 Cross section of filled TSVs using an accelerator derivatized surface method showing the fine grain suppressed growth along sidewalls and large grain accelerated bottoms-up growth in the center.

Experiments with a 5 micron by 40 micron deep TSV structure using this type of "derivatized" surface and accelerator free fill indicates the CEAC phenomena can also be made to strongly occur in the much larger TSV structure[50] as shown in Figure 5.5. In these experiments the filling bath contained suppressor and leveler additives.

5.3.2.4 Combined Transport and Adsorption Theory

Alkokar and Landau[34-36] have developed a combined transport-adsorption model that predicts many of the observed Cu-D superfilling phenomena as resulting from the evolution of surface coverage resulting from the different diffusion and adsorption rates of the accelerator (SPS) and suppressor (PEG) or leveler species. Suppressor molecules being faster to absorb cover a high percentage of surface sites outside the via and inhibit deposition. Additive absorption from the fluid in the TSV interior is sufficient only to cover about 10^{-3} of available copper surface sites before the 10–100 ppm fluid concentrations are depleted, this is not enough to inhibit deposition deep in the via, and subsequent surface coverage must be supplied by diffusion, so the smaller accelerator occupies a high proportion of surface sites in the via bottom. SPS eventually displaces the PEG from adsorption sites, reducing the inhibition. This competitive adsorption occurs at time scale of 100 seconds, which is near the fill time for copper damascene size structures, but since it is only a small fraction of the TSV fill time, this indicates the importance of the leveler species for TSV filling: wherein the levelers have the large slow diffusing characteristic of suppressor but are not displaced by the accelerator (or may displace the accelerator.)

5.3.2.5 Reverse pulse deposition and oxygen sparging

Barkey has shown[52] that dissolved oxygen consumes the Cu^+ ion in solution. Along with Kondo[27] they show this can be combined with reverse pulse plating to create an excess Cu^+ deep inside the TSV and therefore a relative excess of the activated accelerator, Cu(I)thiolate, inside the via as described by the following reactions:

$$2Cu^{2+} + 4MPS \longrightarrow 2Cu(I)\text{thiolate} + SPS + 4H^+$$

$$2Cu^+ + \tfrac{1}{2}O_2 + 2H^+ \longrightarrow 2Cu^{2+} + SPS + H_2O$$

Using this technique with SPS accelerator, PEG suppressor, and Janus-Green leveler they show void free filling of 10 × 70 micron TSVs in 60 minutes.

5.3.2.6 Generalized description of TSV superfilling

Distribution of the applied current between the via bottom, sidewalls and the top field surface is controlled by the relative population of surface adsorbed ACC, SUP and LEV species on the different geometric surfaces. Consider that the total surface is made up of different regions each of which has a certain condition of adsorbed additives: for example the flat wafer surface or field, the TSV wall regions at various depths, and the TSV bottom.

A typical PVD seed layer deposition process produces a thick (2000 to 10,000 angstrom) copper layer on the wafer surface, therefore macro-scale potential drop across the wafer is minimal and plating tool features adjust this to be inconsequential. Because the filling process requires very low current densities, typically in the range of 1 to 10 ma/cm^2, although the copper seed in the via may be only 100 angstroms thick due to shadowing during the PVD process, the potential drop from Top to Bottom of via is in the microvolt range, negligable compared to the deposition potentials for which significant differences are in the millivolt range. With these assumptions the copper surface inside the TSV can be considered a constant potential surface at overpotential η, the overpotential being the interphase potential difference associated with current flow. The current density at each depth, k, is given by the Butler-Volmer equation[51] which characterizes the kinetics of the relationship between current and voltage:

$$j_k = -j_{k,0}\exp\left(-\frac{\alpha_k 2F\eta}{RT}\right)$$

where F=Faraday constant, R=gas constant, T=absolute temperature, $j_{k,0}$=exchange current density on region k, a measure of the speed of the kinetics, α_k=transfer coefficient on region k, a measure of the activation energy barrier symmetry. In this formulation all of the surface specific kinetics are characterized by $j_{k,0}$ and α_k.

Bottoms-up requires a large rate differential, or current density differential, which means the operating potential is ideally set to a value which produces the largest Δj, as shown in Fig. 5.6. A simulated i-v curve is shown for each of the top and bottom regions, the left side of the curve showing Butler-Volmer behavior with a saturation due to other limiting effects (such as copper availability). There is an optimum overpotential for which the ratio between deposition rates of accelerated and suppressed regions is maxim. Recalling that the specifics of the i-v curves depend on local additive adsorption, much of the process development optimization for TSV is empirical testing to identify additive ratios that provide a maxim ratio, however, where each additive ratio must be tested across a range of potential (i.e. applied current density) to identify the maxim.

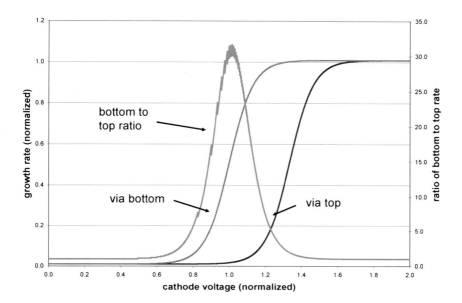

Fig. 5.6 Voltage dependence of the deposition rate for suppressed (via top) and accelerated (via bottom) surfaces, showing the operating regime required to obtain a high ratio of bottom growth rate to top growth rate.

Current is applied and the resulting overpotential η equilibriates such that for each of the different surface conditions k, of given surface area A_k:

$$i_{total} = \sum_k j_k * A_k$$

$$i_{total} = \sum_k -j_{k,0} \exp\left(-\frac{\alpha_k 2 F \eta}{RT}\right) * A_k$$

Consider a simplified example using only two regions: (1) Bottom is the lower 1/3 of the via wall plus the via bottom area, and region (2) Top is the field area outide the TSVs and the remainder of the via sidewall. Evaluate the relative deposition distribution for different superfilling ratios ψ.

$$\psi = \frac{j_{Bottom}}{j_{Top}}$$

$$A = \frac{A_{Bottom}}{A_{Top}}$$

$$i_{total} = j_{Bottom} * A_{Bottom} + j_{Top} * A_{Top}$$

$$\frac{i_{Bottom}}{i_{total}} = \frac{\psi A}{(1 + \psi A)}$$

Figure 5.7 depicts how this simplifed model shows the influence of TSV pitch on the relative current distribution between via and field for various superfilling ratios. Typically TSVs are distributed with a tight pitch and only a partially populated array that depends on device layout, whereas in this simplified calculation a fully populated array is assumed, so the resultant number of TSVs per 6 × 10 mm die is also shown for each different pitch. At high superfilling ratios more of the current flows into the via bottom and therefore if the TSV density is high this becomes a significant fraction of the total current, even though the TSV coverage by area is less than 1%. This effect must be taken into consideration when setting the current density used at various stages of filling.

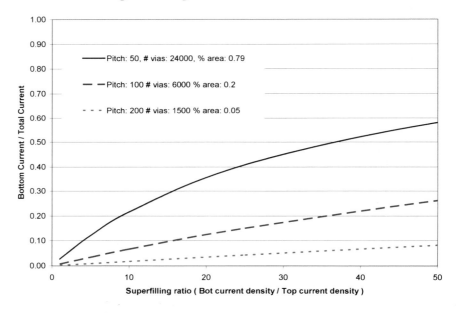

Fig. 5.7 Estimate of initial bottoms-up current as a fraction of total current versus the superfilling ratio ($\psi = j_{Bottom}/j_{Top}$) for various TSV pitches for a 5x50 micron TSV. Number of vias is calculated for fully populated array on 6 × 10 mm die; and % area also for fully populated array at given pitch. Bottom includes lower 1/3 of TSV sidewall, Top includes field and 2/3 of sidewall.

5.4 TSV FILLING PROCESS DEVELOPMENT

To date, theoretical models for TSV superfilling are few so the we have set out general principles based on the available literature for copper damascene. TSV filling process development is now occuring rapidly in many labs around the world and some of these methods and results are discussed in the this section.

TSV process development focuses on many key variables as shown in Table 5.4. As with development of any manufacturing process, a first conservative stage may be obtained which produces good fill with a wide process control window, and subsquent stages are then developed with more economical fill times as the process window is more well defined.

Table 5.4 TSV process development variables.

Input Variables	Output Variables	Process Control
Chemistry type & concentration:	Minimum fill time	Chemical monitoring
Acid	Minimize voids	CVS
Copper	Seam voids	HPLC
Chloride	Sidewall voids	Titration
Suppressor	Missing grain voids	Total Organic Carbon
Leveller	Dimple at the top	Chemistry control method
Accelerator	Minimize overburden	Replenishment by charge
Breakdown products	Uniform thickness	Replenishment by time
Current density vs. Time	Copper grain structure	Amp-hrs/liter lifetime
Number of current steps	Resistivity	Bleed & feed schedule
Current type	Annealing behavior	
DC, Pulse, Pulse reverse		

A design of experiment (DOE) for TSV filling using a commercial chemistry[1] typically involves variation of 2 or 3 levels on each of the main chemical additives, along with several levels for the current density. Especially for high aspect ratio (AR>6) TSVs it is important to first optimize the filling of the first two thirds of the TSV and then to optimize the final filling, otherwise the experimental matrix consumes too many samples and resources. The high cost of both the preparing a TSV structure up to the point of copper filling and the post filling analysis places special constraints on the experimental program.

Chemical additives are present in the 10-100 ppm range, or 2 to 20 milliliters/liter, depending on the specific formulation. Addtive analysis is accomplished with commercial equipment[51] using chemical voltammetry spectroscopy (CVS), high pressure liquid chromatography (HPLC), and titration, typically a combination of techniques is necessary. These are not straightforward measurements. Development of the additive analysis method for a chemical system requires both calibration of the various measurement signatures against a series of known standards, as well as characterization of the measurement spectra against chemical baths operated, or aged, to produce the additive chemical breakdown products typical of a production operation.

Current density used for TSV filling is lower then that used for copper damascene applications, typically in the 0.1 to 5 milli-amp/cm^2 range for TSV applications as compared to 2 to 20 milli-amp/cm^2 range for copper damascene applications. This is indicative of the higher superfilling ratio, ψ, requried for high aspect ratio TSVs as compared to damascene applications, and the fact that larger ψ is acheivable at lower overpotential (i.e. lower applied current density) for most additive systems.

5.4.1 TSV Filling Analysis and Measurement

Presently there is not a cost effective production method for analyzing filled TSV structures to determine the presence of voids. X-Ray tools with computed tomography (X-Ray CT) are capable of identifying voids of several microns in size, but these require long measurement times, on the order of an hour for a handfull of TSVs. An important missing link in the evolution of production TSV fabrication is a reliable and economical in-line measurement tool. Companies such as XRadia, Phoenix-Xray/GE , and Shimadzu are working on higher resolution systems that may eventually have the resolution required for TSV production control.

For development purposes analysis of TSV's to determine the presence of voids can be expensive, requiring either Focused Ion Beam (FIB) analysis or ion-milling and extensive sample preparation as well as scanning electron microscope (SEM) analysis. Typical cost is from $500 to $1000 per sample for analysis suitable to detect sub-micron voids in a 60 micron deep TSV. This technology is also advancing, for example, Oregon Physis and JHT Hyper-FIB have developed an ICP plasma source FIB that is capable of cutting a whole row of vias in the time a standard FIB would cut a single TSV, even for development work this higher speed capability is important for data experimental statistics.   As shown in Figure 5.8 it is useful to cut a row of TSVs such that the analysis is certain to cut at least one via on the centerline, otherwise when looking for a sub-micron void in a single TSV a wrong conclusion of void-free fill may result because the FIB cut was not on the TSV centerline.

A quick turn-around feedback method has been developed[50] to provide feedback during process development. During TSV process development it is important to characterize the growth front at various stages of the filling process and so "partial fill" runs are used wherein for example 4 runs at each of approximately 25%, 50%, 75% and 100% of the total filling charge are run using the same conditions (with exception that more completely filled runs have more charge.) At the end of each run the sample is moved into a nickel plating bath and several microns of nickel are plated over the copper, this stabilizes the growth front during subsequent sample polishing. The sample is then mounted at a slight angle on a polishing block and lapped to expose the TSVs at various heights as shown in Figure 9(a). A close-up image clearly reveals the geometry of the boundary between the filling copper and the protective nickel, as shown in Figure 9(b), where the depth of the cross section is calculated trigonemetrically from the known lateral pitch of the TSV array.

Fig. 5.8 High speed FIB analysis. Left image shows a high speed FIB cut across an array, right image shows a close-up where two TSVs display 0.5 micron voids on the centerline. Other vias likely had similar voids, but appear to be void-free because FIB cut was slightly off-center. Cutting and analysis time was approximately 1 hour. TSV structure courtesy of IMEC.

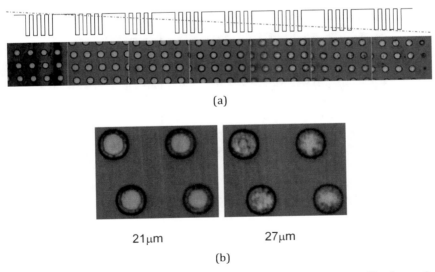

(a)

21μm   27μm

(b)

Fig. 5.9 (a) NEXX-Sectioning™ method of analysis of TSV growth profile shown for 5 × 40 micron sized TSVs. (b) NEXX-Sectioning™ close-up of TSV growth profile showing the boundary between copper and nickel at the 25% partial fill growth front.

The following sections show various features of TSV filling that are optimized or eliminated during process development.

5.4.2 Dimple above the substrate

Following the TSV filling step chemical mechanical polishing (CMP) is used to planarize the substrate and provide a flat surface for subsquent processing, whether it be the BEOL wiring layers or various wiring layers used as part of interposer fabrication.

A potential process defect is a depression in the copper above the TSV, a so called "dimple", as shown in figure 5.10. If the filling process results in this kind of feature it must be optimized so that the dimple is above the oxide layer to which the wafer is polished during the CMP step.

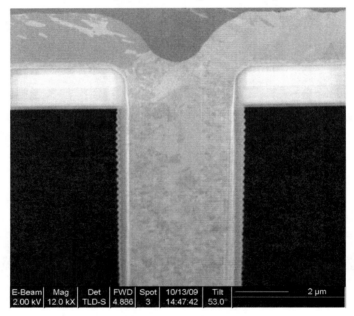

| E-Beam | Mag | Det | FWD | Spot | 10/13/09 | Tilt | 2 µm |
| 2.00 kV | 12.0 kX | TLD-S | 4.886 | 3 | 14:47:42 | 53.0° | |

Fig. 5.10 FIB cross section at the top of a TSV showing that the "dimple" is just above the oxide layer. Note that the overburden is approximately equal to half the 5 micron TSV diameter.

5.4.3 Pinch-off

Early in the TSV filling optimization the "pinch-off" failure mode, shown in Figure 5.11, must be eliminated. This may occur for several reasons, one being insufficient leveller to inhibit growth in the upper corners of the TSV, and a second one being an excessive growth rate (i.e. current density) that exceeds the inhibition capacity of the additive system.

Much of the TSV process development effort is geared to optimizing the ACC, LEV, and SUP concentrations along with the current density versus time profile to generate a robust bottoms-up growth front long before such pinch-off occurs.

Fig. 5.11 FIB/SEM analysis of "pinch-off" void in TSV.

5.4.4 Inadequate wetting of the TSV structure

Particularly for high aspect ratio TSV's the filling process may be corrupted by inadequate wetting of the chemistry to the seed layer inside the TSV. Specialized equipment that includes various types of proprietary vacuum assisted pre-wetting techniques has been developed to minimize this failure mode, see Figure 5.12. For laboratory process experiments, where the specialized hardware may not be available, it is important to distinguish incomplete wetting from early pinch-off (i.e. pinch-off near the bottom of TSV) by careful inspection to identify some plated material at the via bottom.

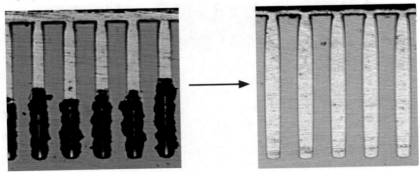

Fig. 5.12 Inadequate wetting of TSVs with AR=10:1 is shown by comparison of using a spray pre-wet (left image) with a vacuum assisted pre-wet (right image). Optical cross section shows no copper plated in the bottom of the poorly pre-wet coupon in contrast to the fully filled vacuum pre-wet TSV.

5.4.5 Eliminating seam voids

Seam voids result from a filling process that is almost conformal: the sidewalls grow toward the center and the copper crystal grains at the converging circular growth front don't merge perfectly, instead sub-micron voids, approximately the size of the copper grain structure, are left along the centerline. Seam voids are most problematic in large TSV structures, such as a 30 micron diameter by 150 micron deep interposer TSV. An open question, is what happens to the plating solution that is trapped in these spaces? Some researchers[54] have reported a spike in the FIB vacuum chamber pressure as the void is cut open, presumably due to sudden sublimation of entrapped fluid. Eliminating seam voids requires experimental vigilance because it requires extreme care to ensure the cross-sectioning (FIB or lap & polish) stops on the centerline of the TSV.

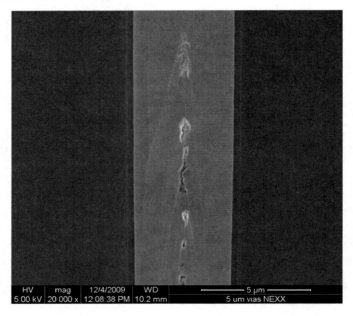

Fig. 5.13 Seam voids along centerline of TSV. Via structure and FIB analysis courtesy of IMEC, filling process by NEXX.

5.4.6 TSV copper grain structure

Soon after electrodeposition the grain structure of the copper within and above the TSV exhibits quite a diverse number of patterns of size, shape, and distribution as shown in Figure 5.14. Copper is well known to self-anneal at room temperature and the rate of grain growth depends on many factors including both the as-grown grain structure and distribution of impurities as well as the nature of surfaces that suround the copper.

Large grain structure is correlated with low impurity copper which performs better in subsequent annealing and CMP processing steps.

Understanding how these factors interact for TSVs and what is important for eventual process control and product reliability will be a vigorous area of TSV filling research over the next decade.

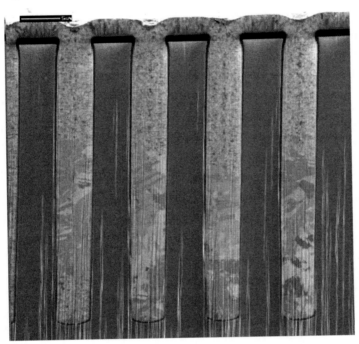

Fig. 5.14 Copper grain structure within TSV shows both large and small grains shortly after plating. Via structure courtesy of IMEC, filling process by NEXX.

5.5 ECONOMICS OF TSV FILLING

TSV filling is a relatively long process, for example from 15 to 60 minutes for a 5 micron diameter by 60 micron deep TSV structure. Fast turnaround, non-destructive post filling analysis is not yet available so the TSV filling will be run at an operating point with high yield confidence. As the chemical management of TSV filling processes for different additive systems are more well understood, and the root causes of voids, dimples, and other defects are well mapped to chemical control parameters, the shorter times will be used. Even at these "short" fill times, plating equipment different from the fountain system used for copper damascene are necessary. For a 15-minute fill time a plating system must have approximately 20 wafers plating in parallel— as opposed to a 3 plating cell system for copper damascene—to produce an output of 60 wafers per hour.

Chemical cost is an important component of the overall COO for TSV filling, it is likely to be about 10–25% of the overall unit process COO. Consumption at the anode—where organic additives could be oxidized--is limited in production tools by various proprietary means of limiting fluid contact of the organic additives to the anode surface. Reactions at the wafer, and in the solution, do however create breakdown products that require either continuous (bleed & feed) or periodic (bath-dumping) refreshing of the chemistry. Estimates of chemical lifetime require a moderate amount of production flow and so are not well known at this time.

5.5.1 Multi-step TSV filling

Most of the development work on TSV filling centers on optimizing a single chemistry mixture for the complete filling process similar to the methods developed for Cu-D applications. A single chemistry approach is less complex but also may be less optimal from a total cost perspective; therefore, we are exploring the potential advantages of using a multi-step filling sequence.

An example of a multi-step TSV filling process is as follows:
1. Pre-wet to ensure no air entrapped in TSV
2. Pre-treatment to prepare Cu surface for additive adsorption
3. Stage-1 plating with strong conformal behavior to add large volume of copper in shortest time
4. Stage-2 plating with high accelerator concentration for strong bottoms-up filling to minimize void risk
5. Stage-3 plating with leveler to minimize overshoot (mound formation) and to provide a bright copper finish in the case of patterned TSV deposition
6. Final clean and dry

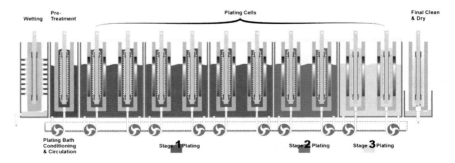

Fig. 5.15 Electrodeposition tool configured for high throughput multi-stage filling of TSVs. 20 wafer plating positions may be configured for either a single stage TSV fill or a multi-stage TSV filling process.

By breaking the TSV fill up into optimized sub-steps, the overall process time and, therefore, cost may be reduced. Because the same family of additives are used for the different steps, this doesn't add significant effort to bath maintenance and control. Figure 5.15 shows an example of a tool with 20 wafer plating positions configured for mulit-step TSV filling. The relative number of process modules for each of the steps is adjusted to match the process times for the particular steps, such that wafers proceed through the tool in a semi-continous manner. A large number of simultaneous wafer plating positions is required, either for a single-step or multi-step filling process, because of the relatively long time of the TSV fill process.

Acknowledgements

The author would like to acknowledge the exerpties and hard work of the NEXX engineering team, Zhen Liu, Johannes Chiu, and Richard Hollman who have done much of the TSV development work for NEXX Systems, and Josh Drexler for providing the figures. Also we would like to thank IMEC and Sematech, as well as many NEXX customers, for providing samples for TSV processing and expert feedback on TSV developments.

References

1. Commercial copper chemistry vendors: Enthone Cookson, Dow Electronic Materials, Atotech, BASF, Sinyang Chemicals

2. Commercial copper plating tool vendors: NEXX, EEJA, Semitool, Ebara, Novellus

3. Andricacos, P. C., Uzoh, C., Dukovic, J. O., Horkans, J., Deligianni, H. (1998). Damascene copper electroplating for chip interconnections, *IBM J. Res. & Dev.*, **42** (5), pp. 567-574.

4. Vereecken, P. M., Binstead, R. A., Deligianni, H., and Andricacos, P. C. (2005). The chemistry of additives in damascene copper plating, *IBM J. Res. & Dev.*, **49** (1), pp. 3-18.

5. Hull, R. O. (1939). Apparatus and process for the study of plating solutions, *U.S. patent 2,149,344.*

6. Beaver, J. F. (1952). Method of plating copper, *U.S. patent 2,602,774.*

7. Fellows, R. A. et.al. (1956). Electrodeposition of copper from an acid bath, *U.S. patent 2,738,318.*

8. Luhn, O., Van Hof, C., Ruythooren, W., Celis, J. -P. (2008). Filling of microvia with an aspect ratio of 5 by copper electrodeposition, *Electrochimica Acta* doi: 10.1016/j.electacta.208.04.002.

9. Dahms, Wolfgang (1990). Aqueous acidic bath for electrochemical deposition of a shiny and tear-free copper coating and method using same, *U.S. patent 4,975,159.*

10. Sonnenberg, W., Fisher, G., Bernards, R. F., Houle, P., Copper electroplating solutions and processes, *U.S. patent 5,252,196.*

11. Hurtubise, Too, Cheng (1998). Copper electrochemical deposition: is it capable of meeting the stringent reliability criteria of the copper damascene process?, *Enthone-OMI*, pp. 1-4.

12. Bonou, L., Eyraud, M., Denoyel, R., Massiani, Y. (2002). Influence of additives on Cu electrodeposition mechanisms in acid solution: direct current study supported by non-electrochemical measurements, *Electrochimica Acta* **47** (2002), 4139-4148.

13. Josell, D., Moffat, T. P., and Wheeler, D. (2007). Superfilling when adsorbed accelerators are mobile, *J. Electrochemical Soc.*, **154** (4), pp. D208-D214.

14. Lee, J. -M. and West, A. C. (2005). Impact of pulse parameters on current distribution in high aspect ratio vias and through-holes, *J. Electrochemical Soc.*, **152** (10) pp. C645-C651.

15. Willey, M. J. and West, A. C. (2006). Microfluidic studies of adsorption and desorption of polyethylene glycol during copper electrodeposition, *J. Electrochemical Soc.*, **153** (10), pp. C728-C734.

16. Willey, M. J. and West, A. C. (2006). A microfluidic device to measure electrode response to changes in electrolyte composition, *Electrochem. and Solid-State Lett.*, **9** (7), pp. E17-E21.

17. Willey, M. J., Reid, J., and West, A. C. (2007). Adsorption kinetics of polyvinylpyrrolidone during copper electrodeposition, *Electrochem. and Solid-State Lett.*, **10** (4), pp. D38-D41.

18. Willey, M. J. and West, A. C. (2007). SPS adsorption and desorption during copper electrodeposition and its impact on PEG adsorption, *J. Electrochemical Soc.*, **154** (3), pp. D156-D162.

19. Gallaway, J. W. and West, A. C. (2008). PEG, PPG, and their triblock copolymers as suppressors in copper electroplating, *J. Electrochemical Soc.* **155** (10), pp. D532-D639.

20. Willey, M. J., Emekli, U. and West, A. C. (2008). Uniformity effects when electrodepositing Cu onto resistive substrates in the presence of organic additives, *J. Electrochemical Soc.* **155** (4), pp. D302-D307.

21. Newman, M. W., Muthukumar, S., Schuelein, M., Dambrauskas, T., Dunaway, P. A., Jordan, J. M., Kulkarni, S., Kinde, C. D., Opheim, T. A., Stingekl, R. A., Worwag, W., Topic, L. A., Swan, J. M. (2006). Fabrication and electrical characterization of 3D vertical interconnects, *IEEE 2006 Elect. Comp. and Tech. Conf.*, pp. 394-398.

22. Moffat, T. P., Wheeler, D., Edelstein, M. D., Josell, D. (2005). Superconformal film growth: mechanism and quantification, *IBM J. Res. & Dev.*, **49** (1), pp. 19-36.

23. Moffat, T. P., Wheeler, D., Kim, S. -K., and Josell, D. (2006). Curvature enhanced adsorbate coverage model for electrodeposition, *J. Electrochemical Soc.,* **153** (2), pp. C127-C132.

24. Kim, S. -K., Josell, D., and Moffat, T. P. (2006). Electrodeposition of Cu in the PEI-PEG-CI-SPS additive system, *J. Electrochemical Soc.,* **153** (9), pp. C616-C622.

25. Kim, S. -K., Josell, D., and Moffat, T. P. (2006). Cationic surfactants for the control of overfill bumps in Cu superfilling, *J. Electrochemical Soc.,* **153** (12), pp. C826-C833.

26. Moffat, T. P., Wheeler, D., Kim, S. -K., Josell, D. (2007). Curvature enhanced adsorbate coverage mechanism for bottom-up superfilling and bump control in damascene processing, *Electrochimica Acta* **53** (2007), pp. 145-154.

27. Kondo, K., Yonezawa, T., Mikami, D., Okubo, T., Taguchi, Y., Takahashi, K., and Barkey, D. P. (2005). High-aspect-ratio copper-via-filling for three-dimensional chip stacking, *J. Electrochemical Soc.* **152** (11), pp. H173-H177.

28. Pasquale, M. A., Barkey, D. P., and Arvia, A. J. (2005). Influence of additives on the growth velocity and morphology of branching copper electrodeposits, *J. Electrochemical Soc.* **152** (3), pp. C149-C157.

29. Forman, R. S. (2007). Advances in wafer plating: meeting the next challenge of through-silicon-via processing, *Chip Scale Rev.,* pp. 61-67.

30. Lee, D. W., Pyn, Sung G., Ko, C. J., Lee, M. H., Kim, Sibum, and Lee, J. G. (2004). Effect of acidity on defectivity and film properties of electroplated copper films, *J. Electrochemical Soc.,* **151** (3) pp. C204-C209.

31. Spiesshoefer, S., Rahman, Z., Vangara, Polamreddy, S., Burkett, S., and Schaper, L. (2005). Process integration for through-silicon vias, *J. Vac. Sci. Technol.* **A23** (4), pp. 824-829.

32. Burkett, S. L., Qiaoi, X., Temple, D., Stoner, B., and McGuire, G (2004). Advanced processing techniques for through-wafer interconnects, *J. Vac., Sci. Technol.* **B221** (1), pp. 248-256.

33. Pasquale, M. A., Gassa, L. M., Arvia, A. J. (2008). Copper electrodeposition from an acidic plating bath containing accelerating and inhibiting organic additives, *Electrochimica Acta* **53** (2008), pp. 5891-5904.

34. Akolkar, R. and Landau, U. (2004). A time-dependent transport-kinetics model for additive interactions in copper interconnect metallization, *J. Electrochemical Soc.* **151** (11), pp. C702-C711.

35. Akolkar, R. and Landau, U. (2009). Mechanistic analysis of the "bottom-up" fill in copper interconnect metallization, *J. Electrochemical Soc.* **156** (9), pp. D351-D359.

36. Mendez, J., Akolkar, R. and Landau, U. (2009). Polyether suppressors enabling copper metallization of high aspect ratio interconnects, *J. Electrochemical Soc.* **156** (11), pp. D474-D479.

37. Li, X., Drews, T. O., Rusli, E., Xue, F., He, Y., Braatz, R. and Alkire, R. (2007). Effect of additives on shape evolution during electrodeposition, I. Multiscale simulation with dynamically coupled kinetic Monte Carlo and moving-boundary finite-volume codes, *J. Electrochemical Soc.* **154** (4), pp. D230-D240.

38. Rusli, E., Xue, F., Drews, T., Vereecken, P. M., Andricacos, P., Deligianni, H., Braatz, R. D., and Alkire, R. C. (2007). Effect of additives on shape evolution during electrodeposition, II. Parameter estimation from roughness evolution experiments, *J. Electrochemical Soc.* **154** (11), pp. D584-D597.

39. Qin, Y., Li, X., Xue, F., Vereecken, P. M., Andricacos, P., Deligianni, Hariklia, Braatz, Richard D., and Alkire, Richard C. (2008). Effect of additives on shape evolution during electrodepsotion, III> Trench infill for on-chip interconnects, *J. Electrochemical Soc.*, **155** (3), pp. D223-D233.

40. Willis, M. and Alkire, R. (2009). Additive-assisted nucleation and growth in electrodeposition, I. Experimental studies with copper seed arrays on gold films, *J. Electrochemical Soc.* **155** (10), pp. D377-D384.

41. Stephens, R. M., Willis, M., and Alkire, R. C. (2009). Additive-assisted nucleation and growth by electrodeposition, II. Mathematical model and comparison with experimental data, *J. Electrochemical Soc.* **156** (10), pp. D385-D394.

42. Huerta Garrido, M. E. and Pritzker, M. D. (2008). Voltammetric study of the inhibition effect of polyethylene glycol and chloride ions on copper deposition, *J. Electrochemical Soc.* **155** (4), pp. D332-D339.

43. Li, Y. -B., Wang, E., and Li, Y. -L. (2009). Adsorption behavior and related mechanisms of Janus Green B during copper via-filling process, *J. Electrochemical Soc.*, **156** (4), pp. D119-D124.

44. Mocoteguy, P., Gabrielli, C., Perrot, H., Zdunck, A., and Sanz, D. Nieto (2006). Influence of the anode and the accelerator on copper bath aging in the damascene process, *J. Electrochemical Soc.*, **153** (12), pp. G1086-G1098.

45. Dixit, P. and Miao, J. (2006). Fabrication of high aspect ratio 35 μm pitch interconnects for next generation 3-D wafer level packaging by through-wafer copper electroplating, *IEEE 2006 Elect. Comp. and Tech. Conf.*, pp. 388-393.

46. Dixit, P., Chen, X., Miao, J., Preisser, R. (2007). Effect of improved wettability of silicon-based materials with electrolyte for void-free copper deposition in high aspect ratio through-vias, *Thin Solid Films* **516** (2008), pp. 5194-5200.

47. Chiu, J. (2009). Angled top-down grind method for improved via filling analysis, *Advanced Packaging*, May 2009.

48. Schultz, A. D., Feng, Z. V., Biggin, M. E., and Gewirth, A. A., *J. Electrochemical Soc.*, **153** (12), pp. G1086-G1098.

49. Barkey, book chapter

50. Chiu, J., Liu, Z., Keigler, A., unpublished (2009)

51. Panovic, M., Schlesinger, M., *Fundamentals of Electrodeposition,* © 2006 Wiley Interscience, p. 77

52. Barkey, D. P., Oberholtzer, F., Wu, Q., *J. Electrochemical Soc.,* **145** (590) 1998.

53. Chemical analysis equipment vendors: ECI www.ecitechnology.com, Ancosys www.ancosys.com, Metrohm www.metrohmusa.com

54. Ruythooren, W., personal communication, IMEC

Chapter 6

3D TECHNOLOGY PLATFORM: TEMPORARY BONDING AND RELEASE

Mark Privett

Brewer Science, Inc.

6.1 BACKGROUND AND INTRODUCTION

Frequently referred to as a "more than Moore approach," stacking of ICs with the use of through-silicon vias (TSVs) was identified as "another direction of improvement" of computing power as early as 1985.[1] From 1985 until the early part of this decade, improvements in lithography and packaging have delivered on the promise of Moore's law, but limitations in lithographic technology and packaging are making costs for further improvements prohibitively high. TSV technology has had limited usage in high-volume production to date because the creation of TSVs in wafers more than 100 µm thick is both time consuming and cost prohibitive.

Thinning a wafer to a thickness much less than 100 µm reduces the cost of TSV creation to a more manageable level. The EMC3D consortium has stated that the cost of TSV creation is now less than $200/wafer. Further, in a testament of potential performance improvements of a package made with TSV technology, Roger Carpenter, of Javelin Design Automation, recently announced modeling software results for its PathFinding software. He stated that "the PathFinding results indicate close to 10 times decrease in dynamic interconnect power of the I/O interface using 3-D interconnect technologies, subsequently allowing the bus-width to increase by 16 times in 3-D implementation without exceeding the power of the original SIP implementation."[2] Modeling software may validate the value of thinning to create TSVs, but to be cost effective, new technology must be developed for handling and processing the thinned wafer.

3D Integration for VLSI Systems
Edited by Chuan Seng Tan, Kuan-Neng Chen and Steven J. Koester
Copyright © 2012 by Pan Stanford Publishing Pte. Ltd.
www.panstanford.com

Stacking wafers by means of a permanent bonding process and then thinning the device wafer already in the permanent stack to expose the previously formed via or creating the via in the thinned wafer already on the permanent device stack has been proposed and is in use for production of 3D stacked packages.[3] Upon exposure of the TSV, or creation of the TSV, several BEOL process steps are still necessary to create the backside redistribution and interconnect, including passivation, photolithography, metallization, and so on. The stacked device, already of significant value because of multiple layers of circuitry, must therefore be exposed to the thermal and chemical processes necessary to create the backside redistribution and interconnects.

Alternatively, several technologies use a temporary carrier, such as a carrier wafer, to allow the thinning and processing of a device wafer. The temporary carrier plus the device wafer may be processed just as if the stack were a wafer of standard thickness. The pair may be handled and processed without risk to the device through completion of the backside redistribution and interconnects. Upon completion of the interconnects, the thinned wafer may be permanently attached to the front side of another wafer and then separated from the carrier wafer for wafer-to-wafer processing (W2W), or separated from the carrier wafer and diced, and then dies may be attached one at a time to the devices. This flexibility allows for W2W bonding, die-to-wafer bonding (D2W), or die-to-die (D2D) bonding and therefore permits many alternatives in processing TSV-containing wafers.

In the following pages, we will explore temporary bonding and TSV creation using several bonding and release (debonding) technologies currently in production on thinned wafers. Although backside processes will vary significantly from company to company, the handling and alteration of a thinned wafer on a carrier, including not only a thin device wafer and a thick carrier wafer but also the bonding material, must remain intact through high-temperature, high-energy-flow, and highly corrosive environments during backside processing. Each bonding and release technology has been developed to address these challenges in a slightly different way. The release of the completed "two-sided device" from the carrier is a major consideration and will involve significant review.

6.2 TEMPORARY BONDING/TSV PROCESS FLOWS

Several process flows have been suggested for TSV creation/completion on thinned wafers supported by carriers. The flows have been divided into two major categories, "vias first" and "vias last," for simplification by Phil Garrou.[4]

In the vias-first process, polysilicon vias as deep as the final, thinned device thickness are formed and filled either prior to device creation, i.e., a true vias-first process, or after device creation in areas left in the design for via creation, i.e., a "vias-middle" process. After the device wafer and via creation are complete, the device wafer is attached to a carrier, facing toward the carrier, and thinned to the appropriate thickness to expose vias prior to backside processing to create a redistribution and interconnect layer (Fig. 6.1).

Fig. 6.1 Vias-first (-middle) process. Vias are created on a full-thickness device wafer after completion of CMOS processes. The device wafer is bonded to a carrier wafer, thinned to expose vias, and then released.

In the vias-last process, a device wafer is created with a "blind" interconnect on the backside of the device, buried in the silicon as the device is built up. After the device buildup is complete, the device wafer is attached face first to a carrier for thinning and backside processing. The device wafer is thinned, vias are etched through the silicon to the "blind" interconnect, the device wafer undergoes passivation, and conductors are deposited on the device to create a functional exposed backside interconnect and redistribution (Fig. 6.2). One such process flow contains eight major steps and has been fully described by the Electronics and Information Technology Laboratory of the French Atomic Energy Commission (CEA Leti).[5] Given that the vias-last process will, by definition, involve more steps after wafer thinning, including via etching, via wall passivation, and via fill or plating, a review of such a process will include all steps included in the vias-first processes.

Fig. 6.2 Vias-last process. CMOS processes are completed on the device wafer with no via formation. The device wafer is mounted onto a carrier wafer and is thinned, and vias are etched into the device wafer while it is on the carrier wafer. The device wafer is then separated from carrier wafer.

6.2.1 Step 1: Bonding

Bonding one wafer to another has been reviewed in previous publications.[6,7] For temporary bonding schemes, this bonding requires coating a device with an adhesive that must later be removed and then bonding the device wafer facedown onto a carrier upon completion of all CMOS processes on the device wafer. The device face or the carrier is first coated with a bondable coating. The device and carrier are brought into contact in a vacuum to ensure that no voids remain during bonding, and the coating is frozen or cured to complete the bonding process. Material choices vary significantly and will be briefly reviewed in section 6.3, but several important characteristics must be shared by all materials and technology choices.

6.2.1.1 Device preparation

After CMOS processing is complete, the device wafer must be prepared for bonding to the carrier wafer, including cleaning, edge preparation, and moisture elimination. First, care should be taken to ensure that the device wafer is free of all particulates. Particulates on the surface may create voids in the bond line or damage the surface during bonding. Second, thinning the device wafer will create a very fragile, knifelike edge on the wafer. Several alternatives have been explored to prevent chipping the wafer edge and are explored more thoroughly in section 6.2.2. A DISCO Corporation process for pre-trimming wafers to eliminate this knife edge gave the most robust wafer edge and is currently recommended. Finally, devices are commonly finished with a passivation step that covers areas of the wafer with oxides of silicon or with polyimide. These surfaces are known to absorb atmospheric moisture that could evolve during subsequent high-temperature processes, and devices should be baked just prior to coating and bonding to eliminate this moisture and prevent subsequent delamination.

6.2.1.2 Coating

Spin-coating the adhesive is the most common method of adhesive application. The spin-coating technology generally yields very good thickness control across the entire wafer, that is, it yields low total thickness variation (TTV). Adhesives currently on the market include both solvent-based adhesives that are baked to release the solvent and 100% solids pre-polymers that are reacted on the wafer to form polymer adhesives.

For solvent-based adhesive usage, bake processes are developed to ensure that no solvent is retained in the adhesive. Bake processes will differ on the basis of not only the adhesive but also the thickness of the adhesive. For thin coatings, baking may be a simple one- or two-step process, but for thicker coatings, multiple steps may be required. Solvent evolution rate in thick coatings, >30 μm, must be slow enough to avoid large blisters in the film but rapid enough to maintain throughput. The TTV of the film may be measured after baking is complete. Either glass or silicon wafers may be used as carriers.

For an adhesive material composed of 100% solids, solidifying the material may involve linearly building up the molecular weight or building it up by cross-linking or "curing." Curing may be completed in the bond line by using acrylic cross-linking activated through exposure to ultraviolet (UV) light or by using thermally activated curing. For the UV-activated material, the use of glass carriers is standard because the carrier must be transparent in the UV range. For the thermally cured materials, the carrier may be either glass or another silicon wafer.

6.2.1.3 No Backside Contamination

Regardless of which wafer, device, or carrier, is coated, the backside of the wafers must always be clean. For the device wafer, the backside will be ground away during the thinning process, but the polymer bonding material would likely contaminate grind wheels. Therefore, the backside of the device wafer must be free of any bonding material. The carrier backside is likely to be in contact with electrostatic and vacuum chucks in several steps during backside processing, so it should be free of any polymer to prevent sticking to or contaminating these chucks.

6.2.1.4 Uniform Bond Line

Whether the process forms vias first or vias last, it is important to create and maintain a uniform bond-line thickness throughout the process. In a vias-first process, the device will be thinned down to the desired thickness to expose the polysilicon or metal (W or Cu) vias. A uniform bond line is necessary to eliminate wafer bowing risks and the potential for some vias to be exposed before others and thus damaged during thinning or other subsequent processes. For example, with a non-uniform bond line, for instance, a bond line several microns thicker at the wafer's edge, a pre-filled copper via might be exposed during the thinning process, grinding and smearing copper across the surface of the device. In a vias-last process, maintaining a uniform bond line is necessary to ensure that process requirements used to etch, insulate, and fill vias will be uniform across the wafer.

Bonding layer coatings are most frequently created with spin-coating processes, which may easily be optimized and monitored using standard profilometers or interferometer technology. However, coating thickness is not the only factor contributing to bond-line thickness. The device wafer and the carrier are brought together with this bonding layer in liquid form, the material being either a low-molecular-weight material that is subsequently cross-linked or a viscous high-temperature thermoplastic material that will solidify upon cooling. Tooling, e.g., vacuum chucks, used to bring the device and carrier together must be flat and parallel. One rigid surface is being brought into contact with a viscous liquid surface. This method reduces the risk of void formation (to be discussed in section 6.2.1.5) but increases the risk of a non-uniform bond line. Direct silicon bonding and Cu/Cu diffusion bonding have zero bond line and therefore do not have the risks of a non-uniform bond line. Benzocylcobutene (BCB) bonding, with a layer of viscous liquid B-staged BCB in between wafers, is very similar to bonding with a thermoplastic or liquid layer that becomes a temporary bonding agent.

Ultimately, the thinned wafer thickness uniformity is the greatest concern. Coating uniformity, bonding, and even wafer thinning may contribute to thinned wafer TTV. In the end, thinned wafer thickness in the bonded pair may be confirmed using infrared (IR) interferometry available from Frontier Semiconductor or by the spectral coherence technology offered by ISIS

Sentronics, or by micrometer measurement of the wafer after separation from the carrier. While this last technique is the simplest and least expensive, it unfortunately has the highest risk. Any handling of the final device, now a silicon wafer that is between 20 μm and 70 μm thick, is very likely to damage the fragile device.

6.2.1.5 Void-Free Bond Line

It is critical to backside processes that no void, which may contain air, be trapped in the bond line. Subsequent processes, including thinning, but more importantly, any process done under vacuum after thinning, could cause a void to expand and blister or burst the wafer. Processes have been developed to ensure a void-free bonding process. Bonding is typically done under vacuum to ensure that no air bubbles are trapped in the bonding material. Inspection of the bonding may be done by scanning acoustic microscopy.[8] J. McCutcheon of Brewer Science has also shown that through-wafer inspection via IR light and an IR camera may be used to detect voids.[9]

6.2.2 Step 2: Thinning

Once bonded to the carrier wafer, the device wafer must be thinned to the appropriate thickness using traditional wafer-thinning technologies. Coarse grinding is used to thin the wafer to within approximately 20 μm of the desired thickness, and then fine grinding and polishing are used to complete the process. In the case of a vias-first process, care must be taken not to grind into the preexisting via-fill material if it is a metal; therefore, tight TTV control is required throughout the entire process.

During development of these thinning processes, it has been noted that the wafer-thinning process creates a knifelike edge around the circumference of the wafer (Fig. 6.3). Although the edge may remain intact during the thinning process, it remains very fragile, and subsequent handling risks chipping the wafer. Developers have proposed several methods to protect the wafer edge, including encapsulation of the wafer edge, pre-trimming of the device wafer to eliminate the knife edge, and even trimming the knife edge off the thinned wafer while it is still on the carrier. Dr. D. Bai investigated a number of these proposals (Fig. 6.4) for protection and reported that the most effective way to prevent chipping of the thinned wafer edge is pre-trimming the device wafer, as proposed by DISCO Corporation.[10] The technique essentially accomplishes two parallel tasks to prevent damage to the wafer edge. First, the wafer is bonded to the carrier without the risk of the extreme edge of the wafer being exposed. Second, the knife edge that would be created by thinning at the bevel of the wafer is blunted. Implementing both preventative measures in one process step greatly reduces the risks of device wafer chipping and breakage (Fig. 6.5).

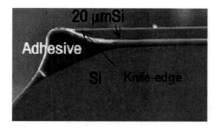

Fig. 6.3 SEM cross-section of the bonded pair edge. Photo courtesy of CEA-LETI.

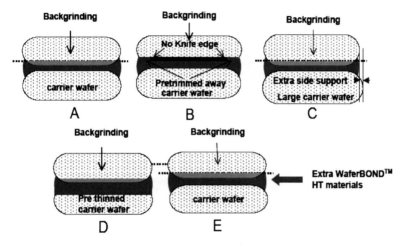

Fig. 6.4 Different edge proposal schemes tested by D. Bai. (A) Standard bonding that would create knife edge. (B) Pre-trimming (notching) of a device wafer to eliminate the knife edge. (C) Use of a large-diameter carrier to support more adhesive at edge and better encapsulate knife edge. (D) Use of a pre-thinned carrier to emulate a larger-diameter carrier, as in C. (E) Use of additional bonding agent at outer circumference of bonded pair to better encapsulate device edge. Only pre-trimming eliminates the knife edge of the thinned device wafer.

Actual inspection and testing of the direct performance of the bonding material essentially stop at this point in the process and become important again upon debonding of the completed thinned TSV-containing device. However, the carrier wafer, and just as importantly, the bond layer holding the device wafer to the carrier wafer, must provide structural support to the thinned wafer through a number of additional steps used to create vias and make them electrically conductive.

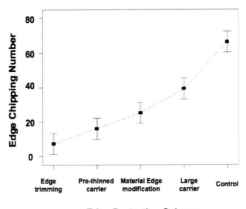

Edge Protection Scheme

Fig. 6.5 Statistical analysis of the chipping data. Pre-trimming device wafers gives better protection than any additional encapsulation of the knife edge.

As previously cited by Henry *et al.*, six major additional steps are required to create an electrically functional via.[5] Some of these steps require exposure to harsh chemicals, including acidic, alkaline, or strong solvent chemistry, high temperatures, vacuum at high temperature, and so forth. The bonding material used must survive each of these different environments without any degradation, support stress caused by coefficient of thermal expansion (CTE) mismatches, provide for a relatively easy separation of a thinned wafer from the carrier, and allow easy removal. Supporting stress under any of these conditions by itself is a tall order for most materials, but these bonding materials must have all these characteristics.

6.2.3 Step 3: Etching Vias

Via etching, for the vias-last process, is commonly the first of these tests of the material and the bonded pair. Via creation may be achieved using wet chemistry as described by Trichur and Shao,[11] laser drilling,[12] or, more commonly, deep reactive ion etching (DRIE). Wet etching of vias is generally limited to low-density vias with a 1:1 aspect ratio. Laser drilling of vias may be cost effective if fewer than 10,000 vias are to be made per wafer and the diameter of the vias is greater than 10 µm. Otherwise, DRIE processing may be the method of choice. DRIE via creation has been extensively reviewed elsewhere in the past,[13] and no further review will be given here.

6.2.4 Step 4: Via Insulation

In order to make vias electrically conductive, it is necessary to fill or plate the via, frequently with copper. Before filling a via, it is necessary to insulate the via to prevent metal migration into the Si. SiO_2 is commonly deposited using chemical vapor deposition (CVD) processes to create this insulating layer. Although several technologies exist for depositing SiO_2 insulating layers, only plasma-enhanced CVD (PECVD) oxides and PECVD tetraethyl orthosilicate (TEOS) films are likely to be suitable for use with temporary bonding for via creation. Other processes, including thermal oxide deposition, low-pressure CVD TEOS, and sub-atmospheric CVD processes, require process temperatures in excess of 400°C, which exceeds the thermal stability of most organic materials.

Although PECVD oxide deposition and PECVD TEOS may be used with temporary bonding, the bonded pair is very sensitive to these high-temperature processes. PECVD requires processing temperatures ranging from 150°C to 400°C. PECVD TEOS requires processing temperatures ranging from 250°C to 400°C. Higher-temperature processes provide better step coverage and generally higher-quality insulation, but these high-temperature processes pose other risks. Organic bonding materials soften significantly or liquefy, and silicone bonding materials expand greatly at temperatures higher than 200°C. Stress created by CTE mismatches within CMOS circuitry and thin Si wafer may cause the thinned, now nearly perforated, wafer to exhibit localized buckling and possibly even delamination within the soft silicone or organic adhesive.

A cooled chuck below the device is used to control device temperature, but surface temperatures on the top of the device may be significantly higher than chuck temperature. Plasma bombardment of the surface may significantly raise the temperature of the device, increase the local stress within the wafer as described above, and even create a thermal gradient through the thickness of the wafer and carrier, causing bowing. Given that polymeric materials have lower temperature limitations than Si, it is very important to test and characterize the temperature of the bond line during PECVD or PECVD TEOS processes.

6.2.5 Step 5: Via Metallization

Once the via is insulated to prevent migration of the conductive material into surrounding silicon and isolate the via, the via must be plated or filled to complete the conductive path. Initially, a seed layer is placed with either a CVD or a sputtering process, and via plating or filling continues with electroplating in highly corrosive wet chemistry. Although temperatures during plating are near room temperature, the corrosive nature of the fluids may be damaging to bonding materials or interfaces between the wafers (carrier or device) and the bonding material. The bonding material must fully encapsulate the device surface during the plating process and prevent delamination and damage to the device surface during plating.

New chemistries and processes are being developed to optimize the via-fill process by minimizing via-fill time and eliminating risks of void formation during via filling. Bonding materials must withstand any of these new chemistries without compromising encapsulation of the device and without softening and thereby allowing movement of the device.

6.2.6 Step 6: Metal Protection

Metal surfaces are then protected from damage and degradation in subsequent processes and environments by using a polymer coating for passivation. BCB 4024 is a common passivation polymer and requires either a 10-hour, 200°C or 1-hour, 250°C curing process. Again the bonding material must not degrade during such high-temperature processes. Degradation could result in volatile formation and blistering or polymer cross-linking or reaction in the bond line that result in particles or residues that are very difficult to remove.

6.2.7 Step 7: Under-Bump Metallization and Interconnect Formation

Typical interconnect formation is completed by solder bumping. Tin/silver/copper (SAC) solders are the most common metal interconnect for use in flip chip and packaging technologies and are evolving from these technologies for use in 3D packaging with TSVs. A three-metal stack may be used for under-bump metallization, including 0.2 μm of titanium, 0.6 μm of nickel to act as a copper diffusion barrier, and 0.1 μm of gold to prevent nickel oxidation. Finally, SAC solder balls are placed on the device and reflowed to complete intermetallic formation and electrical connection. Reflow typically requires exposure to a temperature of 260°C for 45 to 90 seconds. Again, the bonding material must withstand this high temperature without degradation or other change. Alternatively, solder or copper may be electroplated on the devices.

The processes described above may vary significantly from application to application. Times, temperatures, and chemistries differ on the basis of foundry experience and device performance and reliability requirements. PECVD conditions (temperatures higher than 250°C in a vacuum for up to 10 minutes), polymer cure cycles (1 hour at 250°C, in one example), and plating and cleaning chemistries all represent specific requirements for the bonding material and handling of the thinned wafer.

6.3 TEMPORARY BONDING/DEBONDING: MATERIALS OPTIONS

Semiconductor front-end-of-line (FEOL) processing involves hundreds of steps over several weeks to complete devices. Wafers used in these processing steps are typically 650 μm to 750 μm thick so that they will be robust enough to handle all processing steps without breakage. Electronic components packaging is, however, required to be as thin as possible in the completed device not only to save space but also to enable efficient removal of heat generated within the device during operation. Wafers have been thinned to thicknesses of 200 μm to 250 μm prior to final packaging for many years.

As electronic performance requirements have increased over the years, thinning of devices has continued to evolve to a point where wafers are typically thinned to 125 μm to 150 μm thicknesses. For these thinning processes, wafers have been supported on tapes or on rigid carriers with wax bonding materials holding the thinned device. After thinning, the tape may be peeled off with the thinned wafer on a dicing tape on a vacuum chuck. Waxes, coated onto a rigid carrier, provide another option. The rigid carrier may allow the production of thinner wafers, but wafer thickness may not be optimal because the wax coating on the carrier does not provide satisfactory uniformity. Thinning to thicknesses of 100 μm to 150 μm on tape may result in flexing of the device wafer and, ultimately, breakage because the wafers are not rigidly supported. Also, on carriers bonded by wax, TTV may be too great.

6.3.1 Thermoset Materials: UV Cured

One option for temporary wafer bonding for backside processing is to use a transparent carrier with a UV-curable polymer to act as the bonding agent. C. Kessel introduced this material in his paper at the 2004 International Wafer Level Packaging Conference (IWLPC)e.[14] This thinned wafer handling system has two key elements that are required to make the system work: the bonding layer itself is a UV-curable polymeric material, and a light-to-heat conversion (LTHC) layer is included to allow for separation of the carrier from the device. This system is represented graphically in Fig. 6.6.

Spin-coating adhesive on the device allows complete fill and planarization of the device surface. The coated wafer is then bonded to the support glass, previously coated with the LTHC layer. The bonding process is completed using UV light to cure the bonding resin to a semi-rigid elastomeric layer. Support glass diameter is slightly larger than that of the device wafer to ensure adhesive support at the edge of the wafer. The proper adhesive thickness allows for full encapsulation of the device topography, including bumps.

After bonding, the wafer may be thinned to the desired thickness, typically 50 μm. To complete TSV and backside redistribution, the thinned device wafer and the carrier may be handled through the processes described in section 6.1 as if the pair were a full-thickness silicon wafer.

With processing complete, the wafer is attached to a dicing tape, and a 1064 nm YAG laser is rastered over the carrier surface to activate the LTHC layer. This layer heats to a temperature of more than 1000°C, and the integrity of the layer is destroyed by rapid void formation. The glass support is then lifted from the wafer. Finally the elastomeric layer is peeled from the wafer much like tape was pulled from thinned wafers in the past.

The resulting wafer is thinned and mounted to dicing tape for easing dicing. Die may be subsequently assembled into packages using the standard pick-and-place technology.

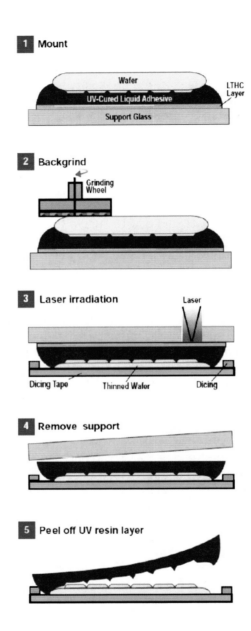

Fig. 6.6 Wafer support system illustration from 3M depicting bonding, thinning, glass removal, and then elastomer removal.[14]

6.3.2 Thermoplastic Bonding Materials: Thermomechanical Slide Debonding

A second process proposed for thinned wafer handling uses solvent-cast spin-applied thermoplastic materials in place of the UV-curable elastomeric adhesive.[15] The basics of this process are depicted in Fig. 6.7.

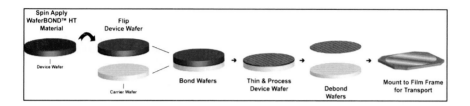

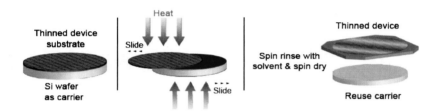

Fig. 6.7 Illustration of a wafer support system using bonding materials from Brewer Science, Inc. The system includes steps for coating, bonding, backside processing, debonding, and cleaning.

As with the UV-curable polymer system, the coating is spin-coated onto the device wafer to planarize topography. Baking of the device wafer is required to eliminate any solvent retained in the solvent-cast polymer. The pair is then bonded under heat and vacuum. Because the polymer is solid at room temperature and becomes a highly viscous liquid at high temperatures, the temperature is brought up well beyond the melting point of the polymer, and two platens bring the device and carrier together with a controlled amount of force in a vacuum environment. The device wafer may then be thinned and processed as necessary on the carrier.

Upon TSV and redistribution layer completion, the pair is heated once again to above the melting temperature of the polymer, and, using two vacuum chucks, the two wafers are slid apart. Each piece is solvent-cleaned in a step similar to a develop process on a full-size vacuum chuck. Finally the thinned device wafer is placed on dicing tape and is ready for packaging.

6.3.3 Release Layer Usage: Thermoset and Nonstick Materials

Thin Materials AG has recently announced the development and commercialization of a new material for temporarily bonding a wafer to a carrier. The process flow is similar to the two previously cited examples, with one preceding step. Initially, a thin "release" layer 100 nm to 150 nm thick is deposited by PECVD on the device wafer. A spin-coatable silicone elastomeric material is then used to coat the carrier wafer, which is then bonded to the device. The silicone material, between 20 μm and 200 μm thick, is thermally cured at temperatures between 150°C and 200°C. Bonding between the adhesive layer and the nonstick covered device is sufficient for thinning and backside processing. When completed, the device is held in place, either by a vacuum chuck or by tape on a vacuum chuck, and the carrier is simply peeled from the device.

Brewer Science is developing a room-temperature debonding scheme with an alternative release layer. For this material the release layer is spin-coated onto the carrier wafer. A solvent-cast spin-coatable thermoplastic polymer is coated onto the device wafer, which is then bonded to the carrier. When completed, the device wafer is held in place and the carrier is peeled away. Unlike with the Thin Materials AG solution, the thermoplastic is then stripped off the device wafer.

6.3.4 Thermoplastic Bonding Materials with Ablation Release

HD Microsystems, a joint venture between Hitachi Chemical and Dupont, has developed a high-temperature polyimide that may be used for temporary bonding. As with the others, the material is spin-coated and bonded to a carrier. The extremely high thermal stability of the polyimide then allows wafer thinning, complete backside processing of the wafer, and finally thermocompression bonding to a permanent device. Lastly, the glass carrier is separated from the die by using laser ablation through the glass or by using solvent debonding through a perforated glass.

6.4 PERFORMANCE EVALUATION

Requirements for the bonding materials proposed in section 6.2 remain largely defined only by their performance in processing. J. Hermanowski reviews these traditional properties for commercially available temporary bonding materials, but in the end acknowledges that actual process testing must be completed prior to technology selection.[16]

Although the electronics industry has for many years been defining the properties of polymeric materials for use in permanent bonding, underfilling, and encapsulation in terms of traditionally measured bulk materials properties, temporary bonding materials require some alternative materials properties that are, by definition, very different from those of permanent materials. The CTE, modulus, and thermal conductivity, for example, may be used to model the performance of an underfill material in use and in thermal cycle testing. However, the use of a silicone elastomer to enable the adhesive peeling process dictates the use of a material with a very low glass transition temperature (T_g) and a very high CTE. Because the thermoplastic material must melt to a high-viscosity liquid for debonding by thermomechanical slide-off debonding, the selected material must have a low T_g. Although the CTE is very high when the temperature is above the T_g of the material in both cases, the modulus is very low, and the material itself does not put stress on the device.

For traditional bonding materials, the bond performance may be tested either by plug pull or by die push-off at room temperature, up to approximately 125°C, but operating temperatures do not generally rise above this. In a bonded wafer, the risk to the device is at temperatures greater than 150°C and up to 300°C. Testing systems may be modified to test bond strength at these high temperatures, but such results may still not reflect performance. Currently, putting device wafers through full processing is the best way to test the performance of these materials.

This required processing presents another set of challenges. Bonding and debonding equipment is not readily available, and even via drilling, insulation, metallization, etc., are frequently not co-located. Further, test vehicles are not widely available. This challenging situation leads to long evaluation times for materials manufacturers and fewer development cycles in a year.

Several potential failure modes may present themselves in backside processing, and each material option mentioned above must be tested through device processing. These failure modes include edge chipping during wafer transfer from boat to boat and within process equipment and delamination due to high-temperature processing and/or chemical attack. More recently, the use of SAC solder bumps for interconnect formation is being tested. This new trend creates the potential for solder bump failure during debonding by peel or slide separation from the device or failure through contamination on the bump.

Once bonded, a device/carrier pair is first thinned, usually by using mechanical polishing technology. Early in development, there was some concern that this thinning process would generate a significant failure rate. Bond energies of materials were tested using the Maszara razor blade test.[15] Thinning of the wafers has, however, not resulted in the expected significant catastrophic failure mechanism. Rather, the major failure that has been identified is chipping at the edge of the wafer. This chipping has been reviewed previously and is addressed in section 6.2.2.

After thinning and polishing, vias may be created in wafers using DRIE, laser drilling, or other techniques. The vias must then be insulated to ensure that once they are filled, the metal filler will not migrate into the silicon and electrically isolate the vias. Insulation of vias may be accomplished with PECVD oxide or PECVD TEOS processes. These processes use a fixed-temperature chuck to hold the wafer. Although chuck temperatures are held constant, the plasma energy imparted to the device may cause the temperature to rise significantly higher than the set point. This high-temperature process imparts significant stress on the device wafer and affects the bonding to the carrier. Localized or large-area delamination during the PECVD process may result.

Several interacting factors may contribute to this delamination. Exposure to high temperatures and a rapid temperature rise during PECVD create stress in the device wafer because of CTE mismatches between features on the device and the surrounding Si. Metals have CTEs three or more times that of Si. With >100 µm of Si, the stress caused by these CTE mismatches may create a slight bowing in the wafer, and a <50 µm device may buckle. Bonding materials that are used in temporary bonding soften at high temperatures. As these materials soften, or even become liquid in the case of thermoplastic materials, they do not support as much stress. Device or bonding materials may have retained solvent or adsorbed moisture that is evolved at these temperatures. Lastly, if glass carriers are used, the CTE mismatch between the carrier wafer and the device creates stress that may result in bowing and delamination.

After backside processing is completed, thinned two-sided devices must be separated from their carrier wafer. For wafer-on-wafer bonding, debonding may take place before removal from the carrier wafer, but, in general, debonding will be completed before permanent bonding occurs. This debonding process also has risks that may damage the device. Both the low-temperature peel process and the elevated-temperature slide process stress the device wafer. Inspection and electrical testing may be used to assess which device debonding process is applicable.

Once debonded from the carrier wafer, any remaining adhesive or wafer treatment that could interfere with interconnect formation must be removed from the device wafer. Removal may be completed by using a peel mechanism for elastomeric adhesives or by using a stripping process similar to a develop process for the solvent-based systems. In each case, the device manufacturer must assess the cleanliness of the device and confirm sufficient cleaning using appropriate measurement tools.

Finally, the device, either in die form or in a whole wafer, must be permanently bonded for packaging. Electrical and product performance testing will complete the stacked 3D package with TSVs.

6.5 SUMMARY AND CONCLUSIONS

Although many 3D devices have been completed, development is ongoing and many challenges will arise as 3D packaging moves forward into high-volume mass production. The technology offers great promise to reduce form factor and greatly improve performance. Market research firms agree that 3D packaging through the use of TSVs will become a widely used commercial solution, although no one knows precisely when. Some market analysts believe that the technology will be in very high volume usage as early as 2013, while others project that such production will not occur until ~2016.

Many corporations are actively pursuing thinned wafer handling for the production of TSVs. Each application has its own processing and performance requirements. In the future, some standard performance requirements may be identified for the temporary bonding adhesives, but at this time, as mentioned by Hermanowski, the best way to choose the right adhesive and thin wafer handling technology is to run devices through complete processing.

References

1. Feynman, R. P. (2000) *The Pleasure of Finding Things Out*, Perseus Publishing, Cambridge, p. 28.

2. *Advanced Packaging*, February 17, 2009.

3. Garrou, P., Bower, C., and Ramm, P. (eds.) (2008), *Handbook of 3D Integration*, Wiley-VCH GmbH & Co. KGaA, Weinheim, pp. 463–486.

4. Garrou, P., Bower, C., and Ramm, P. (eds.) (2008), *Handbook of 3D Integration*, Wiley-VCH GmbH & Co. KGaA, Weinheim, pp. 26–27.

5. Henry, D., Jacquet, F., Neyret, M., Baillin, X., Enot, T., Lapras, V., Brunet Manquat, C., Charbonnier, J., Aventurier, B., and Sillon, N. (2008), Through silicon vias technology for CMOS image sensors packaging, in *Proceedings, 58th Electronic Components and Technology Conference (IEEE ECTC 2008)*, pp. 556–562.

6. Tan, C.S., and Farrens, S. (2008), Wafer-bonding technologies and strategie for 3D ICs, in *Wafer Level 3-D ICs Process Technology*, Springer Science+Business Media, New York, pp. 69–72.

7. Garrou, P., Bower, C., and Ramm, P. (eds.) (2008), *Handbook of 3D Integration*, Wiley-VCH GmbH & Co. KGaA, Weinheim, pp. 27–28.

8. http://www.sonoscan.com/resources/publications.html.

9. McCutcheon, J., McCarthy, L., and Dachsteiner, J. (2008), NIR imaging of bond integrity for wafer bonding applications, in *International Wafer Level Packaging Conference (IWLPC) Proceedings*, pp. 113–116.

10. Bai, D., Zhong, X.-F., Puligadda, R., Burggraf, J., Burgstaller, D., Lypka, C., and Verzosa, J. (2009), Edge protection of temporarily bonded wafers during backgrinding, *ECS Trans.*, 18(1), 757–762.

11. Trichur, R. K., and Shao, X. (2008), A photosensitive, spin-applied masking material for through-silicon via formation for wafer-level packaging, in *Proceedings of SMTA International Conference and Exhibition*, August 17–21, 2008, unnumbered.

12. Lo, W. C, Chen, Y. H., Ko, J. D., Kuo, T. Y., Chien, C. W., Chen, Y. C., Chen, W. Y., Leu, F.-J., and Hu, H.-T. (2005), Development and characterization of low cost

ultrathin 3d interconnect, in *Proceedings, 55th Electronic Components and Technology Conference (IEEE ECTC 2005)*, pp. 337- 342.

13. Garrou, P., Bower, C., Ramm, P. (eds.) (2008), *Handbook of 3D Integration*, Wiley-VCH GmbH & Co. KGaA, Weinheim, pp. 62–78.

14. Kessel, C. R., Behr, A. H., Hoffend, T. R., Jr., Saito, K., Akiyama, R., Takamore, K., and Iwasawa, M. (2004), Wafer thinning with the 3M wafer support system, in *International Wafer-Level Packaging Conference (IWLPC) Proceedings*.

15. Puligadda, R., Pillalamarri, S., Hong, W., Brubaker, C., Wimplinger, M., and Pargfrieder, S. (2007), High-performance temporary adhesives for wafer bonding applications, in *Materials Research Society Symposium Proceedings, vol. 970*.

16. Hermanowski, J. (2009), Thin wafer handling: study of temporary wafer bonding materials and processes, in *IEEE International Conference on 3D System Integration (3D IC)*, September 28–30, 2009, San Francisco, California, USA.

Chapter 7

3D TECHNOLOGY PLATFORM – WAFER THINNING, STRESS RELIEF, AND THIN WAFER HANDLING

Scott Sullivan
DISCO

7.1 INTRODUCTION

Throughout the history of the semiconductor industry new methods for integrating devices have been created. Most if not all methods have been to improve performance and reduce costs. What is particularly exciting about so called 3D integration is that it involves both front end and backend processes. This paper assumes that there is no clearly defined process flow as of yet leaving the way open for each process to change position in the process flow. The two processes circled below in Figure 7.1 is addressed separately.

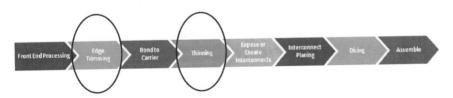

Fig. 7.1

7.2 EDGE TRIMMING

As wafers are thinned, the shape of the wafer edge changes. For 3D target thicknesses of 50 µm and below the wafer edge becomes a knife edge, making

3D Integration for VLSI Systems
Edited by Chuan Seng Tan, Kuan-Neng Chen and Steven J. Koester
Copyright © 2012 by Pan Stanford Publishing Pte. Ltd.
www.panstanford.com

the edge more susceptible to damage and leaving an overhang which can collect particles during grinding. Removing a section of the edge bevel or the entire edge prior to thinning reduces the likelihood of damage and eliminates the pocket for particles to become trapped.

A wafer is beveled on the edge for a number of reasons. Two common reasons are to allow for easier handling and to keep the edge of the wafer clean during the deposition steps. By the time the wafer reaches the back end thinning step the shape of the wafer edge is not as important.

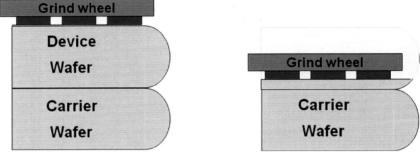

Fig. 7.2

A simple notch of the wafer edge just a bit deeper than the final thickness of the device wafer is enough to dramatically improve the quality and yield of the thinned wafer. As shown in Figure 7.3 below, wafer edge bevels taper at a smooth angle.

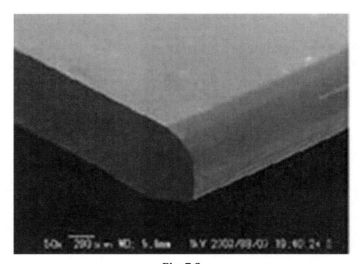

Fig. 7.3

When such wafer is thinned the tapered edge goes to a point as shown in Figure 7.4.

Fig. 7.4

By removing a section of the edge bevel results in a new edge profile, Figure 7.5.

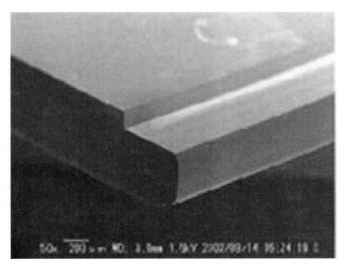

Fig. 7.5

The squared off edge becomes the new wafer edge when thinned, Figure 7.6.

Fig. 7.6

Squared off the edge is thicker and much stronger which should lead to a reduction in chipping.

7.2.1 Experiment

In order to determine the effectiveness of trimming the edges 90 test 200 mm wafers were used. All wafers were bonded on 200 mm silicon carrier wafers and thinned. 45 wafers were thinned to 50 μm and 45 wafers were thinned to 25 μm. Each thickness was divided into three groups. 15 wafers were used as a control, no edge trimming. Of the remaining 30 wafers 15 were edge trimmed at a speed of 10 degrees/second and 15 were edge trimmed at a speed of 6 degrees/second.

Figure 7.7 shows the basic steps. First the edge of the wafer was trimmed. First, using a dicing blade on a DISCO DFD6340 dicing saw, the trimming was done by grinding away a section of the bevel. The depth of the trimming was investigated. From a number of tests the depth of the trim did not seem to matter much as long as any rounding due to wear of the blade was removed during grinding. Next the wafer is flipped over and the device side was bonded to the carrier wafer. The final step is to thin the wafer.

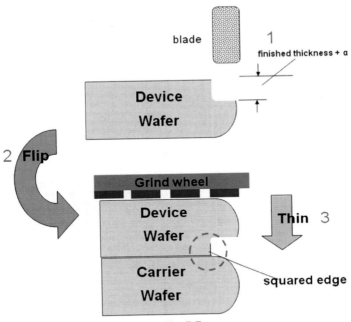

Fig. 7.7

Also examined in this study was the cleanliness of the area of the edge between the two wafers. Figure 7.8 shows where particles tended to accumulate. It is thought that once the device wafer is ground to below the edge trim it is exposed to the grind and cleaning water. Accumulated particles should be more easily flushed away on the wafers that were edge trimmed. Because of the additional processing steps for 3D packaging following the thinning step particles are more of a concern than for traditional thinning.

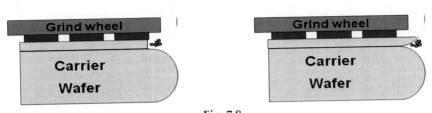

Fig. 7.8

7.2.2 Results and Discussion

There was an assumption prior to this test that trimming the edge would reduce the size of the edge chips. Our results showed that trimming the edge did not have an effect on the size of edge chips only the number. Chart 7.1 below shows a grouping of the results.

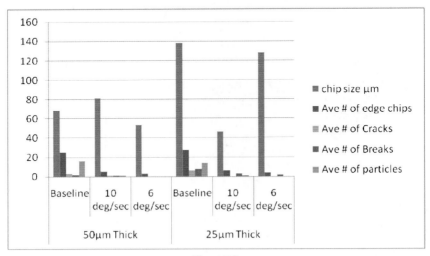

Chart 7.1

Removing the chip size values shows a strong response to the edge quality and wafer yield. Chart 7.2 has chip size removed.

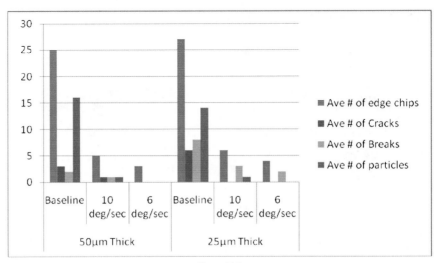

Chart 7.2

Trimming the edge of the wafer reduced the occurrences of edge chips. It is assumed that the reduction in edge chips has led to the reduction in edge cracks and breaks originating from the edge. It is also possible that edge cracks and breaks originating from the edge are independent from edge chips. We saw evidence of this by finding edge cracks and breaks originating from the edge that did not seem to be associated with edge chips. There are plans to investigate this at a future date. In either case there is a strong correlation between edge trimming and wafer edge finish and yield.

7.3 WAFER THINNING

Thinning of wafers is quite possibly the most well understood step in the 3D interconnect process. Wafers have been thinned prior to assembly for many years. The added challenge from the TSVs is that they act as crack initiation points. This issue is conveniently addressed by using a silicon or glass carrier. Targets have been set for via diameters which lead to wafer thicknesses and quality requirements beyond what is currently in production.

Today the dominate form of back end wafer thinning is fixed abrasive in-feed grinding. Two other methods were at different times the predominate methods and are still in limited use. Free-abrasive machining on a lap, often referred to as lapping was the first method used in high volume production, followed by creep-feed grinding.

- Lapping

According to a fairly recent study abrasive lapping was first used sometime between 4000 and 3500 BC. A brittle material such as a stone ax head was rubbed against another softer stone, the lap, with an abrasive. This action produces microscopic conchoidal or clamshell fractures on both the stone being lapped and the lap as the abrasive rolls or is slid between the two surfaces. This loose abrasive method is considered a three-part or three-body mechanism.

Abrasive is continuously added in order to replace the abrasive that is dulled or expelled. Abrasive grit or grains have irregular shapes and sharp corners. As the grit is pressed between the two surfaces the sharp points create the conchoidal fractures. Over time the abrasive becomes more regular shaped loosing the sharp edges becoming dull. Adding abrasive renews the abrasive film between the part and the lap. Uniformity of the abrasive film is one aspect that defines the surface being lapped.

Another determining factor in the uniformity of the surface being lapped is the pressure being applied to the three bodies. Da Vinci envisioned a way to automate the lapping process thereby applying consistent, uniform pressure, Figure 7.9. His drawings show methods for holding the material to be lapped on a rotating lap. The process is essentially the same as today. Two additional improvements made over time have been to control the abrasive and to control the uniformity of the abrasive film.

Properties of the abrasive film are controlled by suspending the abrasive in an aqueous or oil based liquid depending upon the material being lapped.

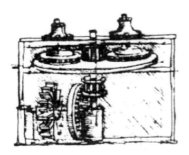 

Fig. 7.9

- Creep feed

The wafer is ground by passing it under a cup wheel. The wafer does not rotate. If the finished thickness is 300 μm, an 8-inch wafer with a starting thickness of 725 μm will need 425 μm ground away. With this much material to remove, the grinding is not done in a single pass. Most grinders have three spindles (Z1, Z2 and Z3-axis) for a rough grind, a medium grind, and a fine grind. As wafers have grown in size, so too have grinders. Creep feed grinding was still used with 6-inch wafers, but with the 8-inch generation, it was almost entirely phased out in favor of in-feed grinders.

- In-feed

In this method, the wafer rotates, as shown in Figure 7.2, and is ground as the cup wheel is lowered. The thickness of the wafer is measured during grinding by a contact probe called an in-process gauge. When the desired thickness is achieved, spindle descent stops and dwells for few seconds then moves onto escape cut, ending the grind. This setup allows grinding to be completed in two stages, rough grinding (Z1-axis) and fine grinding (Z2-axis), meaning that the equipment can be smaller than a comparable creep feed grinder. This is currently the mainstream grinder. Naturally, it is used for 300 mm wafers as well. Hereafter, our discussion assumes the grinder uses the in-feed system.

7.3.1 Carrier

Prior to thinning wafers can be bonded to a temporary carrier. Carriers are most often glass or silicon. There are a number of reasons for the carrier. For the thinning process the carrier supports the wafer allowing for easier handling. Wafer thickness is determined by estimating how strong the wafer needs to be to survive the various handling and thermal, and other steps in the front end. For packaging the thinner a device is the better. This leads to having very thin and weak wafers. Add to this TSVs which act as crack initiating points it is easy to see why carrier wafers are being used.

Along with supporting the wafer, bonded wafers bring a host of challenges with the combination of device wafer, adhesive bond and carrier. As wafers become thinner TTV (Total Thickness Variation) becomes more critical. A 2µm TTV for a wafer that is 200 µm thick may or may not affect the final product. The same 2µ TTV in a wafer that is 20 µm thick will have a profound effect on the final product. For TSV wafers the challenge is compounded. 2008 ITRS stated a via exit diameter of 2 µm in 2011 with a depth to diameter ratio of 10:1.

Following this ITRS roadmap leads us to a device thickness plus silicon thickness of 20µm minus the exposure amount of the TSV interconnect nail, from here on referred to as copper or Cu nail for convenience. Ignoring any variation due to etch depth uniformity and evaluating the error on the thick side of any TTV we start at a thickness of the device layer and silicon of 20µm. At a wafer thickness of 20 µm plus 2 µm TTV some of the Cu nails are exposed. The areas of the wafer with the additional 2µm of silicon thickness because of the TTV do not have exposed Cu nails. If we want to have the Cu nails exposed above the surface of the silicon on the backside of the wafer by a minimum of 2 µm then the target wafer thickness to account for 2 µm of TTV is 16 µm. Working in the opposite direction with a target wafer thickness of 16µm and a TTV of -2 µm there will be some wafers with areas 14 µm thick with Cu nails exposed 6 µm above the surface of the silicon as shown in Table 7.1.

Table 7.1

Target Thickness (µm)	2 µm Positive TTV		2 µm Negative TTV	
	Max Thickness (µm)	Nail Exposure (µm)	Min Thickness (µm)	Nail Exposure (µm)
20	22	0	18	0–2
18	20	0–2	16	2–4
16	18	2–4	14	4–6

From this we know that variations in the amount of Cu nail exposure is 2 x wafer TTV. Wafer TTV is determined by the thinning process capabilities plus the TTV of the carrier and the adhesive bond. For a Cu nail exposure of 2.5 µm +/- 1µm wafer TTV needs to be 0.5 µm or less.

Current backside grinders such as the DISCO DFG8761 are capable of attaining this type of TTV on wafers or carriers alone. The challenge moving forward will be to find ways to either compensate for or reduce the incoming thickness variation of the bonded wafer set.

To quickly grasp the challenge at hand one just has to think of the bond and the variation in surface figure that exists. Each surface of the wafer and carrier can be plano, concave, convex or any combination of the three. Added to the four random surface configurations is the shape of the spun on, sprayed on, or deposited temporary adhesive.

Making the stack coplanar is relatively straight forward when compared with making only one of the three coplanar independent of the other two. Methods for shaping one element of a stack do exist. At this time they are relatively slow taking days to process one element with the above mentioned tolerances.

7.3.2 Stress Relief

With a reduction of form factor and the increase in die stacking, post wafer thinning stress relief became more common. Following the abrasive grind there is a remaining layer of damage to the silicon. This layer contains microcracks which expand the surface creating a tensile stress. Removing this damaged layer with polishing or etching removes the microcracks, removing the stress. For thin wafers and die it is often the microcracks and not the resultant stress that are a problem.

A very thin unsupported silicon wafer can be flexed to terrifying angles. When this happens microcracks will fully propagate through the bulk cracking the wafer. Because 3D wafers are supported by a carrier microcracks and wafer breakage are not the main concern. The issue for 3D bonded pair wafers is the residual stress in the thinned wafer. While reducing the thickness of the silicon the wafer becomes less stiff. Compressive and tensile forces from the materials deposited on the front side of the wafer and the TSVs can overcome the stiffness of the thinned silicon substrate warping the wafer and die. Warped wafers can debond from the carrier and warped die can lead to interconnect failures.

For a first approximation of the stress induced into the wafer by grinding and then and then polished to remove the stress a simplified version of the formula derived by Stoney is used. With this comes the use of some assumptions. First is that that the stress in the wafer is only at the surface that was thinned. Second, that the stressed surface is a film of negligible thickness. Third, that Young's modulus is uniform and independent of crystal direction.

$$\Delta \sigma \; = \; \frac{E t^2}{6R(1-v)}$$

Surface stress, $\Delta\sigma$ is a function of, E the Young's modulus, R the radius of curvature of the substrate and v, Poisson's ratio. Because absolute stress is not of interest at this point we can simply relate an induced stress to the curvature of the substrate. Future work will involve determining the actual stress on the wafer pair and relating that to the bond material, bond method, and downstream processing.

A 200 mm wafer was used to study the amount of material that needed to be removed by a stress relief polish from the ground surface of a wafer. Wafers were thinned to 50 µm using standard grind methods and then a varying amount of material was removed by MCP (Mechano-chemical polishing). At a removal amount of between 1.5 and 2.0 µm the curvature of the wafer and therefore the residual stress from grind goes to zero, Chart 7.3.

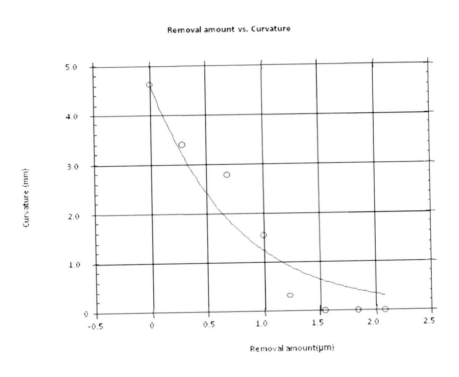

Chart 7.3

• Adhesive voids

Wafers bonded to carriers with a number of different adhesives after thinning were thermal-cycled up to 100°C. This is well below the published operating temperature of each of the adhesives tested. In every case voids were created. The assumption was that residual stress from the grinding step was the cause.

To test this theory the above information was used. Wafers were bonded to glass carriers using the same variety of adhesives and thinned to 50 µm. Half of the wafers were then polished, removing 2 µm. The other half of the wafers were unpolished leaving the residual stress.

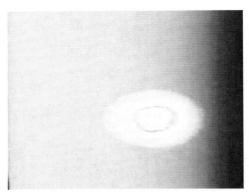

Fig. 7.10

Wafers that were polished and wafers that were only ground exhibited voids. Size, shape, distribution, and number of voids were statistically the same for stress relieved wafers and for wafers with residual stress. Figure 7.10 shows a typical void. Diameter of the voids ranged from 4–3200 µm.

Disproving the theory that residual grind stress leads to voids leaves one to look for other sources. A possibility is adhesive outgassing.

• Warp

In order to understand the magnitude of the stress and corresponding warp in a TSV wafer a number of wafers were thinned. Again the wafers were 200 mm, ground to 50 µm thick and then 2 µm of silicon was removed using MCP. The added variable was that these wafers contained TSVs. The diameter of the TSVs ranged from 10–50 m with pitches ranging from 125–250 µm. For this test it was necessary to not use a carrier in order to measure the curvature and thus calculate the stress.

Figure 7.11 shows the results of measuring the warp of one wafer that is typical of the 13 wafers tested. Fifteen wafers were thinned. Two were broken in handling. Using a TENCORE FLX2320 and the Stoney formula a stress in the neighborhood of 200MPa is calculated. Due to the assumptions, some of which are known to be incorrect, the stress value is seen as only a reference value.

What is understood from this experiment is that stress exists in a TSV wafer. This stress cannot be removed by using a post grind stress relief. In fact it may be beneficial to add a stress to counter the stress from the TSVs.

7.4 CONCLUSION: ECONOMIES OF WAFER SIZE

Over the life of the semiconductor industry there has been a push to larger and larger wafer diameters. Correct or not the belief has been that increasing wafer size will reduce costs. This may hold true with 3D assembly techniques or it may not. This uncertainty makes it necessary to honestly evaluate the costs associated with manufacturing and packaging different wafer sizes.

In the front end the thickness of a wafer is in relationship to the diameter. For simplicity I will use 1/3. Increasing wafer diameter by 1/3 will make the wafer roughly 1/3 weaker. The wafer thickness is then increased by 1/3 to keep the wafer strength constant from size to size. This allows the front end machines to more easily handle wafers and control unwanted effects such as warp during thermal processing.

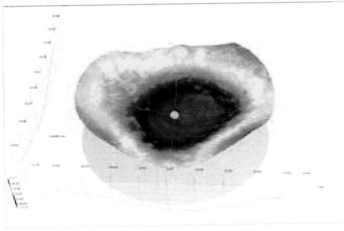

Fig. 7.11

In the back end, the final application determines die thickness which determines wafer thickness. Using the same simple logic as above if wafer diameter and starting thickness are increased by 1/3 this does not change the final thickness. The final thickness of TSV wafers will be determined not by strength but by application. Therefore the final thickness of the wafer remains constant and the strength of the wafer after thinning is reduced by 1/3. Handling or thinning will be the process throughput limiting steps. To compensate for the 1/3 reduction in wafer strength handling will need to be 1/3 slower. Because the wafer started 1/3 thicker, 1/3 more material will need to be removed during thinning, increasing the time to grind by 1/3. To hold throughput constant 1/3 more grinders will need to be utilized.

This very simplistic calculation demonstrated that increasing wafer diameter increases per wafer capitol costs and WIP (Work In Process) costs. These increased costs may or may not be compensated for by the few batch process steps used in the manufacture of a device. As semiconductor manufacturing has evolved batch processing has given way to single wafer and stepping. Assumptions of economies of scale should be challenged.

Much of the development work being conducted is on 200 mm wafers. The industry assumption is that all 3D and TSV development will be transferred to 300 mm wafers. Would it in fact be lower cost to stay at 200 mm?

In conclusion we need to keep in mind the purpose of the TSV. It is an interconnect, and in some cases a thermal pathway, nothing more or less. By keeping this in mind we can engineer elegant, simple and low cost solutions.

Chapter 8

ADVANCED DIE-TO-WAFER 3D INTEGRATION PLATFORM: SELF-ASSEMBLY TECHNOLOGY

Takafumi Fukushima, Kang-Wook Lee, Tetsu Tanaka and Mitsumasa Koyanagi

Tohoku University

In three-dimensional (3D) integration based on die-to-wafer stacking, high-precision chip alignment and low-temperature chip bonding are major two key technologies. However, traditional robotic pick-and-place chip assembly for the die-to-wafer 3D integration has a great disadvantage in production throughput due to the sequential one-by-one process. To overcome the issue, a new multichip self-assembly technology using liquid surface tension as a driving force has been proposed. In the advanced die-to-wafer 3D integration using the self-assembly technology, by the surface tension of liquid droplets, a large number of known good dies (KGDs) with/without through-Si vias (TSVs) can be simultaneously, quickly, and precisely aligned to pre-determined areas photolithographically formed on Si wafers, and consequently, the self-assembled KGDs can be tightly bonded to the wafers at room temperature. Here, an overview of the self-assembly technology for the advanced die-to-wafer 3D integration using the self-assembly technology is introduced and the potential applications of the self-assembly-based 3D integration are described.

3D Integration for VLSI Systems
Edited by Chuan Seng Tan, Kuan-Neng Chen and Steven J. Koester
Copyright © 2012 by Pan Stanford Publishing Pte. Ltd.
www.panstanford.com

8.1 INTRODUCTION—WHY SELF-ASSEMBLY IS REQUIRED?

In the past few years, development of three-dimensionally stacked integrated circuits (3D ICs) has increasingly activated towards a technological breakthrough overcoming limitations in downscaling electron devices by Moore's law. The most striking feature of the 3D ICs is that a huge number of very short TSVs with a length of several tens microns can vertically connect multiple stacked thin chips, as shown in Figure 8.1. Therefore, the 3D integration technologies can not only miniaturize the chip size and reduce the length of long global wirings used in system on a chip (SoC) and long bonding wires used in system in packaging (SiP), but also significantly increase the signal processing speed and decrease the power consumption.[1-3] In addition to the TSV formation, several key technologies are required for the 3D IC fabrication: microbump formation, chip/wafer thinning, adhesive injection (unique underfilling under vacuum[4]), and chip/wafer alignment and bonding. As shown in Figure 8.2, these key technologies can provide vertically stacked thin wafers (or chips) in which TSVs with 30 μm in length and metal microbumps with 5 μm in width can electrically connect between upper and lower wafers (or chips).

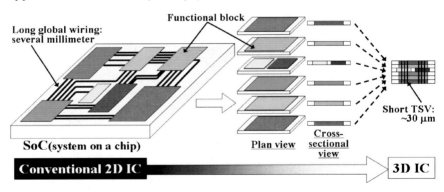

Fig. 8.1 Background on 3D IC development.

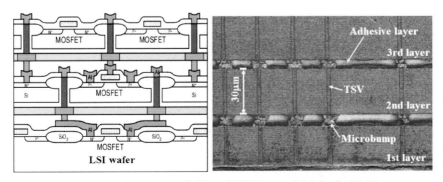

Fig. 8.2 Cross-sectional structure (left) and SEM image (right) of a 3D IC.

Stacking methods for 3D integration is mainly classified into three categories: die-to-die, wafer-to-wafer, and die-to-wafer, as shown in Table 8.1. The die-to-die 3D integration approach is employed for packaging applications. The production yield is high due to the use of KGDs only although the production throughput is extremely low due to the one-by-one processes at the chip level. In contrast, the production throughput in the wafer-to-wafer 3D integration approach is very high due to the batch processes at the wafer level. However, the total production yield exponentially decreases with an increase in the number of stacked layers because defective dies can not be removed from wafers to be stacked. Thus, the wafer-to-wafer 3D integration is useful for high-yield products such as dynamic random access memory (DRAM). On the other hands, the 3D integration based on die-to-wafer stacking can dramatically improve the production yield because KGDs can be selectively used and wafer-level processes can be applied after chip assembly. Therefore, the die-to-wafer 3D integration approach is thought to be the best solution to increase both the production throughput and the production yield. In the die-to-wafer 3D integration, traditional robotic pick-and-place chip assembly is a candidate for an alignment and bonding method of upper and lower chips. However, the pick-and-place method has a serious trade-off problem between assembly throughput and alignment accuracy. The alignment accuracy decreases with the reduction of assembly time, as shown in Figure 8.3.

Table 8.1 Comparison between stacking methods for 3D integration.

	Die-to-die	Die-to-wafer	Wafer-to-wafer	Advanced die-to-wafer
Production throughput	Extremely Low	Low (pick &place)	High	High (self assembly)
Production yield	High (use of KGD)	High (use of KGD)	Low	High (use of KGD)
Applications	Packaging	CIS, logic, memory, MEMS, etc.	DRAM (high-yield products)	CIS, logic, memory, MEMS, etc.

Single chip
pick-up tool

Chip tray   Wafer

	Equipment A	Equipment B	Equipment C
Assembly throughput:	450 chips/hour (8 sec/chip)	~3,000 chips/hour (~1.2 sec/chip)	~7,000 chips/hour (~0.5 sec/chip)
Alignment accuracy:	± 2 µm	± 10 µm	± 10 µm
Bonding type	Flip-chip bonding	Flip-chip bonding	Die bonding

Fig. 8.3 Traditional pick-and-place chip assembly.

By the way, the diameter and pitch of TSVs have been decreased year by year. M. Koyanagi *et al.* (Tohoku University) has already developed a fine TSV technology using W buried interconnections with a thin polycrystalline-Si as an adhesive layer. Figure 8.4(a) shows SEM cross-section of the W buried interconnections with a diameter of 0.6 µm.[2] It is also obvious that the TSV diameter and pitch can be further reduced to less than 1 and 2 µm, respectively. On the other hand, Figure 8.4(b) shows a photomicrograph and an SEM image of metal microbump array with a size of 2 µm. M. Motoyoshi (Zycube) and M. Koyanagi (Tohoku University) *et al.* have reported the fine-pitch InAu microbumps formed by a special lift-off technique using evaporation.[5,6] Thus, chip stacking using ultra-fine TSVs and metal microbumps requires higher alignment accuracy within 1 µm. However, the conventional robotic pick-and-place chip assembly takes an immense amount of time to precisely align and tightly bond a large number of chips to three-dimensionally stack in layer, as described above. Surface-tension-driven chip self-assembly is a promising technology in order to satisfy both the requirements of high assembly throughput and high alignment accuracy.

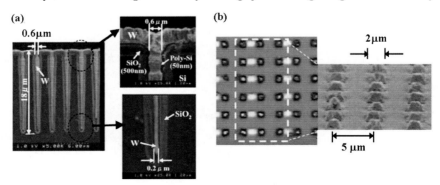

Fig. 8.4 SEM images of fine-pitch W-TSV (a) and InAu microbumps (b).

8.2 OVERVIEW OF SELF-ASSEMBLY TECHNOLOGY

8.2.1 Conventional self-assembly technologies

Self-alignment of KGDs manipulated and positioned by traditional robotic pick-and-place methods is widely known to be a flip-chip packaging technology. The self-alignment largely differs from self-assembly featured in this chapter. As shown in Figure 8.5, solder bumps formed on the KGDs are reflowed in the self-alignment process after flip-chip bonding. As a result, the KGDs are precisely aligned to a printed circuit board (PCB) by the surface tension of the molten solder.[7] On the other hand, fluidic self-assembly is well known to be just a handling method for microstructures such as optical devices that are too small to pick them up. The first fluidic self-assembly technology has been reported in 1994 by H.-J. J. Yeh *et al.* at the University of California, Berkeley, as shown in Figure 8.6. In their fluidic process, trapezoidal GaAs chips are assembled by shape recognition to a substrate on which the corresponding inverted trapezoidal pockets are formed.[8] U. Srinivasan *et al.* (the University of California, Berkeley) has described the self-assembly of Si micro-mirrors by the capillary force of an epoxy resin in 2001, as shown in Figure 8.7.[9] In addition, several groups have reported the self-assembly of Light Emitting Diodes (LEDs)[10] and Si parts[11] by the shape recognition and the surface tension of low-melting-point solder alloy. In the latter paper, the chip self-assembly technology is proposed for wafer-level packaging applications. However, these self-assembly technologies are not targeting toward 3D IC fabrication. In addition, generally, epoxy resins as adhesive materials involve much ionic contaminants and low-melting-point solder alloys as metal bump materials are thermally unstable at above 200°C. Therefore, these processes will surely be difficult to apply in large scale integration (LSI) manufacturing including 3D IC fabrication. Recent 3D integration trends are directed to die-to-wafer stacking by which many KGDs can be sequentially assembled to wafers in high yield. However, the classical pick-and-place methods for chip alignment and bonding are major obstacle for industrial use due to the serious trade-off problem described in the previous session.

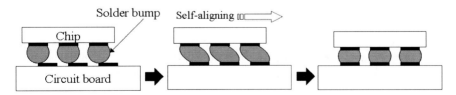

Fig. 8.5 Self-alignment using surface tension of molten solders.[7]

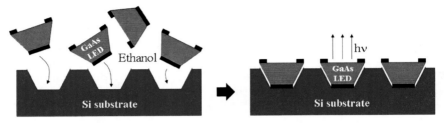

Fig. 8.6 Fluidic self-assembly of trapezoidal LED by shape recognition.[8]

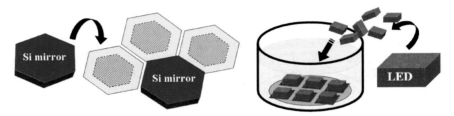

Fig. 8.7 Self-assembly of Si micro-mirrors by surface tension of an epoxy resin (left)[9] and LED chips by surface tension of a low-melting-point solder (right).[10]

8.2.2 Super chip integration technology

To fulfill strong demands for parallel chip assembly, T. Fukushima and M. Koyanagi *et al.* (Tohoku University) have proposed 3D integration technology using a self-assembly method, and demonstrated for the first time that a multichip self-assembly technique can be applied to 3D integration of LSI chips in 2005.[12] A 3D memory having P-doped polycrystalline-Si as a conductive material for TSVs has been successfully fabricated. They call the self-assembly-based 3D integration "Super Chip Integration" that is conceptually illustrated in Figure 8.8, where various kinds of chips with different chip sizes, different materials, and different devices are stacked to produce the *3D Super Chip* consisting of processors, memories, sensors, and microelectromechanical system (MEMS) devices and so on. As shown in Figure 8.9, a multichip pick-up tool enables simultaneously carry a large number of KGDs from a chip tray to a wafer. The many KGDs are quickly and precisely aligned and then bonded to the wafer. The driving force of the parallel assembly is the surface tension of small droplets of liquid. At the Tohoku University, reconfigured wafer-to-wafer 3D integration as a new self-assembly-based 3D integration has continuously introduced in 2007.[13] In the reconfigured wafer-to-wafer 3D integration, described in the last session in detail, many KGDs are temporarily self-assembled to a carrier wafer named the reconfigured wafer, followed by TSV and microbump formation, and then the KGDs are transferred to another LSI wafer. In 2008, they have presented a heterogeneous multi-chip module (MCM) integration technology, by which LSI chips, passive device chips, and MEMS chips are self-assembled to a flexible polyimide substrate.[14] In addition, their die-to-wafer 3D researches have demonstrated that chips having metal microbumps can be directly self-assembled to LSI wafers in a flip-chip bonding manner.[14,15]

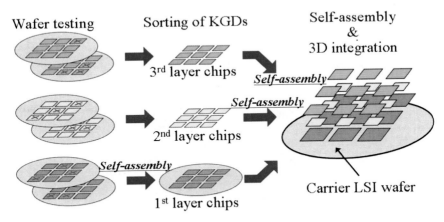

Fig. 8.8 Conceptual viewgraph of Super Chip Integration.

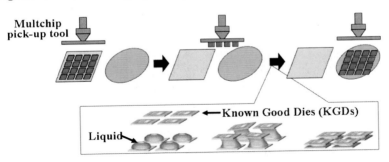

Fig. 8.9 Surface-tension-driven multichip self-assembly in batch.

Many researchers and product vendors have also started developing surface-tension-driven self-assembly technologies for advanced packaging and 3D integration. From 2006, the HYDROMEL project has started by a big European consortium in which DATACON as a big Austrian company for die/flip-chip bonders, Helsinki University of Technology, and so forth participate.[16] The HYDROMEL means hybrid self-assembly that combines traditional robotic chip transfer by capillary self-assembly. M. Mastrangeli and E. Beyne *et al.* (IMEC) have presented their self-assembly researches for 3D integration early in 2008.[17] They employed SAM (Self-Assembled Monolayer) technology to realize their die-to-wafer 3D integration using self-assembly. F. Grossi and L. Clavelier *et al.* (LETI) have reported their self-assembly results for 3D integration in late 2008.[18] They introduce a direct bonding technique for their die-to-wafer 3D integration using self-assembly.

8.3 SELF-ASSEMBLY TECHNOLOGY FOR 3D INTEGRATION

The self-assembly technology for 3D integration includes high-precision chip alignment and low-temperature chip bonding. The following is described in detail of both two key technologies. In this session, self-assembly of various kinds of chips having cavity structure and metal microbumps is also introduced.

8.3.1 High-precision chip alignment

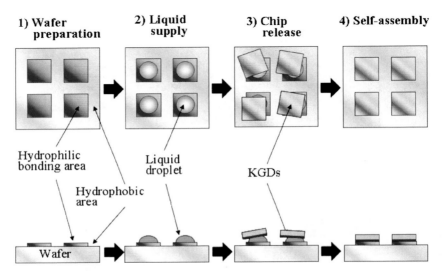

Fig. 8.10 Multichip self-assembly process for 3D integration.

Figure 8.10 shows a process sequence of multichip self-assembly for advanced die-to-wafer 3D integration. First, hydrophilic bonding areas are formed on a wafer and the surrounding areas are rendered hydrophobic. Then, many chips are roughly pre-aligned to the hydrophilic bonding areas. After chip release, the chips are precisely aligned to the bonding areas at once. Finally, the chips are tightly bonded to the hydrophilic areas after the liquid droplets are evaporated. Figure 8.11 shows video frames from a movie of a multichip self-assembly event. X, Y, and θ directions are largely misaligned on purpose immediately before the chips fall off. However, the chips are precisely aligned to the hydrophilic bonding areas within 0.1 sec. Figure 8.12 shows magnified snapshots of a chip self-assembly event taken by a high-speed camera and viewed from the lateral side. As seen from this figure, a horizontally-vibrated chip is precisely aligned to a bonding area by the surface tension of liquid in split seconds. High wettability contrast between hydrophilic bonding areas and the surrounding hydrophobic areas is an important parameter to realize high alignment accuracy in the chip self-assembly process. As shown in Figure 8.13, even a large volume of water is confined to the hydrophilic bonding areas by high hydrophobic property of the surrounding areas. Figure 8.14 shows dependence of alignment accuracy on liquid volume and chip size. The left figure shows liquid volume effects and the right figure shows chip size effects. Liquid volume hardly effects on alignment accuracy, but excess liquid volume tends to largely decrease alignment accuracy. When chips ranging in size from 1 mm to 5 mm are employed, alignment accuracy is very high in any case. Figure 8.15 shows self-assembly events of a larger chip and a rectangular chip. As shown in this figure, the 20-mm-square chip can be precisely self-assembled to another lower 20-mm-square chip within 0.5 sec. The 4 × 9-mm chip can be also self-assembled to another lower 4 × 9-mm chip in a short time although these chips are perpendicularly-placed each other before chip release. From these results, large chips such as microprocessors and rectangular chips such as memories turn out to be applicable in the self-assembly-based 3D integration. Figure 8.16 shows self-assembly of different-size chips. In the top of this figure, a 2-mm-square chip is successfully self-assembled onto a larger 3-mm-square chip. On the other hand, in the bottom of this figure, a 4-mm-square chip can be also self-assembled onto a smaller 3-mm-square chip. These results indicate that various sizes of chips are available for the 3D chip stacking by self-assembly. Figure 8.17 shows the result of self-assembly with chips having different chip height. 280-μm- and 500-μm-thick chips can be self-assembled to a wafer at one time. Thinner chips less than 50 μm can be also self-assembled in a similar manner. After self-assembly, the chips are thinned down to several tens microns from the top side to give planarized chips having uniform chip height. KGDs can be self-assembled to not only Si wafers but also various kinds of substrates. Figure 8.18 shows a photomicrograph of 3-mm-square chip array self-assembled on a flexible polyimide substrate on which hydrophilic bonding areas are formed with SiO_2 by plasma-enhanced CVD.

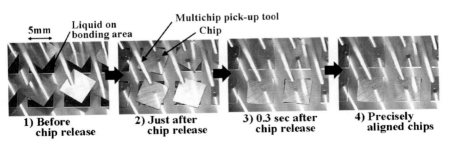

Fig. 8.11 Video frames from a movie of a multichip self-assembly event.

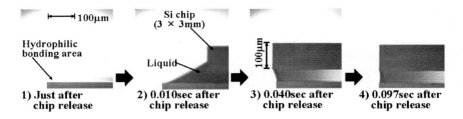

Fig. 8.12 Magnified cross-sectional view of a self-assembling chip to a bonding area.

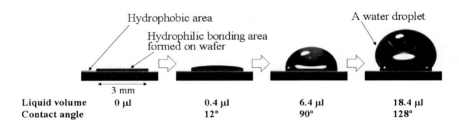

Fig. 8.13 Relationship between liquid volume and water contact angle.

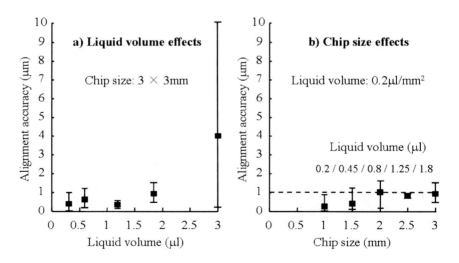

Fig. 8.14 Dependence of alignment accuracy on liquid volume (left) and chip size (right).

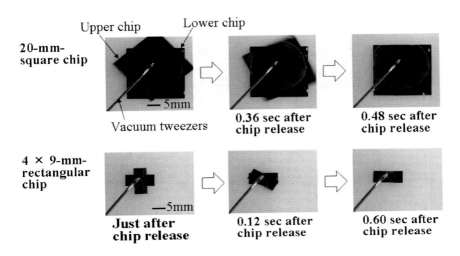

Fig. 8.15 Self-assembly of large-size (top) and rectangular (bottom) chips.

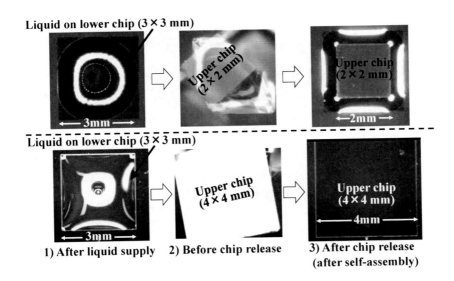

Fig. 8.16 Self-assembly of chips with different chip size.

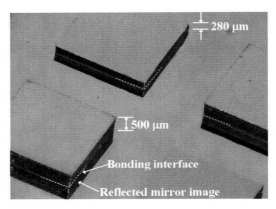

Fig. 8.17 Self-assembly of chips with different chip height.

Fig. 8.18 Self-assembly of chips on a flexible polyimide substrate.

Figure 8.19 introduces self-assembly of various chips: for example, chips with cavity structure and chips with metal microbumps. Both chips can be self-assembled to substrates with high alignment accuracy. The self-assembly of cavity chips are targeting for MEMS and passive device chips such as coil. The results of self-assembly with the cavity chips were presented in 2008 IEDM (IEEE International Electron Devices Meeting).[14,19] On the other hand, the results of the bump chips will be described in detail in 2009 IEDM.[15] In the former self-assembly, microchannel structures are formed in the hydrophilic bonding areas because liquid used in the self-assembly can be effectively removed from the cavity. In the latter self-assembly, in addition to high-precision chip alignment, high-density interconnections through microbump-to-microbump bonding can be obtained by just heating up to melting or eutectic temperatures of the microbump materials.

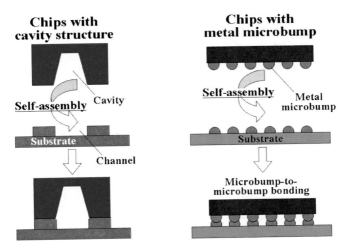

Fig. 8.19 Self-assembly of chips having cavity structure (left) and metal microbumps (right).

8.3.2 Room-temperature chip bonding

Here, low-temperature direct chip bonding without thermal compression is introduced to the multichip self-assembly technology for 3D integration. Direct bonding is an innovative method to provide adhesive-free bonding interface between Si/Si, Si/SiO$_2$, and SiO$_2$/SiO$_2$, and so forth.[20] T. Fukushima and M. Koyanagi *et al.* (Tohoku University) have first demonstrated both alignment and bonding of millimeter-scale large chips to Si wafers at room temperature without applying mechanical force by using self-assembly with an aqueous solution.[12] Chip alignment by the liquid surface tension and the subsequent liquid evaporation within several minutes leads to chip bonding to hydrophilic bonding areas formed on the wafers. The mechanism of the room-temperature and load-free chip bonding is thought to be SiO$_2$-SiO$_2$ direct bonding without adhesive layers. An additive dissolved in the aqueous solution used in self-assembly can promote SiO$_2$/SiO$_2$ bonding between chips and wafers.[21] The bonding strength of self-assembled chips to wafers is measured. Figure 8.20 shows the measurement results of shear bonding strength between a wafer having thermal oxide and the self-assembled chips having various types of oxide: these oxides are employed as hydrophilic bonding materials for self-assembly. The chips having thermal oxide can provide high bonding strength, whereas bonding failure is resulted from chips having as deposited plasma-TEOS oxide and sputtered oxide. In general, the surface of as deposited plasma-TEOS oxide is not smooth, so the resulting bonding strength is low. However, CMP process more than 1 min can provide highly smooth surface, as shown in Figs. 8.21 and 8.22. Thus, high bonding strength is obtained from chips having the polished plasma-TEOS oxide. As shown in Figure 8.23, a self-assembled chip is destructed after the pull bonding test because the chip is tightly bonded to the wafer by the room-temperature direct bonding without thermal compression.

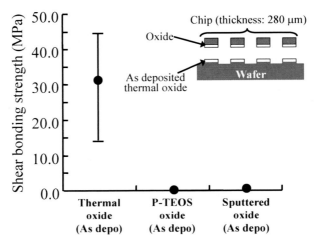

SiO$_2$ types on the bonding surface of chips

Fig. 8.20 Relationship between oxide type and bonding strength.

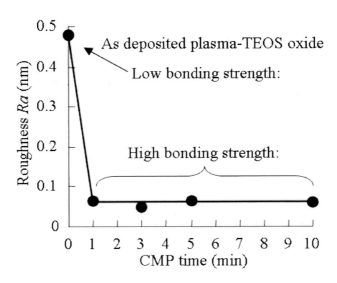

Fig. 8.21 Relationship between CMP time and surface roughness.

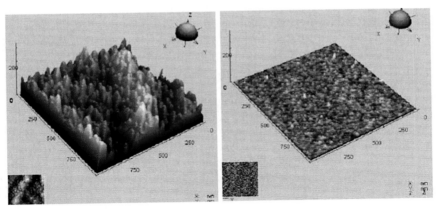

Fig. 8.22 AFM images of plasma-TEOS oxide before (left) and after (right) CMP.

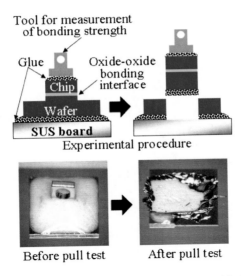

Fig. 8.23 Pull bonding strength measurement of a self-assembled chip.

8.4 3D & HETERO INTEGRATION BY SELF-ASSEMBLY

Self-assembly-based 3D integration approaches are simply divided into two categories, as shown in Figure 8.24. One is self-assembly of KGDs having metal microbumps, which is shown in the left side of this figure. Thin or thick KGDs with/without TSVs are directly self-assembled upside down to an LSI wafer through metal microbump-to-microbump bonding to give high-density interconnections between the KGDs and the wafer. The other is reconfigured wafer-to-wafer 3D integration,[13,22] which is shown in the right side of this figure. A large number of KGDs for the first layer are self-assembled on a carrier wafer in the reconfigured wafer-to-wafer 3D integration. The KGDs with/without metal microbumps on their top surface are just placed or temporarily bonded to the carrier wafer in face-up (like die bonding) or face-down (like flip-chip bonding) manners. After that, TSVs and/or metal microbumps are formed to the self-assembled KGDs. The resulting reconfigured wafer consisting of the first KGDs and the carrier wafer is aligned and bonded to another faced LSI wafer. Consequently, the many KGDs are removed from the carrier wafer and transferred to the LSI wafer. The reconfigured wafer is transformed to a carrier wafer and reused again in the subsequent stacking process of many different KGDs for the second and the third layers. After underfilling between upper and lower chips and the subsequent chip thinning at the wafer level, 3D ICs are fabricated in batch by repeating the processes listed above.

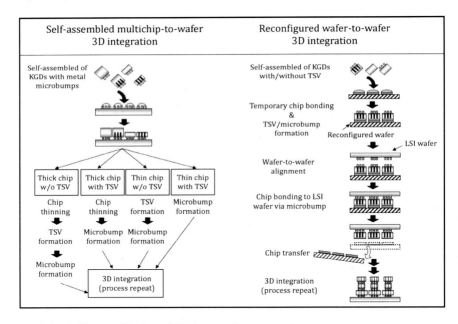

Fig. 8.24 Self-assembly-based 3D integration categories.

By using the self-assembly technique, various sizes and various kinds of many chips can be stacked (or mounted) to wafers (or substrates), as shown in Figs. 8.25–8.27. Three-layer chip stacking with different-size chips is shown in Figure25, where the largest 7-mm-square chip for the top layer can be self-assembled on smaller 5-mm- and 6-mm-square chips for the lower layers. Figure 8.26 shows a technical challenge for 38-layer chip stacking by self-assembly. The total chip height is about 1.3 mm and each chip thickness is approximately 34 µm. Further chip stacking more than 40 layers is technologically possible by self-assembly. Figure 8.27 shows a photomicrograph of a heterogeneously integrated system consisting of a capacitor chip, an inductor chip, and two LSI chips, which are self-assembled to a polyimide substrate in a face-up bonding manner. These chips are then electrically connected with high-aspect-ratio Cu sidewall interconnections that are crossing over both bottom and top corners of the chips with 100 μm in height.[14]

Fig. 8.25 Three-layer chip stack with different sizes (5, 6, and 7-mm square).

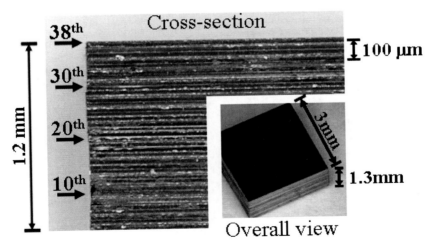

Fig. 8.26 38-layer chip stack with 3-mm-square dies.

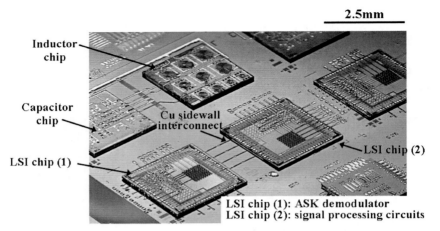

Fig. 8.27 Hetero integration of an inductor chip, a capacitor chip, and two LSI chips by self-assembly.

In 2008, a new alignment and bonding equipment called self-assembled multichip bonder has been developed at Tohoku University to practically use the multichip self-assembly technology for 3D IC fabrication, as shown in Figure 8.28.[14] By using the prototype self-assembled multichip bonder, a large number of chips can be aligned with high alignment accuracy and tightly bonded to an 8-inch wafer, as shown in Figure 8.29. More than 500 chips can be quickly (< 0.1 sec) and precisely aligned in batch and bonded onto the entire 8-inch wafer at room temperature by the multichip self-assembly technology, as shown in Figure 8.30. The alignment accuracy was found to be approximately 400 nm on average.

Fig. 8.28 Overview of a self-assembled multichip bonder.

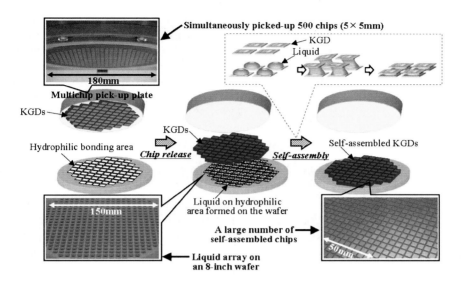

Fig. 8.29 Self-assembly of 500 chips to an 8-inch wafer using a self-assembled multichip bonder.

Fig. 8.30 A magnified photograph of the self-assembled 5-mm-square chip array to the 8-inch wafer.

Acknowledgments

This work was performed in the Micro/Nano-machining research and education Center (MNC) and Jun-ichi Nishizawa Research Center at Tohoku University. This work was supported from 2006 to 2008 by NEDO (New Energy and Industrial Technology Development Organization) as a part of the "Highly Integrated, Complex MEMS Production Technology Development Project". This research was partially supported in 2009 by Grant-in-Aid for Scientific Research (Grant-in-Aid for Young Scientists (A)), the Ministry of Education, Culture, Sports, Science and Technology, No. 19686022. This research was also supported from 2010 to 2011 by Japan Society for the Promotion of Science (JSPS), Grant-in-Aid for Scientific Research "Grant-in-Aid for Scientific Research (S)", No. 21226009.

References

1. Koyanagi, M., Kurino, H., Lee, K.-W., Sakuma, K., Miyakawa, N. and Itani, H. (1998). Future system-on-silicon LSI chips, *IEEE Micro*, **18**, pp. 17–22.

2. Koyanagi, M., Nakamura, T., Yamada, Y., Kikuchi, H., Fukushima, T., Tanaka, T. and Kurino, H. (2006). Three-dimensional integration technology based on wafer bonding with vertical buried interconnections, *IEEE Trans. Electron Devices,* **53**, pp. 2799–2808.

3. Koyanagi, M., Fukushima, T. and Tanaka, T. (2009). High-density through silicon vias for 3-D LSIs, *Proc. IEEE*, **97**, pp. 49–59.

4. Matsumoto, T., Satoh, M., Sakuma, K., Kurino, H., Miyakawa, N., Itani, H. and Koyanagi, M. (1997). New three-dimensional wafer bonding technology using adhesive injection method, *Proc. 29th Int. Conf. Solid State Devices and Mater.,* SSDM, pp. 460–461.

5. Matsumoto, T., Kudoh, Y., Tahara, M., Yu, K.-H., Miyakawa, N., Itani, H., Ichikizaki, T., Fujiwara, A., Tsukamoto, H. and Koyanagi, M. (1995). Three-dimensional integration technology based on wafer bonding technique using micro-bumps, *Proc. 27th Int. Conf. Solid State Devices and Mater.,* SSDM, pp. 1073–1074.

6. Motoyoshi, M., Kamibayashi, K., Koyanagi, M. and Bonkohara, M. (2007). Current and future 3-dimensional LSI technologies, *Tech. Dig. 1st 3D System Integration Conf.,* 3D-SIC, pp. 8.1–8.14.

7. Hayashi, T. (1992). An innovative bonding technique for optical chips using solder bumps that eliminate chip positioning adjustments, *IEEE Trans. Compon., Hybrids, and Manuf. Technol.*, **15**, pp. 225–230.

8. Yeh, H. J. and Smith, J. S. (1994). Fluidic self-assembly for the integration of GaAs light-emitting diodes on Si substrate, *IEEE Photonics Technol. Lett.*, **6**, pp. 706–708.

9. Srinivasan, U., Liepmann, D. and Howe, R. T. (2001). Microstructure to substrate self-assembly using capillary forces, *J. Microelectromech. Syst.*, **10**, pp. 17–24.

10. Zheng, W. and Jacobs, H. O. (2004). Shape-and-solder-directed self-assembly to package semiconductor device segments, *Appl. Phys. Lett.*, **85**, pp. 3635–3637.

11. Fang, J. and Böhringer, K. F. (2006). Wafer-level packaging based on uniquely orienting self-assembly (the DUO-SPASS processes), *J. Microelectromech. Syst.*, **15**, pp. 531–540.

12. Fukushima, T., Yamada, Y., Kikuchi, H. and M. Koyanagi. (2005). New three-dimensional integration technology using self-assembly technique, *Tech. Dig. 51st Int. Electron Devices Meeting*, IEDM, pp. 359–362.

13. Fukushima, T., Yamada, Y., Kikuchi, H., Konno, T., Liang, J., Ali, A. M., Sasaki, K., Inamura, K., Tanaka, T. and Koyanagi, M. (2007). New three-dimensional integration technology based on reconfigured wafer-on-wafer bonding technique, *Tech. Dig. 53rd Int. Electron Devices Meeting*, IEDM, pp. 985–988.

14. Fukushima, T., Konno, T., Kiyoyama, K., Murugesan, M., Sato, K., Jeong, W.-C., Ohara, Y., Noriki, A., Kanno, S., Kaiho, Y., Kino, H., Makita, K., Kobayashi, R., Yin, C.-K., Inamura, K., Lee, K.-W., Bea, J.-C., Tanaka, T. and Koyanagi, M. (2008). New heterogeneous multi-chip module integration technology using self-assembly method, *Tech. Dig. 54th Int. Electron Devices Meeting*, IEDM, pp. 499–502.

15. Fukushima, T., Iwata, E., Ohara, Y., Noriki, A., Inamura, K., Lee, K.-W., Bea, J.-C., Tanaka, T. and Koyanagi, M. (2009). Three-dimensional integration technology based on reconfigured wafer-to-wafer and multichip-to-wafer stacking using self-assembly method, *Tech. Dig. 55th Int. Electron Devices Meeting*, IEDM, pp. 349–352.

16. Zhou, Q. (2009). Hybrid assembly based on robotic assembly and fluidic self-alignment, *Tutorial in 3rd Smart System Integration*.

17. Mastrangeli, M., Ruythooren, W., Hoof, C. V. and Celis, J.-P. (2008). Establishing solder Interconnects in capillary die-to-substrate self-assembly, *Tech. Dig. 2nd 3D System Integration Conference*, 3D-SIC, pp. 45–55.

18. Grossi, F., Cioccio, L. D., Renault, O., Berthier, J. and Clavelier, L. (2008). Self assembly of dies to wafer using direct bonding and capillary forces, *Symposium E: Materials and Technologies for 3-D Integration in Mater. Res. Soc. Fall Meeting*, MRS.

19. Fukushima, T., Konno, T., Iwata, E., Kobayashi, R., Kojima, T., Murugesan, M., Bea, J.-C., Lee, K.-W., Tanaka, T. and Koyanagi, M. (2011). Self-assembly of chip-size components with cavity structures: high-precision alignment and direct bonding without thermal compression for hetero integration, *Micromachines*, **2**, pp. 49–68

20. Tong, Q.-Y., Cha, G., Gafiteanu, R. and Gosele, U. (1994). Low temperature wafer direct bonding, *J. Microelectromech. Syst.*, **3**, pp. 29–35.

21. Fukushima, T., Iwata, E., Konno, T., Bea, J.-C., Lee, K.-W., Tanaka, T. and Koyanagi, M. (2010). Surface tension-driven chip self-assembly with load-free hydrogen fluoride-assisted direct bonding at room temperature for three-dimensional integrated circuits, *Appl. Phys. Lett.*, 96, 154105.

22. Fukushima, T., Iwata, E., Ohara, Y., Murugesan, M., Bea, J., Lee, K., Tanaka, T., and Koyanagi, M. (2011). A multi-chip self-assembly technologyfor advanced die-to-wafer 3D integration to precisely align known good diesin batch processing, *IEEE Trans. Electron. Packag. Manuf.*, accepted.

Chapter 9

ADVANCED DIRECT BOND TECHNOLOGY

Paul Enquist

Ziptronix, Inc.

9.1 INTRODUCTION

New technology platforms are adopted as application requirements exceed the capabilities of installed technology platforms. The inherent scalability of conventional planar semiconductor manufacturing technology[1] has allowed continuous improvements in conventional technology platforms like lithography, front end processes, and interconnect to meet customer demands for over 40 years. The interconnect technology platform has become an impediment in allowing continuation of this improvement due to fundamental limitations that arise from increasing parasitics that result in increased power consumption and signal delay.[2] Increasing market demand for mobile devices that require minimum footprint, long battery life, and "more than Moore"[2] functionality has further exacerbated these limitations and resulted in the adoption of 3D technology to meet customer requirements.

3D technology consists of the bonding and interconnecting of stacked layers of 2D die or wafers, preferably to achieve higher functionality in a smaller footprint at lower cost to facilitate adoption. It consists of essentially three die or wafer fabrication technology platforms; bonding, thinning, and interconnecting. The requirements of each of these platforms generally depend on the other platforms and the sequence in which they are implemented in a 3D process flow. For example, thinning before bonding may require a thin die or wafer handling capability that thinning after bonding does not require. The lack of a self-consistent set of 3D standards for these platforms or process flows has contributed to an ad hoc investigation of a wide range of possibilities for these platforms throughout universities, research organizations and the semiconductor industry.[3]

3D Integration for VLSI Systems
Edited by Chuan Seng Tan, Kuan-Neng Chen and Steven J. Koester
Copyright © 2012 by Pan Stanford Publishing Pte. Ltd.
www.panstanford.com

The manufacturing adoption of 3D technology has principally leveraged the improvement of existing technology platforms like adhesives, wirebonding, and bumping to achieve rapid adoption at low cost. Examples of this include die stacking with tiered wirebonding[4, 5], and package-on-package[4, 6] assemblies. Relentless demand for smaller form factors and increased functional density has resulted in the development of a through-silicon-via (TSV) interconnect technology platform suitable for implementation in the Back End.[7] The deep vias that are characteristic of this technology platform increase cost and interconnect pitch resulting in limited scalability and preventing realization of the full cost and performance potential benefit of 3D technology.

3D integration will enable the full potential cost and performance benefits of 3D technology to be realized by combining the highest lateral density of 3D interconnects and vertical density of heterogeneous stacking with wafer scale economics. This capability results in some specific requirements for the three 3D technology platforms. First, the interconnect technology platform will require a TSV to propagate an interconnection between stacked die or wafers to eliminate parasitics that result from other interconnect approaches like wirebonding. However, it is important that the TSV not intrude into the Back-End-of-Line (BEOL) multi-level metal interconnect stack to avoid complicating or compromising the BEOL interconnect routing or increasing the die size. Building this type of TSV in a wafer foundry has a number of advantages with regard to cost and process integration optimization.[8] Availability of this technology platform is imminent as indicated by wafer foundry roadmap announcements regarding this type of TSV integrated into their wafer fabrication[8] and process design kits. Second, the thinning technology platform will need to be compatible with very thin 2D layers, less than ten microns for some applications. A 3D process flow that thins 2D layers before bonding will require a thin wafer handling or temporary wafer bonding technology that has yet to be proven suitable for high volume semiconductor manufacturing. Alternatively, a bond technology that allows thinning after bonding allows use of an existing wafer thinning technology platform. Third, the bond technology platform needs to be compatible with a scalable areal density of vertical electrical interconnections to support pixilated applications like image sensors and displays that rely on reductions in pixel pitch to achieve reduced cost, footprint, or increased performance. This requires an electrical interconnection integral to the bond to avoid a TSV-based inter 2D layer interconnect that increases cost and limits scalability with a TSV etch and fill exclusion extending through the BEOL interconnect of one of the 2D layers.[8] Another requirement for this technology platform is adequate bond strength and uniformity to meet the minimum 2D layer thickness requirements.[8] It is further preferred for the bond strength and uniformity to be adequate to allow thinning after bonding to avoid thin die or wafer handling requirements. Furthermore, achieving this bond strength and uniformity with a low thermal budget is required for heterogeneous applications involving 2D layers with significant coefficient of thermal expansion (CTE) mismatch. A fourth bond technology platform capability requirement is post-bond compatibility with BEOL and Back End fabrication process including lithography, backgrinding, CMP, via etch and fill, and thermal cycling that are required for a number of 3D process flows.

This chapter provides a review of advanced direct bond technology and evaluates its suitability as a bond technology platform for 3D integration. It is shown with the aid of comparison to other bond technologies that advanced direct bond technology delivers optimum 3D interconnect lateral scaling, vertical scaling, bond strength and uniformity, thermal budget and post-bond fabrication capability with a low cost-of-ownership and supply chain synergy resulting in an ideal solution for the requirements of a bond technology platform for 3D integration.

9.2 DIRECT BONDING

Direct bonding refers to forming a bond directly between the surfaces of two materials without an intervening third material, for example an adhesive, to realize the bond[10]. The surface roughness of the materials is preferably sufficiently smooth to allow the formation of a high density of bonds between atoms on the respective surfaces. The resulting bond strength is a function of the density and type of inter-atomic bonds. The types of inter-atomic bonds can be one or a combination of ionic, covalent, or metallic and hence result in a bond with strength comparable to the materials that are bonded. For applications like semiconductors wherein the material has a form factor characterized by a wafer with a specified thickness, bow and warp; the values of these parameters are preferably sufficiently low to allow the bond strength to adequately deform the wafers to bring the material surfaces into intimate contact. High temperatures are typically required after the materials are placed into contact to achieve bond strengths adequate for semiconductor applications.[11] High temperatures and pressures may also be used, for example to accommodate excessive surface roughness with fusion of the direct bonded materials. High bond strength can also be achieved without the use of high temperatures or pressures by placing sufficiently smooth and clean surfaces together in a non-contaminating vacuum environment.[12, 13] These temperatures, pressures, and/or vacuum environments are generally not compatible with the requirements for a bond technology platform suitable for 3D integration. Furthermore, these environments would increase the cost-of-ownership (CoO) of this platform.

9.3 ADVANCED DIRECT BONDING

Advanced direct bonding, as used in this chapter, refers to a bond technology platform that is suitable for 3D integration. A key requirement is high bond strength at low thermal budget to ensure compatibility with temperature-sensitive CMOS and other devices, for example III/Vs, that benefit from 3D integration. It is further desirable to be able to execute the bond process in ambient manufacturing conditions to avoid the high cost-of-ownership associated with specialized, vacuum, or low throughput bond tools. An advanced direct bond technology that meets these and other requirements is described below. This advanced direct bond technology has been trademarked as ZiBond™ and is covered by a number of Ziptronix patents and patent applications.[14–20]

9.3.1 Low Thermal Budget

Conventional direct bond technology requires a specialized vacuum environment to clean bond surfaces and keep them clean until they are brought together in order to achieve a high bond strength at low thermal budget [12, 13]. Advanced direct bond technology resolves this limitation, and satisfies a key requirement for a 3D integration bond technology platform. This section describes how this limitation is resolved with a surface preparation that includes terminating the bond surface with a species that enables the formation of high strength chemical or molecular bonds after bond surfaces are brought into contact in an ambient manufacturing environment within a low thermal budget. Characterization of this bond and its capability to support very aggressive 3D vertical scaling is also discussed.

9.3.1.1 Surface preparation

A number of surface preparations are possible within this advanced direct bond technology. This chapter will focus on some of the surface preparations that are most suitable for 3D integration. Given that CMOS will most likely be the largest constituent for 3D integration, and CMOS is comprised mostly of silicon-based transistors and silicon oxide-based dielectrics in the BEOL interconnect, this section will focus on the use of advanced direct bonding to bond wafers with a silicon oxide coated surface to each other or silicon wafers. The surface preparation consists of a mechanical and chemical specification portion as described below.

Mechanical Specification
Achieving high bond strength at low temperatures requires a sufficient contact area between the surfaces being directly bonded. Microscopically, this requires that the surfaces are smooth on the order of an atomic dimension to enable the highest areal density of atomic bonds at the bond interface. Given that the inter-atomic distance in silicon and silicon dioxide

is about 0.2 nm, the maximum RMS surface roughness should be less than 1nm and preferably less than 0.5 nm. Macroscopically, this may require the bond surface bow, warp, and non-planarity to be sufficiently low to allow contact across the entire bond surface. This is typically achieved with the SEMI Standards for wafer thickness, bow, and warp and a non-planarity of less than 5-10 nm per 100 microns when bonding silicon wafers.

Both silicon and silicon oxide RMS of less than 0.5 nm can be readily achieved with chemo-mechanical polish (CMP) using industry standard tools, pads and slurries. The ability of this advanced direct bond technology to leverage industry standard CMP technology minimizes the adoption barrier for the technology.

The surfaces of conventional CMOS wafers are typically not suitably planar for the highest possible bond strength due to final BEOL fabrication steps including pad metal, passivation, and pad cut. This can be accommodated by eliminating these process steps with the added benefit of process cost savings, or by planarizing the non-planar surface with oxide deposition and CMP.

The appropriate elimination of process steps depends on the type of CMOS BEOL, typically aluminum or copper. A typical aluminum BEOL fabrication sequence includes metal deposition and etch, oxide deposition, via cut to metal, tungsten CVD and tungsten CMP. Stopping the aluminum BEOL fabrication sequence after tungsten CMP can result in a surface within or close to the planarity and roughness specification for advanced direct oxide bonding. A typical copper BEOL fabrication sequence includes dielectric deposition, via and trench etch, conductive copper diffusion barrier liner, copper electroplated or PVD fill, copper CMP, and silicon nitride non-conductive copper diffusion barrier. Stopping the copper BEOL fabrication sequence after copper CMP or silicon nitride diffusion barrier can result in a surface within or close to the planarity and roughness specification for advanced direct oxide bonding. On either of these intermediate aluminum or copper BEOL surfaces, if the planarity and roughness is not suitable, a subsequent planarization, for example submicron silicon oxide deposition and CMP, is typically adequate to comply with the specification and provide a surface suitable for bonding to either silicon or silicon oxide.

Chemical Specification

Achieving high bond strength at low temperatures also requires constituent atoms on respective surfaces to be able to form strong bonds with each other without requiring high temperatures to drive chemical reactions at the direct bond interface. This advanced direct bond technology can achieve this objective by terminating at least one of the direct bond surfaces with a suitable atomic, chemical or molecular species. Subsequent contact of this first terminated surface with a second terminated or un-terminated surface results in a spontaneous direct bond with high strength. This bond strength, or increase in bond strength with time after contact, may be increased or accelerated with a nominal temperature cycle, for example less than 200°C, after contact. This type of bond may also be referred to as a chemical or molecular bond due to the use of chemical reactions or a molecular surface termination.

Surface activation may be required in order for a surface termination to be effective. Surface activation may include the removal of organic contaminants resulting from CMP, and/or the breaking of atomic bonds to increase the availability of atomic bonds for termination and/or the absorption increase of surface termination reaction byproducts after direct bond surface contact. Activation may be accomplished with a very slight etch of the surface, for example with a plasma-based process, that substantially maintains or does not materially increase the surface roughness.

The effectiveness of a specific surface termination depends on the materials comprising the direct bond surfaces. An effective termination is defined as one which facilitates high bond strength within a low thermal budget. Nitrogen or a nitrogen-based atomic or molecular species are an effective surface termination when direct bonding silicon oxide to silicon or silicon oxide. The surface termination may be accomplished with exposure to a dry process, for example a plasma-based process, a wet process, for example a spin-rinse-dry process, or a gas, for example a vapor. The termination may be in-situ with or after activation and a second termination may augment a first termination.

9.3.1.2 Bond Process

Surfaces that comply with the mechanical and chemical specifications may be bonded by simply aligning and placing them together in an ambient environment. The tool required for this process depends on the alignment accuracy required. For example, bonding to an un-patterned wafer can typically be accomplished with a simple flat or notch alignment. Alternatively, bonding to a patterned wafer will typically require a tool with an alignment accuracy specified by the patterns on the wafers to be bonded. The primary placement requirement is to bring the wafers together so that they contact with the required alignment accuracy and distortion. If the initial wafer contact does not extend over the entire wafer surface, the spontaneous bond strength over the initial area of contact will pull the remaining area into contact, subject to the control of the placement tool and the mechanical properties of the wafers.

The high bond strength at low temperature characteristic of this advanced direct bond technology alleviates the alignment/placement tool from requiring application of a temperature and/or pressure bond cycle after placement. This enables increased tool throughput and lower tool cost, significantly reducing tool CoO. The surface termination can also be effective at preventing ambient contamination that would otherwise require a specialized environment, e.g., vacuum, to be implemented that would decrease tool throughput and increase tool cost and CoO.

9.3.1.3 Post-Contact Kinetics

After the wafers are brought into contact, the bond strength increases spontaneously depending upon the kinetics associated with the chemical nature of the termination. This preferably occurs outside the alignment/placement tool in batch to not decrease tool throughput or increase tool CoO. This can result in a bond strength that is a significant fraction or multiple of the silicon fracture strength, which can be enabling for 3D integration applications requiring the bonding of materials with high CTE mismatch. 3D integration applications without high CTE mismatch, for example CMOS-to-CMOS, can achieve an accelerated bond strength that is a multiple of the silicon fracture strength, with a moderate 100-200°C post-contact bake as shown below.

9.3.1.4 Bond Characterization

Three different types of bond characterization and representative data are used to demonstrate the efficacy and suitability of this advanced direct bond technology for 3D integration. The measurement of interfacial bond energy and its dependence on a variety of bond surfaces and bond surface termination processes will first demonstrate the high bond strength at low thermal budget capability of the technology. An analysis of the bond interface with secondary ion mass spectroscopy (SIMS) will then provide evidence of the type of surface termination than can be used to achieve an effective termination. Finally, an evaluation of low distortion capability required for 3D integration applications wherein high accuracy post-bond lithography alignment is essential will be given.

Interfacial Bond Energy
The interfacial bond energy is a measurement of the strength of the direct bond. It can be characterized with mechanical loading of and observing failure at the bond interface. The bond interface can be loaded by normal, shear, or a combination of normal and shear forces. The two most common types of bond energy measurement techniques reported are the so-called four point bend (FPB)[21] and crack-opening (CO)[22, 23] techniques. The FPB technique bends the bonded wafers about a fulcrum and results in both normal and shear loading at the bond interface. The CO technique inserts a wedge between the bonded wafers resulting in normal loading at the bond interface. The FPB technique results in higher bond energy values due to the inclusion of shear in the measurement. The CO technique is a more direct indication of the bond interfacial energy between the bonded wafers but is a more difficult measurement to repeat and reproduce. All bond energy measurements reported in this chapter were obtained with the CO technique.

The CO technique relies on insertion of a wedge to separate bonded surfaces that results in a crack whose length can be measured with IR microscopy when bonding lightly doped, un-patterned silicon wafers or acoustic microscopy when bonding non-IR transparent materials. The interfacial bond energy can be calculated from eq. 1 [10],

$$(3h^2 E_1 t_{w1}^3 E_2 t_{w2}^3)/(32L^4(E_1 t_{w1}^3 + E_2 t_{w2}^3)) \tag{9.1}$$

where h is the wedge thickness at the point of contact with the bonded wafer pair, E_1 and E_2 are the Young's modulus of wafers 1 and 2, respectively, t_{w1} and t_{w2} are the thickness of wafers 1 and 2, respectively, and L is the observed crack length. The bond energy as a function of crack length resulting from this relation for 0.7mm thick silicon wafers and a 0.1mm thick wedge is given in Fig. 9.1. Note that for a fixed accuracy of crack length measurement, the accuracy of this technique decreases significantly as the bond energy increases. Proper design of a wedge insertion tool allows gauge repeatable and reproducible interfacial bond energy measurements well in excess of the silicon fracture strength for the wafer and wedge thickness used in Fig. 9.1. If the wedge has a sharpened edge to facilitate insertion at the bond interface, e.g., a typical razor blade, it is necessary to add the length of this taper to the IR observed crack length to get the true crack length and avoid calculating an erroneously high bond energy. This error can be as high as 20% for a 0.5 mm taper and 5 J/m² bond energy.

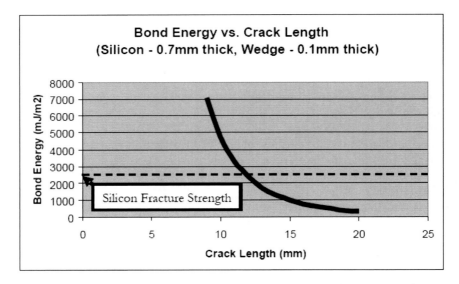

Fig. 9.1 Interfacial bond energy vs. crack length for the Crack-Opening technique using 0.7mm thick silicon wafers and a 0.1 micron thick wedge.

The bond energy depends on the materials being bonded, the surface chemistry before, during, and after bond surfaces are placed into contact and the time and temperature after bond surfaces are contacted. The surface chemistry can be controlled with activation and termination of the bond surfaces enabled by this advanced direct bond technology. Interfacial bond energies have been characterized for a wide variety of surfaces including silicon with native oxide bonded to silicon oxide and silicon oxide bonded to silicon oxide for a wide variety of types of silicon oxides, surface activation and termination processes, and post-contact temperature cycles. Silicon oxides evaluated have included thermally grown, silane-based plasma enhanced chemical vapor deposition (PECVD), tetraethyl ortosilicate (TEOS) PECVD, high density plasma PECVD, and physical vapor deposition. A wide variety of activation and termination processes including combinations of Ar, N_2, and O_2 plasma reactive ion etch (RIE), aqueous ammonium hydroxide at various concentrations, vapor ammonium hydroxide, and de-ionized (DI) water have been characterized. Effective surface terminations have been established using any of these plasma processes, with or without further processing before the wafers are placed into contact. These processes do not result in an adverse increase in surface roughness as verified by atomic force microscopy. A surface roughness less than 0.5 nm is typically adequate to achieve good results.

One of the more effective surface activation and terminations can be achieved with a nitrogen plasma RIE process. An example of the effectiveness of this process is given in Table 9.1 where a range of interfacial bond energies after various post-contact temperatures as measured by the CO technique for processes with a nitrogen plasma RIE surface activation and termination process are compared to processes without a plasma surface termination process. This table includes results either with or without subsequent aqueous water or ammonia hydroxide post-plasma processes prior to surface contact, silicon or silicon oxide direct bonded to silicon oxide, and various types of silicon oxide.

Table 9.1 Interfacial bond energies in mJ/m^2 after various post-contact temperatures for processes with a nitrogen plasma RIE surface termination process compared to processes without a plasma surface termination process. Results include either with or without subsequent aqueous water or ammonia hydroxide post-plasma processes prior to surface contact, silicon or silicon oxide direct bonded silicon oxide, and various types of silicon oxide.

Plasma RIE Prosess	None	N$_2$
Post-Plasma Process	None, H$_2$O, or NH$_4$OH	None, H$_2$O, or NH$_4$OH
300°C Bake	< 1,000	3,200 – 4,700
200°C Bake	< 500	1,700 – 4,700
100°C Bake	< 300	1,000 – 2,200
RT (No Bake)	< 300	300 – 1,000

As shown in Table 9.1, a conventional silicon or silicon oxide to silicon oxide direct bond without plasma surface activation and termination results in low interfacial bond energies even after post-contact temperatures as high as 300°C. In contrast, a nitrogen plasma RIE surface activation and termination is effective at achieving interfacial bond energies in excess of the silicon fracture strength at post-contact temperatures as low as 100°C. Moreover, high interfacial bond energies can be obtained at room temperature which may be enabling for heterogeneous integration applications requiring bonding of materials with a high CTE mismatch. [24] Comparable bond energies are obtained when direct bonding silicon to silicon oxide or silicon oxide to silicon oxide when exposing only the silicon oxide surfaces to the plasma RIE process. Bond energies in the lower portion of the indicated range are typically obtained when direct bonding silicon oxide to silicon oxide and exposing only one compared to both of the silicon oxide surfaces to the plasma RIE process.

Plasma RIE activation and termination processes using gases other than nitrogen can also be used. For example, Table 9.2 provides a range of interfacial bond energy results after various post-contact temperatures as measured by the CO technique for surface activation and termination processes incorporating a nitrogen plasma RIE processes compared to oxygen or argon plasma RIE processes. Table 9.2 further partitions these results into use of a subsequent post-plasma aqueous ammonia hydroxide rinse vs. water or no rinse. Results include silicon or silicon oxide direct bonded to silicon oxide and various types of silicon oxide.

Table 9.2 Interfacial bond energies in mJ/m^2 after various post-contact temperatures for various surface activation and termination processes using a nitrogen plasma RIE process compared to an argon or oxygen plasma RIE processes and subsequent aqueous ammonia hydroxide rinse compared to aqueous water or no rinse. Results include silicon or silicon oxide direct bonded to silicon oxide and various types of silicon oxide.

Plasma RIE Prosess	N$_2$		Ar or O$_2$	
Post-Plasma Process	NH$_4$OH	None or H$_2$O	NH$_4$OH	None or H$_2$O
300°C Bake	3,200 – 4,700		1,900 – 2,700	1,300 – 2,200
200°C Bake	1,700 – 4,700		1,400 – 2,700	500 – 1,900
100°C Bake	1,000 – 2,200		500 – 1,400	400– 600
RT (No Bake)	300 – 1,000		< 500	< 300

The results in Table 9.2 indicate that high interfacial bond energies after low post-contact temperatures can be obtained with surface activation and termination processes using plasma RIE processes other than nitrogen; however, higher bond energies are typically obtained at lower post-contact temperatures with a nitrogen plasma RIE process. Furthermore, when using non-nitrogen plasma RIE processes, higher interfacial bond energies are typically observed with subsequent aqueous ammonia hydroxide rinse compared to water or no rinse while this improvement is not typically observed when using a nitrogen plasma RIE process.

These results indicate that termination of the bond surface with a nitrogen-species is an effective method to obtain high interfacial bond energies after low post-contact temperatures. The nitrogen-based termination can be implemented with a nitrogen plasma RIE process or exposure to ammonia hydroxide. An ammonia hydroxide rinse prior to bond surface contact can also help remove surface particulates generated during the plasma process, but may be difficult to implement in production bond tools. A surface activation and termination process comprising a nitrogen plasma RIE process and a water rinse is a favorable process for achieving high interfacial bond energy after low post-contact temperature, robust particulate control, simple production bond tool compatibility, and low CoO. An adequate bond energy is often achieved after exposing only one of the direct bond surfaces prior to contact allowing a reduced tool cost and CoO compared to exposing both bond surfaces.

Bond Interface Analysis

Evaluation of surface termination resulting in high interfacial bond energy after low post-contact temperatures can be facilitated with an analysis of the bond interface. Secondary Ion Mass Spectroscopy (SIMS) can provide good spatial resolution and sensitivity of a terminating species when looking for elements at the bond interface not present in large concentrations in the vicinity of the bond interface. For example, when bonding silicon or silicon oxide to silicon oxide, profiling of nitrogen can be used to provide evidence of surface termination with a nitrogen species. This example is of particular interest given the correlation of high bond energy with nitrogen-based processes noted above.

SIMS profiles of nitrogen, boron, and oxygen through an advanced direct bond interface of silicon oxide to silicon using a single step nitrogen plasma RIE activation and termination process on the silicon oxide surface prior to bonding is given in Fig. 9.2. Oxygen and boron were profiled as markers of the bond interface and nitrogen was profiled for evidence of nitrogen-based surface termination. A boron peak is observed at the silicon oxide to silicon direct bond interface as expected for air exposed silicon-based surfaces.[25] A nitrogen peak is also observed at the direct bond interface as expected for a silicon oxide surface terminated with a nitrogen species prior to bonding. The boron and nitrogen profile widths are consistent with that expected for monolayer-scale boron contamination and nitrogen termination of bond surfaces prior to bonding given the depth of the profile and the material profiled. The nitrogen dose is about half the silicon oxide areal density and comparable to the silicon areal density of the silicon oxide. This, in conjunction with Table 1, indicates that a nitrogen surface termination can result in very high interfacial bond energies at low post-contact temperatures.

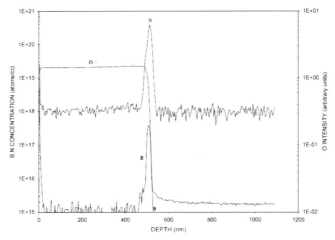

Fig.9.2 SIMS profile of boron, nitrogen, and oxygen through a direct bond interface of silicon oxide to silicon with nitrogen plasma RIE surface activation and termination of the silicon oxide prior to bonding.

As shown in Table 9.2, oxygen and argon plasma RIE processes on silicon oxide surfaces can also result in high interfacial bond energy at low post-contact temperature. SIMS bond interface analysis of silicon or silicon oxide direct bonded to silicon oxide using a surface activation and termination process incorporating oxygen or argon plasma RIE has also been performed to evaluate the nature of surface termination with these processes. Figure 9.3 provides the SIMS profiles of nitrogen, boron, and oxygen through an advanced direct bond interface of silicon oxide to silicon oxide using an oxygen plasma RIE activation and termination process on both silicon oxide surfaces prior to bonding. The boron peak marks the silicon oxide to silicon oxide advanced direct bond interface as noted previously. A nitrogen peak is also observed at the bond interface, although at a lower dose than observed for the nitrogen plasma RIE termination. This nitrogen peak is attributed to residual nitrogen in the plasma RIE chamber after evacuation to the base pressure. This was verified by observation of a plasma emission spectroscopy transient at plasma RIE initiation and a reduction of this transient and the SIMS nitrogen dose with increased plasma RIE chamber evacuation. A reduced interfacial bond energy at low post-contact temperatures was also observed with increased plasma chamber evacuation and reduced SIMS nitrogen dose and is consistent with surface termination by a nitrogen species using plasma RIE as an effective mechanism for achieving high direct bond energy at low post-contact temperature.

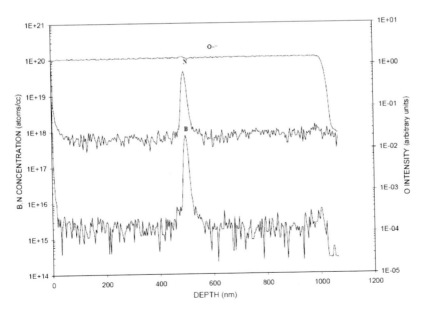

Fig. 9.3 SIMS profile of boron, nitrogen, and oxygen through a direct bond interface of silicon oxide to silicon oxide using oxygen plasma RIE prior to bonding.

Bond Distortion

Distortion is a measure of variance of a specific pattern on a wafer. Distortion on CMOS wafers results from the cumulative deposition and etch of patterned materials with different CTE at various temperatures. It is thus dependent on the wafer fabrication process and can exceed the ability of a stepper or scanner to step and repeat a pattern across the wafer. Distortion in 2D semiconductor wafer fabrication is typically not an issue because the alignment accuracy requirements are typically relaxed as the geometries are increased as the wafer is fabricated from the front-end-of-line through the BEOL. A typical distortion for 200 mm CMOS wafers at the 130 nm node is 20-30 nm.

Bond distortion can be induced in a bonded wafer as a result of stresses induced during alignment, placement into contact, and/or bonding of wafers. This distortion can be greater than 500 nm depending on the bond process and details of the alignment and wafer contact process.[26] The maximum allowed distortion needs to be within post-bond wafer fabrication alignment requirements which is dependent on the application. For example, a maximum distortion spec for backside illuminated image sensors maybe less than 100 nm.

Bond distortion can be measured by bonding the BEOL face of a CMOS wafer to another wafer, thinning the CMOS substrate, revealing alignment marks, then measuring these alignment marks with a stepper or scanner. This advanced direct bond technology has a fundamental advantage over other bond technologies in that the inherent high bond strength at low post-contact temperatures precludes distortion from bond interface slip that can occur with other bond technologies during post-contact temperature and/or pressure cycles required to achieve high bond strength. The bond distortion resulting from this advanced direct bond technology is thus limited to alignment and contact placement effects which can be controlled to < 100 nm.[26]

9.3.3.5 Bond Scalability

Bond scalability refers to the minimum wafer thickness that can be bonded onto another wafer. This is typically accomplished with thinning of the bonded wafer, either before or after bonding. A bond technology that supports thinning after bonding is desirable for a 3D integration bond technology platform because it eliminates the need for thin wafer handling tools and processes that have yet to be proven in volume manufacturing.

The mechanical and chemical specification of the advanced direct bond technology results in excellent uniformity of the interfacial bond strength after contact at low temperature. This uniformity allows thinning of a bonded wafer to micron scale thickness. This is distinct from other bond technologies like aligned patterned copper pillar or copper thermo-compression that bond at aligned copper regions but not between these aligned copper regions. The lack of a uniform bond over the entire wafer typically limits the minimum thickness that can be achieved. This limitation is exacerbated by the CTE mismatch for applications requiring the bonding of heterogeneous materials.

An example of the benefit of the uniform high bond strength at low post-contact temperature capability in a 3D integration application is the use of this advanced direct bond technology to fabricate symmetric intrinsic heterojunction bipolar transistors.[27, 28] The process flow for this fabrication involves frontside heterogeneous device fabrication followed by planarization, advanced direct bonding to a handle or CMOS wafer, device substrate removal leaving micron scale device thickness, and backside device fabrication symmetric to frontside device fabrication, resulting in minimizing extrinsic device parasitics and allowing device performance to approach intrinsic limits. The uniform, high bond strength at low post-contact temperatures was required to accommodate the materials bonded and their CTE mismatch for this application. Furthermore, submicron alignment with low distortion was enabled resulting in high device fabrication yield required for manufacturability.

Another example of bond scalability is the fabrication and stacking of multi-layer SOI structures. A typical thin film SOI wafer has < 500nm device and buried oxide layer thicknesses. This advanced direct bond technology is well suited to the fabrication of these thin film SOI wafers and the stacking of submicron thick, thin film SOI layers with an iterative process of advanced direct bonding and SOI handle substrate removal. This process may also be augmented with device and interconnect fabrication before and after bonding, subject to respective thermal budget requirements, of which the advanced direct bonding is not the limiting process.

9.4 INTEGRAL INTERCONNECT

A primary requirement for a 3D integration bond technology platform is the ability to realize vertical 3D interconnects with scalable areal density between bonded layers of CMOS. This can be achieved with an advanced direct bond technology wherein heterogeneous surfaces comprising conductors surrounded by an insulating matrix are aligned and bonded with a direct bond component. This advanced direct bond technology has been registered as direct bond interconnect (DBI®) and is covered by a number of Ziptronix patents and patent applications[29–30]. Details of the DBI® device structure, fabrication methods, interconnection mechanisms, and practical examples are given below.

9.4.1 DBI® Device Structure

A cross-sectional schematic of a DBI® device structure example is given in Fig. 9.4. This example is comprised of two bonded DBI® heterogeneous surfaces, each comprised of electrically conducting and electrically isolating components, to form a DBI® interface comprised of a direct bond between the electrically isolating components and a conductive bond between the electrically conducting components.

Insulator	Conductor	Insulator	Conductor	Insulator
Insulator	Conductor	Insulator	Conductor	Insulator

DBI® Interface

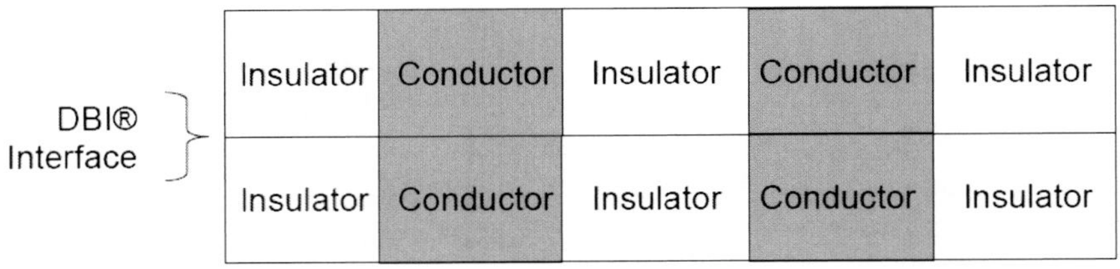

Fig. 9.4 Schematic cross-section example of Direct Bond Interconnect (DBI®) device showing DBI® interface after aligning and bonding two DBI® heterogeneous surfaces comprised of electrically conducting and electrically insulating components.

These components may be perfectly aligned as shown in Fig. 9.4 or may be misaligned. Misalignment may be unintentional, for example as a result of alignment limitations of tools used to place the heterogeneous surfaces together, or intentional, for example if the conductors on respective surfaces have differing sizes or shapes. Misalignment may result in an insulating bond component between a conducting element on one surface and an insulating component on the other surface.

A plan view schematic of a DBI® heterogeneous surface example is given in Fig. 9.5. This example illustrates a high density of closely packed electrically conducting components electrically isolated by a single electrically isolating component. Perfect alignment and bonding of two of these DBI® surfaces can result in a cross-section shown schematically in Fig. 9.4. It is also possible for the DBI® surface to comprise a single electrically conducting component perforated by a number of electrically isolating components. The alignment and bonding of two of these surfaces may be useful for certain applications, for example construction of a perforated ground plane at the DBI® interface.

The electrically insulating component can be comprised of a direct or non-adhesive bond with an insulating material. Examples of insulating materials include silicon oxides or nitrides including oxynitrides, low-k dielectrics, and other insulators used in CMOS BEOL fabrication.

The electrically conducting DBI® component can be metallic or non-metallic. An example of a conducting non-metallic component is polysilicon. Examples of metallic components include elements like aluminum, copper, gold, nickel, etc., alloys of copper, nickel, titanium, etc., and solders like eutectic (63% Tin, 37% Lead), high lead (95% Lead, 5% Tin), or lead-free (97.5% Tin, 2.5% Silver). The conducting bond component can be direct or indirect, and with or without fusion, significant mass transport, grain growth or reflow across the DBI® interface.

The heterogeneous DBI® surface topography is preferably nominally planar to facilitate the bonding. The precise planarity required across the bonding surface can be categorized into long range and short range considerations. Long range considerations include parameters like wafer bow, warp, and flatness that are dependent on properties of the bonded materials like thickness and modulus. Short range considerations include parameters like surface roughness and planarity in the vicinity of the transition between the electrically conducting and electrically insulating components. For example, the surface of the electrically conducting component can be coplanar, elevated, or recessed with respect to the electrically insulating component as shown in Fig. 9.6. Elevation is typically referred to as bumping and recession is typically referred to as dishing. The amount of bumping or dishing that can be accommodated depends on a number of factors including the CTE and modulus of the electrically conducting and insulating components and the relative topography of the DBI® surfaces bonded, e.g., dished to dished, planar to dished, etc., but are typically on the order of a few nanometers. These considerations are discussed in more detail below.

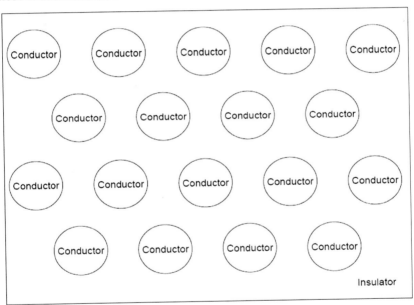

Fig. 9.5 Schematic plan view example of Direct Bond Interconnect (DBI®) surface prior to bonding showing electrically conducting components electrically isolated by an electrically insulating component.

9.4.2 DBI® Fabrication Methods

A number of different fabrication methods are possible to build the DBI® structure. The details of these methods depend on the specific materials used for a particular type of DBI®. Examples of these methods are described according to fabrication of the heterogeneous DBI® surface prior to bonding and bonding of heterogeneous DBI® surfaces to achieve the DBI® device structure.

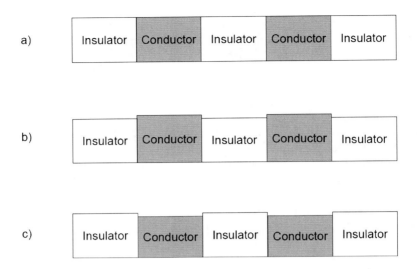

Fig. 9.6 Schematic cross-section example of Direct Bond Interconnect (DBI®) surface showing electrically conducting component surface a) coplanar, b) elevated or bumped, and c) recessed or dished relative to insulating component surface.

9.4.2.1 Pre-Bond DBI® Fabrication

The DBI® pre-bond fabrication requires formation of a heterogeneous surface with requisite planarity and surface roughness suitable for DBI® bonding. These requirements can be met with a variety of process flows that utilize standard semiconductor industry process tools that have been used in volume CMOS BEOL production for a number of years. Three different types of these process flows are described below.

Damascene DBI® Process Flow
Damascene fabrication is pervasive in the fabrication of CMOS wafers. It involves etching of vias and/or trenches followed by filling of these features with metal then removing excess metal outside of these features with chemo-mechanical polishing (CMP). There are basically two types of CMOS wafers in volume semiconductor manufacturing today, those with aluminum or copper BEOL multi-level interconnect to route electrical interconnections within the IC. An aluminum BEOL typically uses tungsten filled vias or plugs to electrically interconnect adjacent levels of interconnect separated by an inter-level dielectric (ILD). A copper BEOL uses copper filled vias to electrically interconnect adjacent interconnect levels. The tungsten plugs are fabricated with a single damascene process wherein a single etch step is used to etch vias through an ILD deposited on top of the previous interconnect level. CVD Tungsten is then deposited on the wafer that fills the vias followed by tungsten CMP to remove excess tungsten outside of the vias. The next

interconnect level is then fabricated on top of and electrically interconnected to the tungsten plugs. The copper vias are fabricated with a dual damascene process wherein two etch steps, each less than the ILD thickness but the combination through the ILD thickness, are used to etch vias that electrically interconnect adjacent interconnect levels and a trench that defines the upper interconnect level. A barrier layer and copper CVD or electroplating fills the vias and trenches followed by copper CMP to remove excess copper outside of the trenches. A cross-sectional schematic of a single and dual damascene process after CMP is given in Fig. 9.7.

Adhesion and barrier layers are typically incorporated in the single and dual damascene processes to improve the adhesion obtainable between aluminum, tungsten or copper to the ILD and the diffusion of these metals into the surrounding ILD. These layers are used prior to or after interconnect metal deposition and can be electrically conducting, e.g., TiN, TiW, or Ta/TaN, or electrically insulating, e.g., SiC or SiN, depending on electrical interconnection requirements to adjacent materials.

| Single Damascene | Dual Damascene |

Fig. 9.7 Cross-sectional schematic after single and dual damascene processes.

Both single and dual damascene processes shown schematically in Fig. 9.7 can result in a surface suitable for DBI® bonding using any metal compatible with these processes. This is due to the capability of CMP to produce a heterogeneous surface with adequate ILD surface roughness, metal surface roughness, and planarity between the ILD and metal surface components, for example as shown schematically in Fig. 9.6. Reliable, high yield 3D electrical interconnections can be built with various values of these parameters. The DBI® interconnection mechanism depends on the specific combination of these parameters as described in more detail below.

Planarized Stud DBI® Process Flow
Some 3D integration applications may require a DBI® metal that is not conducive to a damascene process. An alternate process flow that may be suitable for these metals is to first form DBI® studs, then planarize the surface with an ILD deposition and CMP process that reveals the studs. The DBI® studs may be formed with a variety of methods, for example using electroplating on a seed layer through a photoresist mask followed by isolation with a masked or unmasked seed etch prior to ILD deposition.

Post-planarization DBI® Metallization Process Flow
The damascene and stud process flows described above require CMP of the DBI® metal. Some applications may call for a DBI® metal that is not amenable to CMP. For these DBI® metals, a pre-bond DBI® process flow that

deposits DBI® metal after CMP may be used. In this process flow, a planar surface is first prepared, for example by a damascene or stud process flow. A very thin layer of DBI® metal is then deposited on top of and electrically connected to the conductive planarized surface portions to which DBI® connections are desired. Although this deposition results in a non-planarity of the DBI®surface, this non-planarity may be accommodated by high direct bond energy between planar portions of this DBI® surface. The degree of non-planarity that can be accommodated is dependent on a number of factors, for example the pitch and extent of the DBI® connections.

9.4.2.2 DBI® Bonding

The DBI® bonding primarily consists of aligning and placing two DBI® surfaces into contact. It may also include an advanced direct bonding surface activation and termination process as described above prior to alignment or a heat treatment after bonding to facilitate 3D electrical interconnections.

Alignment and Placement
A key advantage of DBI® is the ability to bond by simply aligning and placing two DBI® surfaces into contact. The bond tool required to implement this process does not require a temperature and/or pressure bond cycle resulting in significant CoO savings from improved tool cost and throughput compared to bond processes that require a bond cycle after placement.[31, 32]

The primary bond tool specification is the accuracy with which wafers or die can be aligned and placed. This accuracy can be characterized according to average and variable components across the bond surface. The average alignment accuracy is what is typically referred to as alignment accuracy and quoted by bond tool manufacturers. The variable alignment accuracy is the variance of alignment across the bond surface. The variable component is typically much smaller than the average component and not an issue for a number of 3D applications. However some 3D applications, for example those that require accurate overlay of post-bond fabrication features to pre-bond fabrication features, have a much tighter specification for variable than average alignment. The bond distortion described above is an example of this variable component.

Submicron average alignment accuracy can be achieved by the use of high resolution optics, minimizing wafer or die movement between alignment and contact, and vibration isolation. A number of tools with this accuracy capability designed for volume production are available on the market.

The variable alignment accuracy can be minimized by minimizing the distortion of die or wafers that are bonded and controlling how the bond surfaces are brought in to contact. Control of bond surface contact includes control of die or wafer bow and warp with a combination of starting material specification and the design and use of the tooling used to bring the bond surfaces into contact.

A significant advantage of DBI® compared to other bond technologies is high bond energy after bond surface contact without post-contact temperature or pressure that can result in slip at the bond interface and compromise in average or variable alignment accuracy. This advantage can be optimized with incorporation of the advanced direct bonding surface activation and termination process described above.

Bond Configurations Suitable for CMOS BEOL Process Integration
The integration of a 3D IC bond process into the CMOS BEOL process flow with the minimum number of additional process steps is desirable to minimize the 3D IC CoO. This can be accomplished with a bond technology that can bond a surface typically presented in the BEOL process flow. Two of these surfaces are shown schematically in Fig. 9.7 wherein the single damascene is an example of a typical CMOS aluminum BEOL wafer after a tungsten plug CMP process and the dual damascene is an example of a typically CMOS copper BEOL wafer after copper CMP. These two surface types are suitable candidates for DBI® process integration in a volume manufacturing environment given their pervasive adoption in mainstream CMOS fabrication and compatibility with DBI® pre-bond fabrication requirements. These types of surfaces can be bonded to themselves or each other, resulting in the three possible configurations for DBI® bonding shown schematically in Fig. 9.8.

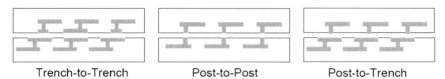

Trench-to-Trench   Post-to-Post   Post-to-Trench

Fig. 9.8 Cross-sectional schematics of possible bonding configurations with single and dual damascene pre-bond DBI® fabrication.

Regardless of bond configuration there will typically be overlap of DBI® metal to ILD resulting from mismatched patterns as shown in Fig. 9.8 or misalignment. This can be a reliability issue for some combinations of DBI®metal and ILD, for example copper and silicon oxide, respectively. This issue can be resolved with incorporation of an ILD barrier layer at the DBI®surface, for example silicon nitride when using copper as DBI® metal and silicon oxide as ILD. Cross-sectional schematic examples of this variation before and after DBI® bonding are given in Fig. 9.9. The pre-bond schematic shows the ILD barrier layer at the DBI® surface and a similar ILD barrier layer on top of the subcutaneous last metal with these two ILD barrier layers separated by non-barrier ILD layer through which a via is etched to the last metal, followed by conductive barrier layer and DBI® metal deposition via fill, and CMP planarization. The post-bond schematic shows misalignment of a post-to-post bonding configuration as shown in Fig. 9.8 and protection of ILD from DBI® metal by ILD barrier layer.

Before DBI® Bonding   After DBI® Bonding

Fig. 9.9 Cross-sectional DBI® schematics of damascene process including barrier ILD to accommodate bond misalignment before and after DBI® bonding.

Post-Bond Thermal Cycle

After DBI® surfaces are aligned and placed together, the wafers may be heated to facilitate the bond or 3D interconnect. This can be done after contact as a separate step or in a downstream process step required for another purpose. The DBI® interfacial bond energy after contact can be sufficient to allow downstream heating in batch without externally applied pressure to minimize CoO.

9.4.3 DBI® Interconnection Mechanisms

As indicated above, many different types of conductors, surface planarities and post-bond thermal cycles are possible for DBI®. These variations can result in different mechanisms of 3D interconnect formation. This section describes some of these mechanisms that can be realized with and without post-contact heating.

9.4.3.1 Without Post-Contact Heating

DBI® interconnections can be realized without post-contact heating. This can be accomplished with a direct bond between both insulating and conducting DBI® components after contact of two aligned DBI® surfaces. This is possible if the surfaces of the insulating and conducting components are sufficiently smooth and planar. Sufficiently planar includes the possibility that the conducting component is slightly non-planar, for example bumped relative to the insulating component as shown schematically in Fig. 9.6. The degree of non-planarity that can be tolerated depends on a number of parameters including the specific conductor or insulator and the relative extent of conductor or insulator. The required smoothness or RMS is also dependent on a number of parameters including the degree of non-planarity and relative extent of the conductor. For example, a higher RMS on a bumped conductor may be tolerated relative to the insulator if the insulator extent is sufficient to provide enough interfacial bond energy to apply adequate pressure on the conductor to accommodate the higher conductor RMS.

9.4.3.2 With Post-Contact Heating

DBI® interconnections can also be realized with post-contact heating. Heating may be used to facilitating interconnections if a solder or alloy is used for the conductive component that would benefit from reflow. Heating may also be used to make a direct bond without fusing if the conductive component is dished or has a native surface oxide that can be reduced or compromised with pressure. Heating may be further useful if fusion of an elemental DBI® conductive component is desired. These scenarios are described in more detail below.

Reflow
DBI® 3D interconnections can be made using solders and alloys as conductive components. However, these materials typically require heating or reflow after positioning to make a reliable electrical interconnection. An advantage of the reflow process is the conductive material can be reshaped to improve interconnection yield or reliability. For example, if there is a gap between aligned conductive solder or alloy components due to dishing on both DBI® surfaces or a larger dishing on one DBI® surface compared to bumping on the opposed surface, the reflow can result in a bridging of this gap resulting in DBI® interconnections.

Internal Thermo-compression
DBI® 3D interconnections can also be made with post-contact heating using conductive materials that do not reflow. These interconnections can be made with a conductive component that is planar, bumped or dished and/or has a higher RMS relative to the adjacent insulating component. The mechanism for this type of DBI® interconnection can be an internal thermo-compression whereby a thermal cycle results in an expansion of the higher CTE conducting component relative to the lower CTE insulating component and the interfacial bond energy of the opposed insulating components is sufficient to compress the opposed conductive components, without application of any external pressure, to form high yield, reliable 3D interconnects. The thermal cycle resulting in 3D interconnections can be significantly less than that required to melt the conductive component, depending on the CTE, planarity, and modulus of the conducting and insulating components. This post-contact expansion can result in increased dishing tolerance of the conductive component than allowed without thermal cycling. A microscopic cross-

section of this type of connection may exhibit a distinct grain structure on each side of the DBI® bond interface.

Fusion
DBI® interconnections can also be made with post-contact fusion. Fusion will typically occur with higher thermal cycles compared to those required for internal thermo-compression without fusion and may occur after an internal thermo-compression 3D interconnection has been made. This type of DBI®interconnection may be desired if sufficient reliability is not realized with internal thermo-compression without fusion. A microscopic cross-section of this type of connection will typically exhibit a continuous grain structure across the DBI® bond interface.

9.4.4 DBI® Examples

A number of different types of DBI® have been implemented. These have focused on the use of silicon oxide as the insulating component and an elemental metal as the conducting component. A family tree of some of these implementations including organizations who have reported results is given in Fig. 9.10. These implementations are described in more detail below.

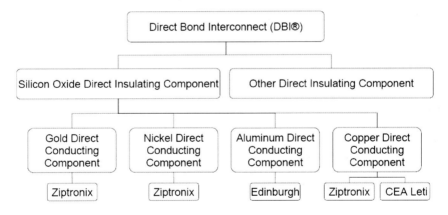

Fig. 9.10 – Family tree of DBI® implementations reported by various organizations.

9.4.4.1 Gold

Gold as a conductive DBI® component is an attractive option because the lack of a stable native oxide, high CTE and low modulus are expected to facilitate a good electrical interconnect with little, if any, post-contact thermal budget. Unfortunately, gold is not compatible with a foundry CMOS BEOL manufacturing environment or conventional CMP planarization. Nevertheless, a gold DBI® process may be desirable for some 3D IC integration scenarios, for example a die-to-wafer implementation.

DBI® pre-bond fabrication with damascene or planarization stud process flows may be problematic due to this CMP incompatibility. However, the post-planarization DBI® metallization process flow described above is suitable for gold. Ziptronix has used this approach to demonstrate direct gold DBI®interconnections with a direct metallic bond using a gold thickness of 20-100 nm above a planarized silicon oxide surface.[33] Although the smallest pitch reported was about a millimeter, 3D interconnections were achieved at room temperature without any post-contact thermal cycle.

9.4.4.2 Nickel

The advantages of nickel as a conductive DBI® component include nominal barrier layer requirement due to low diffusivity in silicon oxide, compatibility with CMOS BEOL manufacturing environments, and low susceptibility to dishing due to high hardness. The disadvantages include low CTE, high modulus and permeability, and novelty in aluminum or copper CMOS BEOL.

Direct nickel DBI® is currently an advanced DBI® processes with regard to interconnect pitch, reliability, CMOS process integration and manufacturing adoption. Ziptronix has reported DBI® interconnection pitch of 1.5 micron as shown in Fig. 9.11. DBI® interconnection yield > 99.999% at a 10 micron pitch and < 3 um² overlapping conductive component area with interconnect resistivity < 0.5 Ohm-um², post-contact thermal budget < 300°C, and reliability exceeding JEDEC standards with a planarized stud process flow have also been reported.[8, 32, 34] This process has been used to build product for a number of 3D IC applications including backside illuminated image sensors[35] and heterogeneously integrated displays[35] and has been non-exclusively licensed for production.[36]

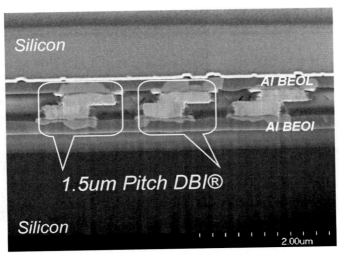

Fig. 9.11 Scanning electron micrograph of 1.5 micron pitch, direct nickel DBI® 3D interconnect without fusion.[32]

9.4.4.3 Aluminum

The advantages of aluminum as a conductive DBI® component include high CTE, low modulus, nominal barrier layer requirement due to low diffusivity in silicon oxide, and utilization in high volume CMOS BEOL manufacturing environments. A primary disadvantage is a stable surface oxide that can impede electrical interconnections.

Edinburgh[37] has reported results with a direct aluminum DBI® 3D interconnect using a damascene process with 20-40 nm of dishing and a 435°C post-contact bake. A contact resistivity of 2.6 Ohm-um^2, and DBI® yield greater than 97% and 99.9% for overlapping conductive component areas of 100um^2 and 196 um^2, respectively, were achieved.

9.4.4.4 Copper

Copper as a conductive DBI® component is attractive due to its high CTE, low modulus, and utilization in high volume CMOS BEOL manufacturing environments. A primary disadvantage is a high diffusivity in silicon oxide that requires a barrier layer to a silicon oxide ILD.

CEA-Leti has reported results with a direct copper DBI® 3D interconnect using a damascene process without a barrier layer. A contact resistivity of 0.98 Ohm-um^2 after a post-contact thermal cycle of 200°C from single connection Kelvin structures was achieved.[38]

Ziptronix has also reported results with a direct copper DBI® 3D interconnect but using a damascene process that incorporated a barrier layer. A yield in excess of 99.9995% and contact resistivity < 0.5 Ohm-um^2 was achieved from 10 micron pitch serial daisy chains with 463,000 connections, cross-sections from which are shown in Fig. 9.12. Direct copper DBI® 3D interconnections were achieved without copper fusing with a 125°C post-contact thermal cycle and preliminary reliability results indicated compliance with JEDEC standards.[39]

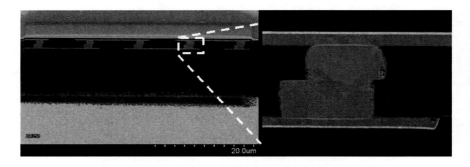

Fig. 9.12 Scanning electron micrographs of 10 micron pitch direct copper DBI® 3D interconnect without copper fusion.

9.5 SUMMARY

This chapter has reviewed the ZiBond™ and DBI® advanced direct bond processes and their capability. Their attributes have been shown to provide an excellent solution to the requirements of a 3D IC bond technology platform. These attributes include a high, uniform bond strength at low thermal budget with an integral interconnect that is scalable to submicron pitch. The technology can be implemented with a variety of material combinations, with the optimum choice dependent on supply chain implementation considerations. Post-bond fabrication capability is typically limited by the CMOS in the 3D IC or the tools used to implement the process and not the fundamental technology. This enables a wide application space possible with these advanced direct bond technologies.

References

1. G. E. Moore, "Cramming more components onto integrated circuits," Electronics, Volume 38, Number 8, April 19, (1965).

2. International Technology Roadmap for Semiconductors, 2007.

3. Handbook of 3D Integration, Vols. 1 and 2, Wiley, 2008.

4. P. Garrou, "Introduction to 3D Integration," Handbook of 3D Integration, Vol 1, Chapter 1, 2008, p.1.

5. Prismark – Stacked Die Report, 2002.

6. M. Dreiza, L. Smith, A. Yoshida, J, Micksch, "Stacked Package on Package (PoP) Design Guidelines," International Wafer Level Packaging Congress, San Jose, CA, November 2005.

7. P. Garrou, C. Bower, "Overview of 3D Integration Process Technology," Handbook of 3D Integration, Vol 1, Chapter 3, 2008, p.25.

8. P. Enquist, "High Density Direct Bond Interconnect (DBI™) Technology for Three Dimensional Integrated Circuit Applications," Fall MRS, 2006.

9. P. Garrou, "Foundry TSVs are a comin' – TSMC makes their play for a bigger portion of the pie," Semiconductor International, May 2, 2008.

10. Q.-Y. Tong and U. Gosele, Semiconductor Wafer Bonding, Wiley (1999).

11. R. Stengl, T. Tan, U. Gosele, "A Model for the Silicon Wafer Bonding Process," Jpn. J. Appl. Phys., 28, 1735-1741, 1989.

12. U. Gösele, H. Stenzel, T. Martini, J. Steinkirchner, D. Conrad, and K. Scheerschmidt, Appl. Phys. Lett. 67, 3614, (1995).

13. H.Takagi, K.Kikuchi, R.Maeda, T.R.Chung, T.Suga, "Surface activated bonding of silicon wafers at room temperature," Appl.Phys.Lett., 68(1996), 2222-2224.

14. Q.-Y. Tong, G. Fountain, and P. Enquist, US patent 6,902,987, "Method for Low Temperature Bonding and Bonded Structure," June 7, 2005.

15. Q.-Y. Tong, G. Fountain, and P. Enquist, US patent 7,041,078, "Method for Low Temperature Bonding and Bonded Structure," May 9, 2006.

16. Q.-Y. Tong, US patent 7,109,092, "Method of Room Temperature Covalent Bonding," September 19, 2006.

17. Q.-Y. Tong, G. Fountain, and P. Enquist, US patent 7,335,572, "Method for Low Temperature Bonding and Bonded Structure," February 26, 2008.

18. Q.-Y. Tong, US patent 7,335,996, "Method of Room Temperature Covalent Bonding," February 26, 2008.

19. Q.-Y. Tong, G. Fountain, and P. Enquist, US patent 7,387,994, "Method for Low Temperature Bonding and Bonded Structure," June 17, 2008.

20. Q.-Y. Tong, G. Fountain, and P. Enquist, US patent 7,553,744, "Method for Low Temperature Bonding and Bonded Structure," June 30 2009.

21. Q. Ma, J. Mat. Res. 12(3), 840-845 (1997).

22. M.S. Metsik, "Splitting of mica crystals and surface energy," J. Adhes., 3, 307, 1972.

23. W.P. Maszara, G. Goetz, A. Cavilia, and J.B. McKitterick, "Bonding of silicon wafers for silicon–on-insualtor," J. Appl. Phys., 64, 4943 (1988).

24. B.P. Abbot, J. Caron, J. Chocola, K. Lin, S. Malocha, N. Naumenko, P. Welsh, "Advances in RF SAW Substrates," 2nd Intl. Symp. On Acoustic Wave Devices for Future Mobile Communication Systems, March 5, 2004.

25. F.A. Stevie, E.P. Martin, Jr., P.M. Kahora, J.T. Cargo, A.K. Nanda, A.S. Harrus, A.J. Muller, and H.W. Krautter, "Boron contamination of surfaces in silicon microelectronics processing: Characterization and causes," J. Vac. Sci. Technol. A9 (5), Sep/Oct 1991, 2813-2816.

26. P. Enquist, G. Fountain, C. Petteway, unpublished.

27. P.M. Enquist, D.B. Slater, US Patent 5,318,916, "Symmetric Self-Aligned Processing," June 7, 1994.

28. P. Enquist, D. Chow, Q.-Y. Tong, F. Reed, G. Fountain, G. Hudson, A. Rose, J. Hancock, and M. Simons, "Symmetric Intrinsic HBT / RTD Technology for Functionally Dense, LSI 100 GHZ Circuits," GOMAC (2000).

29. Q. Y. Tong, P. M. Enquist, A. S. Rose, U.S. Patent 6,962,835, "Method for Room Temperature Metal Direct Bonding," November 8, 2005.

30. Q.-Y. Tong, G. Fountain, and P. Enquist, US patent 7,602,070, "Method for Low Temperature Bonding and Bonded Structure," October 13, 2009.

31. Yole, "3DIC Report", 2007 Edition, Revised August 15, 2008.

32. P. Enquist, "Scalability and Low Cost of Ownership Advantages of Direct Bond Interconnect (DBI®) as Drivers for Volume Commercialization of 3-D Integration Architectures and Applications", Mater. Res. Soc. Symp. Proc. Vol. 1112, 2009, p.33.

33. [33] Q. Y. Tong, "Room Temperature Metal Direct Bonding," Applied Physics Letters, 89, 1, 2006.

34. P. Enquist, "Direct Bonding Processes for 3-D Integration," Handbook of 3D Integration, Vol 2, Chapter 11, 2008, p.487.

35. P. Enquist, G. Fountain, C. Petteway, unpublished.

36. www.ziptronix.com

37. H. Lin, J.T.M. Stevenson, A.M. Gundlach, C.C. Dunare, A.J. Walton, "Direct Al-Al contact using lot temperature wafer bonding for integrating MEMS and CMOS devices," Microelectronics Engineering, 85, (2008), 1059-1061.

38. P. Gueguen, L.D. Cioccia, J.P. Gonchond, P.Gergaud, M. Rivoire, D. Scevola, M. Zussy, D. Lafond, L. Clavelier, "3D Vertical Interconnects by Copper Direct Bonding," Mater. Res. Soc. Symp. Proc. Vol. 1112, 2009, p.81.

39. P. Enquist, G. Fountain, C. Petteway, A. Hollingsworth, and H. Grady, "Low Cost of Ownership Scalable Copper Direct Bond Interconnect 3D IC Technology for Three Dimensional Integrated Circuit Applications," IEEE 3D System Integration, September 30, 2009.

Chapter 10

SURFACE MODIFICATION BONDING AT LOW TEMPERATURE FOR THREE-DIMENSIONAL HETERO-INTEGRATION

Akitsu Shigetou

National Institute for Materials Science (NIMS)

This chapter gives an overview of low-temperature bonding methods based on surface modification technologies. It is important to decrease the process temperature to around room temperature for application-oriented discrete system packaging, which requires three-dimensional interconnections between diverse substrates. It is critical for the contacting surface to have a well-modified chemical structure and topography for high bond quality at such low temperatures. This chapter focuses on surface modification methods induced by the beam irradiation process, and in it, the technical background, principles, important bondability factors, applications, and future tasks are described.

10.1 INTRODUCTION

One of the main focuses in the recent developments in electronics packaging has been the creation of a novel multifunctional system through the mixed integration of diverse materials, as represented in the concept of "More than Moore." For example, as illustrated in Figure 10.1, three-dimensional (3D) hetero-integrated LSI with new components such as non-heat-resistant organic materials is expected to have a high practical value in the near

3D Integration for VLSI Systems
Edited by Chuan Seng Tan, Kuan-Neng Chen and Steven J. Koester
Copyright © 2012 by Pan Stanford Publishing Pte. Ltd.
www.panstanford.com

future; this is because the scale at which these components can be fabricated now overlaps with that of conventional semiconductor microelectronics.[1] However, fabricating such mixed-layer structures using conventional semiconductor microfabrication processes is considered difficult. Severe variations in material properties, bond mechanisms, and process temperatures dramatically increase the process complexity. A simple assembly (bonding) process for discrete substrates is necessary. Naturally, at the microscopic level, the bond strength and mechanical/electrical function at the interface should be seamless from one material to another; therefore, the bonding process has to be designed to enable high alignment accuracy and good control of the interfacial structure. For the former, a low process temperature should be provided to take into account the thermal expansion mismatch between conventional substrate materials and soft materials. The elimination of heating process is ideal, but the considerably low temperature, around 150°C for example, would be tolerant of bondability when the temperatures in other assembly processes in practical packages are considered. In addition, it is effective for the integration of soft materials such as resins and polymers to reduce the temperature to around 150°C. For the latter, the influence of the bonding atmosphere (gas component, pressure, temperature, etc.) on the chemical/physical structure of the contacting surface has to be understood to enable active modification of the interfacial structure.

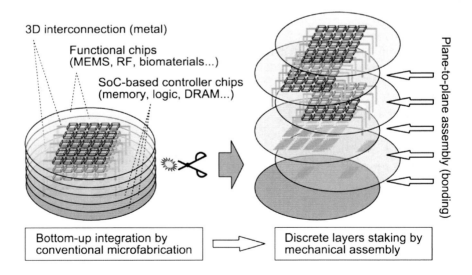

Fig. 10.1 Concept of future 3D interconnection achieved by bonding of discrete materials.

A technical difficulty in satisfying these factors lies in generating sufficient electronic interactions across the interface between materials at such low temperatures. An atomically clean surface has a high binding energy because of the inhomogeneous electrical density distribution. Once the chemisorbed molecules on the active surface are removed and the bonding atmosphere devoid of other adsorption sources (i.e., high vacuum), the highly energetic surface attracts nearby molecules to form new bonds, and thus stabilize the energy status. This implies that, in conventional bonding methods, contact between two materials in ambient air occurs between stable thick adsorbate layers such as oxide and organic contaminants; consequently, sufficient bonding strength cannot be obtained in short experimental times between these layers at low temperatures of around 150°C. Interfacial diffusion is dominant at such low temperatures, and the volume diffusion of inner atoms through the adsorbate layer is not enough to either increase the bonding strength or trigger the tunneling effect at the metal interface that ensures good electrical conductivity. This is the reason why materials being joined together are heated to temperatures near their respective softening points in conventional bonding methods such as diffusion and thermo-compression bonding. Moreover, the low process temperature cannot help in breaking the adsorbate layer and enlarging the contact area through surface deformation at the moment of contact. The flatness of the contacting surface is also an important factor. Therefore, the chemical/physical structure of the flattened surface, including the adsorbate layer, must be modified for high bond quality at temperatures below 150°C.

Here, "surface modification" for low-temperature bonding contains the following procedures: (1) elimination of the initial thick adsorbate covering the active surface (i.e., surface cleaning); and (2) control of the adsorption behavior occurring on the atomically clean surface. Procedur 2 is not compulsory, depending on the bond mechanism of the material, but Procedure 1 is critical to the bond quality of almost all interconnection materials at low temperature. For surface cleaning, it is necessary to apply energy that is higher than the dissociation energy of the adsorbed molecules to the initial surface. The beam-induced dry process, physical etching, and chemical reactions are widely used. Wet processes such as chemical washing are also used, but they are outside the scope of this chapter. Either way, the cleaned surface has to be protected from blind adsorption of gas molecules throughout the bonding process to achieve low-temperature bonding; therefore, the optimization of multiple parameters through surface analysis techniques is necessary for carrying out the bonding process. In this chapter, the typical optimization procedure for the beam-induced surface modification process is described after explaining the fundamental principles and actual bondability factors. Some applications and future challenges are then introduced.

10.2 FUNDAMENTAL BEAM-INDUCED SURFACE MODIFICATION TECHNOLOGIES

10.2.1 Principles

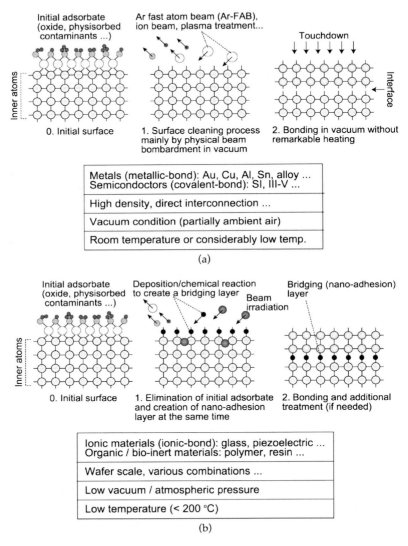

Fig. 10.2 Schematic representations of the principles of surface modification bonding methods. (a) Metallic bond and covalent bond materials. (b) Ionic bond materials and other materials such as polymers. Considerable bond strength can be obtained in ionic bond materials at room temperature, but additional low-temperature heating is highly effective for increasing the bond quality.

Surface-activated bonding (SAB) is a typical example of a low-temperature bonding method. In this, systematic surface modification processes are based on the beam irradiation energy.[2] The SAB method was inspired by observations of the adhesion of space debris on satellites during the early age of space development. It is based on a very simple friction phenomenon: the inert layer covering the surface of the satellite material is physically rubbed off by the collision with debris in vacuum conditions, which automatically creates a tight adhesion on the newly exposed active surface. The "activated" surface condition helps generate a high binding energy between the surfaces, so the term "surface activation" was utilized to characterize this bonding method.

Figure 10.2 shows a schematic representation of the beam irradiation process used in the conventional SAB method. The beam-induced surface modification process for the SAB method is as follows: (1) select a proper beam source, (2) optimize the irradiation condition, and (3) control the adsorption speed after irradiation. These steps are carried out depending on the binding mechanisms of the contacting materials.

As shown in Figure 10.2(a), metallic and covalent bond materials such as Al, Cu, and Si can be bonded with the simplest SAB method, i.e., a dry etching process in vacuum through physical bombardment of the inert gas atoms using Ar fast-atom beam (Ar-FAB), ion beam, or plasma irradiation. On these surfaces, eliminating the initial adsorbate layers results in exposure of the unstable metal surface or dangling bonds. These highly energetic active surfaces can be maintained for certain periods of time if sufficiently high vacuum conditions are provided. Ideally, once these active surfaces come within interatomic distances, surface reconstruction is automatically triggered to stabilize the energetic surfaces stable; tight atomic bonds are created with no additional heating or loading. Some beam-damaged layers with disordered lattice structures and the deformation-induced residual strain can be expected, but in practical use, the bond interface created with this process generally behaves like a new grain boundary.

In contrast, clean surfaces of ionic bond materials such as SiO_2 cannot be bonded with the abovementioned physical beam bombardment process because in-plane relaxation of the surface energy readily occurs on the clean surface. Moreover, electric repulsion between the polarized surfaces obstructs their physical approach to interatomic distances. An effective method for creating high binding energy on the surfaces of ionic bond materials is by forming an ultrathin bridging layer at the same time as the beam irradiation (surface cleaning) process, as shown in Figure 10.2(b). So far, such bridging layers, known as nano-adhesion layers,[3] are typically obtained by the simultaneous deposition of metal ions through sputtering with surface cleaning or by single/multiple reactive ion/plasma beam irradiation, which accelerates the surface cleaning and chemical reaction between the material surface and gas atoms at the same time. Unlike the simple surface activation method mentioned previously, the bond interface created with this method contains a thin interfacial layer whose binding mechanism is different from the surrounding inner atoms.

10.2.2 Bondability factors

Major factors that influence the bond results for the surface modification processes mentioned in Section 10.2.1 include the (a) beam irradiation (surface cleaning) conditions and (b) amount of gas exposure for the cleaned surface before contact. For metals and covalent bond materials, tight bonds are easy to produce if the cleaned surfaces are joined together immediately after the beam bombardment process. However, in general, a certain time is necessary for sample handling and precise alignment in the actual bonding process. Consequently, it is difficult to keep the surface perfectly clean in vacuum chambers for conventional bonding apparatuses with background vacuum pressures of 10^{-2}–10^{-6} Pa (of course, ultrahigh vacuum chambers generate better vacuums). Most actual bonding materials have re-adsorbed molecules on their surfaces before contact. Therefore, to clarify the possible bonding conditions, it is necessary to not only identify the thickness and chemical structure of the initial adsorbate layer but also to understand the growth behavior of the newly created molecular layer. For ionic bond materials, these studies have to be carried out on the nano-adhesion layer as well. The influence of beam irradiation on the topography also has to be considered. Theoretically, a contact load is not necessary for the surface modification bonding method if the contacting surface is atomically flat; however, actual surfaces naturally have roughness. Beam irradiation occasionally increases the surface roughness, which obstructs contact between the surfaces, although the irradiation process is indispensable for improving bondability at low temperatures. In other words, there is usually a trade-off between the topographical and beam irradiation conditions; it is important to estimate the threshold roughness to obtain both full contact and a clean surface. Although other mechanical factors affect the state of contact (sample handling in parallel, etc.), estimating the roughness is still necessary to isolate the topographical influence on bondability from other constraint factors.

Simple estimations for adequate gas adsorption and surface roughness conditions can be performed through both analytical calculation and numerical simulation. First, for gas adsorption onto a clean surface, the increase in the adsorbate thickness is generally considered to follow Gibbs's equation for the initial steps until the formation of a layer covering the clean surface. In this equation, the number of gas atoms that collide with the unit area of a surface for a unit time is approximated by

$$\Gamma = P/(2 \cdot \pi \cdot M \cdot k_B T)^{1/2} \, [\mathrm{m^{-2} s^{-2}}] \qquad (10.1)$$

where P is the gas pressure, M is the molecular or atomic weight of gas, k_B is Boltzman's constant, and T is the temperature. The number of oxide layers in the unit area is obtained by

$$t \cdot \Gamma / N \, [\text{m}^{-2}] \tag{10.2}$$

where t is the exposure time and N is the number of oxygen ions in the monoxide layer of the unit area. This is the most severe adsorption condition since the probability of adsorption is usually much less than 1 for actual materials. Assuming that the bonding material is copper, that the gas molecule is oxygen, that all colliding oxygen molecules are adsorbed onto the clean surface, and that cubic-close-packed copper oxide is generated. For example, the necessary time to generate ideal copper monoxide is around 60 s at a vacuum pressure of 2×10^{-5} Pa. However, longer exposure times are used for conventional bonding because only island-like oxide is believed to exist for the initial adsorption steps; a continuous covering layer is created only after the number of layers increases to around ten.

For the surface roughness, one of the simplest calculation methods is based on Hertz's contact theory, which assumes that the cross-sectional structure of the surface is an elastic sinusoidal body and that the height distribution of the surface asperities approximately follows the normal distribution, i.e., "nominally flat surface".[4, 5] In this method, the threshold roughness is described as the relationship between the wavelength and wave amplitude and is derived from the equation

$$h^2 / L < 32 \cdot \Delta \gamma \cdot (1 - v^2) / (\pi \cdot E) \tag{10.3}$$

where h is the amplitude, L is the wavelength, $\Delta \gamma$ is the surface energy, v is Poisson's ratio, and E is Young's modulus. This expression assumes that the elastic asperities make contact with a rigid flat plane and that full contact between the surfaces is generated automatically without additional load if the elastic energy to recover the compressed asperities is lower than the adhesion energy between the surfaces. When plotting the values of h against L, the sufficient conditions for L and h are located under the curve. For example, the threshold wave amplitude h for a standard Si wafer with an average wavelength L of around 100 nm becomes smaller than 1 nm; for a Cu film highly flattened through the chemical mechanical polishing (CMP) process, the threshold h is around 2 nm for an L of 100 nm. Most conventional SAB results for elastic materials such as Si[6] have been proven to follow this estimation; however, a larger roughness is usually allowed for metals to ensure that bondability owing to the plastic deformation of surface asperities contributes towards enhancing the contact area. For metals and other materials with plastic deformation, it is better to use this roughness estimation to figure out the safest condition.

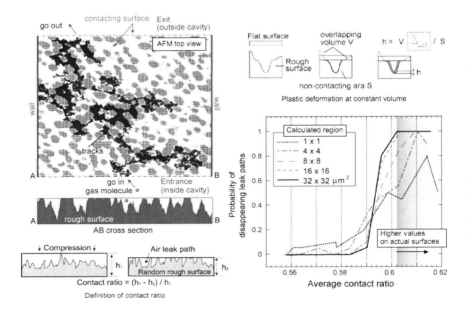

Fig. 10.3 Example of the simulated behavior for the plastic deformation of surface asperities at touchdown, derived from actual AFM measurement data. In this simulation, it was assumed that the plastic surface was mated with a rigid-flat surface and that a "nominally perfect contact" occurred when a gas molecule was not able to pass through the gap between these two surfaces.

Numerical simulations on random rough surfaces have been performed to provide more practical estimations including the plastic deformation derived from actual measurement data. Figure 10.3 shows an overview of an example of a simulation result, where the actual contact ratio between the compressed surfaces was calculated using actual topographical data taken using an atomic force microscope (AFM).[7] This simulation was originally performed to evaluate the capability of vacuum sealing in micro-electro-mechanical-systems (MEMS) packaging, not to determine the threshold roughness for bondability. Therefore, the specific criteria for compression pressure and surface roughness were not considered in the calculations, but the results clarified how much the surface should be compressed in practical materials. The assumptions in the calculations were as follows: (a) AFM top views for different square sizes were prepared as the testing surface; (b) a gas atom was put into the gap between the contact surfaces from one side of the square edge; (c) the atom moved randomly with elastic collisions against the surface asperities; and (d) if the atom reached the other side of the square edge within a finite period of time, it was regarded as the generation of a gas leak path. Deformations of the surface asperities were assumed to occur to maintain a constant volume. The compression ratio at which the gas leak path statistically vanishes (i.e., disappearance probability = 1) was then derived using the Monte Carlo method. For example, an Au film with a mean surface roughness of 1.5 nm reaches its complete disappearance probability at a compression ration of around 60%, regardless of the square sizes. As mentioned later in Section 2.4, nominally flat metal surfaces usually adhere to each other tightly when the relative compression ratio between the surfaces is estimated to be higher than this value.

10.2.3 Optimization of bond parameter

Surface modification bonding methods have some general steps to optimize the bond parameters. These steps are described here using the metal Cu when bonded through basic physical beam bombardment of Ar atoms. The procedure for ionic bond materials is not discussed in this section because there is a wide range in the types of usable chemical compounds for the nano-adhesion layer.

1. The fundamental properties of bonding materials, including their oxide and possible adsorbate, should be understood. In particular, the bond mechanism, lattice structure, mechanical constants related to the fracture strength, thermodynamic constants, etc., should be known in advance.

2. Sufficient bond conditions for surface roughness and gas adsorption need to be estimated to determine the safest (i.e., severest) bond parameters. For gas adsorption, Equations 1 and 2 can be used as a basic derivation of the critical adsorption thickness at which a continuous covering layer is created on the clean surface. If available, numerical estimations such as first-principle calculations are more preferable. For the surface roughness, the simplest way to specify the threshold condition is to use theoretical calculations such as Eq. 3. Although the surface deformation behavior is dramatically changed by the mechanical contact conditions during the actual bonding processes, it is still important to maintain a safe flatness to eliminate unexpected influences on the bondability.

3. The chemical/physical structure of the initial surface should be analyzed. First, surface analysis such as X-ray photoelectron spectroscopy (XPS) and Auger electron spectroscopy (AES) is necessary to understand the structure of the initial adsorbate. This is indispensable for determining the minimum beam irradiation conditions to obtain an atomically clean surface because the removal rate at the surface is different from that for inner atoms. Ideally, the removal rate should be measured using a standard specimen of the specified adsorbate material. For example, Figure 10.4(a)[8] shows the XPS depth profiling results for the chemical composition of a CMP-Cu film corresponding to the etching depth. The presence of C 1s, which mainly originated from organic contaminants, was found at the outmost surface; the copper oxide (and usually

hydroxide) as indicated by O 1s was observed beneath the contaminant with the composition ratio decreasing concomitantly with the etching depth. The inner Cu layer was exposed at an etching depth of around 8 nm, which was regarded as the minimum etching depth necessary to clean the surface. This transition can also be observed in the chemical shifts of Cu LMM spectra that are presented in Figure 10.4(b).[8] The chemical shift is also available from photoelectron binding energy (Cu 2p3/2), but in some materials, it is difficult to isolate the oxide peak (in this case, Cu_2O) from the peak of the pure material because those peaks are too close to each other. When this occurs, the kinetic energy of Auger electrons (Cu LMM) is occasionally more useful. Next, for the surface roughness, precise surface profiles, such as from AFM, are necessary to measure the mean roughness, average wavelength, and height distribution. The horizontal shape information is more accurate if considered together with the power spectral density (PSD). If there is a large periodic waviness in addition to a small surface roughness, AFM measurements should be performed for not only small areas (a few square microns), but also large (more than several tens of square microns) to evaluate the influence on the contact condition.

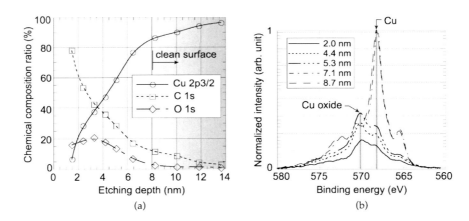

Fig. 10.4 XPS depth profiling results for the initial Cu surface. (a) Changes in the chemical composition ratio. (b) Chemical shift of Cu LMM spectra corresponding to the etching depth.

4. The influence of beam irradiation on the surface roughness should be analyzed to determine the allowable range of surface cleaning conditions. First, the top view, cross-sectional view, Ra or RMS roughness, height distribution, and PSD (if available) should be taken using AFM for each irradiation condition. Especially for surfaces of polycrystalline materials, where the in-plane etching rate is different

due to diverse lattice orientations (Figure 10.5),[8] it is important to know how much the mean roughness increases without significant deviation of the height distribution from the nominally flat surface. In addition, the influence of the beam source and beam irradiation angle on the etching condition should be analyzed if possible. For example, in Figure 10.5, the irradiation angle was fixed at 45°, at which the removal rate was generally considered to be at its maximum. Further, the change in surface energy should be examined if the bonding material indicates a remarkable electrical charge-up induced by the beam irradiation. The safe beam irradiation conditions can then be determined by overlapping (a) the sufficient roughness to ensure perfect contact as calculated in Step 2 and (b) the minimum etching depth for eliminating the initial adsorbate as calculated in Step 3. For the CMP-Cu surface shown in Figure 10.5, the allowable range in etching depth turned out to be 8–15 nm.

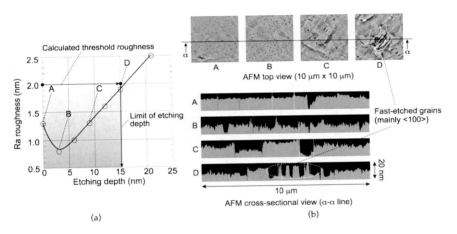

(a)   (b)

Fig. 10.5 Results of AFM analyses on the topographical influence of beam bombardment. (a) Change in the mean surface roughness corresponding to the etching depth. (b) Top views and cross-sectional views of the 10 μm × 10 μm surface, which indicate the increasing roughness due to the in-plane difference in etching speed because of diverse lattice orientations.

5. The relationship between the bonding strength and amount of gas exposure after the surface cleaning process should be clarified to determine the threshold condition for gas adsorption on the clean surface. A simple way to express the amount of exposure is by using the product of gas pressure and exposure time (Pa·s) for the initial steps of adsorption. Figure 10.6(a)[8] shows an example of CMP-Cu film bonding results. In this example, the nominal bond area, which was obtained from the contour of the film fracture on the debonded surface in the die-

shear test, was used as an alternative to the bonding strength because it is difficult to accurately measure the interfacial bonding strength in thin film samples. The amount of exposure changed for both constant vacuum pressure and constant exposure time; the limit to ensure high bondability was 0.1–0.2 Pa·s in either case. XPS observations showed a Cu_2O layer that was around 2 nm thick, which corresponds to about ten molecular layers; it was created at the threshold amounts of exposure. With better exposure conditions, a nominally voidless interface was obtained, as shown in Figure 10.6(b). In addition, the growth speed of the re-created adsorbate for long-range gas exposure should be quantified from the depth profiling results if possible. When good exposure conditions cannot be maintained during the actual bonding process, low-temperature annealing is effective at improving the bonding strength. Understanding the necessary diffusion length for such adsorbates is an effective way to avoid applying an unnecessarily high heating temperature.

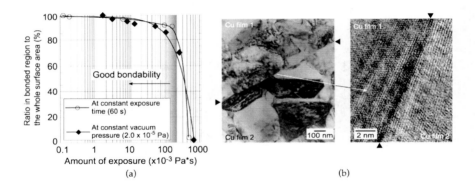

(a)   (b)

Fig. 10.6 Influence of the amount of exposure after the surface cleaning process. (a) Changes in the ratio of the bonded region to the whole surface. (b) TEM images of the nominally voidless interface achieved in low exposure conditions.

6. The bond quality should be evaluated to see if the selected bonding condition was correct. Detailed investigations on the interfacial structure and its transition in long-range storage or harsh aging conditions are necessary, as well as fundamental evaluations of the bonding strength and electrical performance, to discuss the overall mechanical/electrical reliability in accordance with industry standards.

10.2.4 Applications

Conventional applications

Figure 10.7 shows transmission electron microscopy (TEM) images of the fundamental bond interfaces created from a simple surface activation process through physical bombardment with Ar atom or ion beams at room temperature. Figure 10.7(a)[9] shows the bond interface between Al single crystals. The deformation of surface asperities induced the crystal rotation near the interface and resulted in the mismatch of lattice orientations, which was observed as a contrast to the residual strain. However, no readily visible void or intermediate layers were found at the interface. The thickness of such a disordered layer is generally a few nanometers at most, and the residual strain gradually becomes relaxed even at room temperature. Especially for polycrystalline materials, the relaxation is more accelerated compared to single-crystal materials due to the large numbers of defects and grain boundaries penetrating to the contact surface. Consequently, the influence of the interfacial structure on the electrical properties is not serious compared to conventional bonding methods using high heating temperatures. For metal/metal interfaces, a significant decrease in the RC delay is expected for signal transmissions through ultrafine pitch electrodes. Figures 10.7(b) and (c)[10, 11] represent the bond interfaces of Al/diamond and Si/GaAs, respectively. Most metals and covalent bond materials can create hetero-interconnections through the physical beam bombardment process; therefore, many combinations with large coefficients of thermal expansion (CTE) mismatches are enabled at room temperature. Unlike metallic interfaces, the elastic bodies of covalent bond materials often generate the amorphous beam-damaged layer, and the bond interface becomes vague. The thickness of the damaged layer is influenced by the beam energy, but it is generally limited to within a few nanometers. Gas atoms from the beam source are sometimes found trapped in this layer, so it is necessary to control the beam energy to avoid undesirable degassing when an additional heating process is required for the bonded sample.

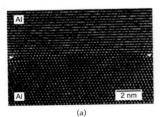
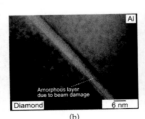
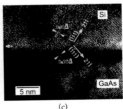

(a)   (b)   (c)

Fig. 10.7 TEM images of typical SAB interfaces between (a) single crystalline Al bulks, (b) Al and diamond, and (c) Si and GaAs.

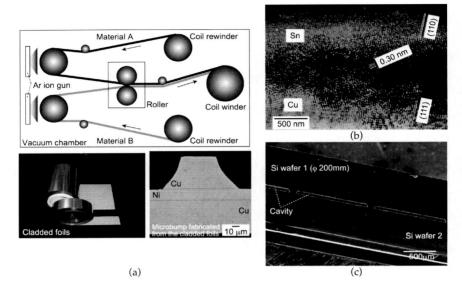

(a)   (c)

Fig. 10.8 Examples of applied SAB interfaces. (a) Cu cladding materials with flat surface and low compression pressure. (b) Sn/Cu interface with an intermetallic compound layer, which is much thinner than that produced by conventional soldering methods. (c) Bonding of cavity wafers for the MEMS device packaging.

Some of the bonding results created by this simple surface cleaning process have been used commercially, as shown in Figure 10.8. Figure 10.8(a)[12] outlines the cold-rolling process for the fabrication of Cu cladding foil. The rolling pressure is dramatically reduced by installing the surface cleaning step before the roller. This cladding process enables the use of flattened foils in place of conventional double-treated foils because a large plastic deformation is not necessary for improving the adhesion strength. Such smooth interfaces contribute towards reducing the stacking thickness of the printed wiring board and the skin effect in high-frequency transmissions. Figure 10.8(b)[13] presents an example of the interface between solder materials, where the structure of the eutectic intermetallic compound became easy to modify using the atomically clean surfaces. The elimination of the surface oxide for solder materials was performed not only with Ar beam irradiation but also with hydrogen-induced deoxidization; these processes have the significant advantage of decreasing the reflow temperature of flux-less solder paste.[14] Figure 10.8(c)[15] shows the bonding results for a Si cavity wafer. As mentioned in the previous section, semiconductor materials can be bonded in high vacuum conditions without visible voids, which is useful in vacuum sealing techniques for MEMS sensor packaging because very little degassing is expected at low bonding temperatures. In this case, a gas leak rate of lower than 1.0×10^{-12} Pa·m^3/s was realized at the same time as the bonding was carried out.

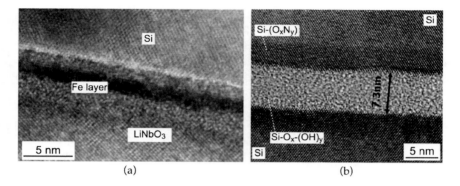

(a)   (b)

Fig. 10.9 TEM images of the bond interfaces created by the nano-adhesion layer method. (a) Si/LiNbO$_3$ interface through the thin Fe layer. (b) Oxynitride bridging layer created by the sequential plasma process. The sequential plasma process was especially effective for Si and glass-related materials.

The bond interfaces presented in Figure 10.9 are typical examples using ultrathin bridging layers (i.e., nano-adhesion layers). Figure 10.9(a)[16] shows the LiNbO$_3$/Si interface used for the acoustic wave filter module. In this case, an ultrathin Fe layer was used as the adhesion layer; it was fabricated by sputter-deposition carried out at the same time as the surface cleaning process by Ar ion beam irradiation at room temperature. XPS and AES interfacial analyses showed that the Fe component was included in the amorphous beam damage layer, not as a continuous film; hence, the interface was considered to partially contain metallic bonds. Although investigations on the influence of such metallic bonds on the electrical characteristics are still ongoing, the use of nano-adhesion layers has increased the capability of optical modules for hetero-integration. In related bonding results, the interconnection between the GaN wafer and wiring metal as well as the Si waveguide substrates were realized using thin metal films at room temperature.[17, 18] Figure 10.9(b)[19] shows another type of bridging layer, which was created by chemical reaction between the gas atoms from the beam source and the surface of the bonding materials. In this example, Si and quartz glass wafers were tightly bonded by a Si oxynitride bridging layer generated using the O$_2$ reactive ion beam etching (RIE) process, followed by microwave N$_2$ radical treatment in low vacuum. Unlike conventional fusion bonding methods that create hydrogen bonds at the interface between hydrophilic surfaces at around 1000°C, this sequential plasma treatment process requires no additional annealing process after contact. For every nano-adhesion layer method, controlling the chemical structure is the next task. The development of a novel bridging layer, which would work not only as the adhesive but also as a specific functional layer, is expected to be feasible in the near future.

Advanced application - Cu bumpless interconnection

The dramatically reduced process temperatures in the surface modification bonding methods especially helps contribute to shrinking the interconnection pitch of area-array flip-chip structures. Conventionally, there has been a severe technological gap in the interconnection pitch between off-chip bonding and on-chip microfabrication, as shown in Figure 10.10. Therefore, a narrow pitch of less than 10 μm is strongly required for bonding technologies to build up the discrete layer structures and ensure seamless signal integrity throughout all of the interconnection levels. One of the most serious bottlenecks for this demand has been the CTE mismatch, which is related to the bump-like electrodes between the layers (substrates). Very small bumps corresponding to a bond pitch of a few microns cannot compensate for the thermal strain between the substrates because plastic deformation is not enough at high heating temperatures used in existing bond techniques. In addition, general self-alignment effect induced by surface tension on the melted metal/solder materials cannot be expected when the necessary alignment accuracy is reduced to about 1/10 of the bond pitch, e.g., less than 1 μm. Therefore, next-generation high-density 3D interconnections should eliminate the bump electrodes, fabricate a highly flattened layer structure where the surfaces of the metal wiring and insulation nominally appear on the same plane, and reduce the bond temperature. This "bumpless interconnect"[20] is schematically represented in Figure 10.11.

The layer structure for the bumpless interconnect, where wiring is formed inside the insulation layer together with the through-hole electrodes, can be fabricated through the damascene process. An excellent example of a bumpless structure was created by a build-up process based on conventional semiconductor microfabrication,[21] but it is outside the scope of this section. For bonding of bumpless structures, a pioneering study realized an 8 μm–pitch interconnection of Cu electrodes through an oxide bonding process on the insulation layer followed by diffusion bonding of the Cu electrodes.[22] A significant benefit of this process was that very tight adhesion was obtained over the entire substrate surface, although a heating process at around 350°C was necessary for several hours to accelerate Cu diffusion through the thick adsorbate at the interface to ensure both the bond strength and electric interconnection. Meanwhile, a 6 μm–pitch bumpless interconnect (pin count was around 50,000 electrodes/mm^2) was achieved without additional heating using physical bombardment of Ar atoms onto the highly flattened Cu electrodes, as shown in Figure 10.12(b).[23] The surfaces adhered tightly without visible voids or intermediate compounds, so the interfacial electrical resistance became as low as the volume resistance of the mating electrodes. Moreover, the voidless interface provided high bonding strength and electrical reliability after high temperature storage testing at 150°C for 1000 h. However, it was impossible to generate a high binding energy on the insulation layer (i.e., ionic bond material) at room temperature in this case. Moreover, the surface of the insulation layer had to be etched back by around 50 nm after the damascene process to ensure Cu-Cu contact; this was because the surface of the Cu electrodes was generally indented due to the CMP process during the last step of the sample fabrication. Consequently, this resulted in a narrow gap between the insulation layers; hence, a Cu frame surrounding the electrodes was fabricated at the same height as the electrodes. This frame structure contributed to not only improving the bonding strength but also providing a high-quality vacuum seal to the electrodes at the same time as the bonding process.

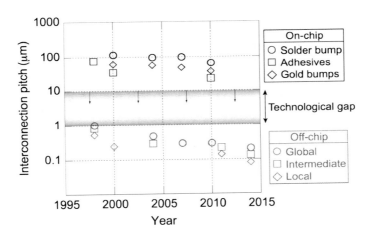

Fig. 10.10 Roadmap of the interconnection pitch off-chip and on-chip.

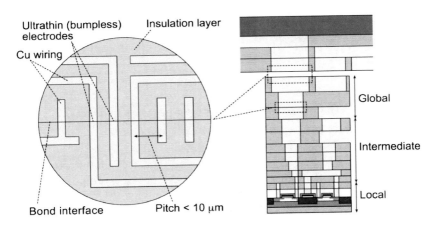

Fig. 10.11 Schematic representation of the bumpless interconnection concept. This structure was originally proposed to alternate with the global interconnection layers.

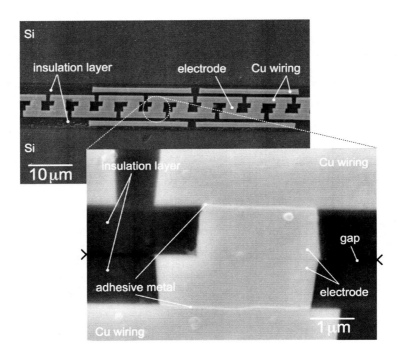

Fig. 10.12 Enlarged images of the interface between 6 μm–pitch bumpless electrodes. A narrow gap of less than 100 nm remained at the interface between the insulation layers.

For either case of bumpless interconnections, the important point is that mixed materials on the same surface could not be bonded simultaneously at low temperature due to the difference in bonding mechanisms for each material. This clearly shows the technical tasks required for future 3D interconnections include the integration with non-heat-resistant soft components. It is necessary to develop a novel bonding process that can be used for different materials at the same time to provide tight adhesion at low process temperatures in an ambient air-like atmosphere.

10.3 NEXT CHALLENGES IN SURFACE MODIFICATION BONDING PROCESSES

Bonding methods based on beam-induced surface modification have extended the field of hetero-integration at low process temperatures, but there are still some technical hurdles for realizing the borderless integration illustrated in Figure 10.1. One pressing requirement, as mentioned in the previous section, is carrying out the low-temperature bonding process at ambient pressure, especially in ambient air. This not only reduces the process complexity but also has a beneficial impact on potential non-heat-resistant components such as organic materials. Although it is difficult to carry out the surface cleaning step at atmospheric pressure, other bonding steps such as the alignment should be conducted in ambient air-like conditions. The main focus needs to be on "how to maintain bondability on a surface exposed to adsorptive conditions."

For Au, tight adhesion can be obtained simply by removing organic contaminants because no stable oxide is generated around room temperature. For example, Au microbumps were successfully interconnected in N_2 gas at atmospheric pressure.[24] Bonding in ambient air was also possible, but this experiment used N_2 gas to prevent readsorption of hydrocarbonaceous contaminants. In contrast, other oxidizing materials such as Si, Al, Cu, etc. have to be bonded on the premise of oxides and other adsorbed molecular layers forming in ambient air conditions. To ensure interconnection at low temperatures for such adsorptive conditions, modifying the chemical structure and thickness of the adsorbate layer that is newly created during the gas exposure is strongly required. This is because sufficient volume diffusion of inner atoms through the interfacial layer is indispensable for improving the bondability and inducing the electric tunneling effect (for metals) at low temperature. Low-temperature bonding in ambient air can be paraphrased as an advanced nano-adhesion layer method, where the adsorbate (interfacial) layer is expected to work as not only a bridging layer but also an electric functional layer. There are various candidates for the ultrathin bridging layer, but assuming normal air as the fundamental bonding atmosphere, water and oxygen molecules are most critical for the chemical binding condition on a clean surface. Therefore, using CMP-Cu films as a basic example, the growth behavior of the adsorbate and the bonding quality are described within the parameters of humidity in oxygen gas at atmospheric pressure.[25]

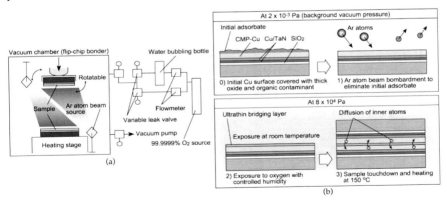

Fig. 10.13 Schematic representations of (a) the bonding apparatus used to create an ultrathin bridging layer on metal surfaces and (b) the bonding process carried out at low temperature and atmospheric pressure. In (a), the upper sample was attached to a rotatable fixture, which was supported with a pair of (*X-Y*) plate springs to compensate for the chip tilt.

The bonding apparatus illustrated in Figure 10.13(a) can be used for above investigations. This machine was fabricated by installing Ar fast atom beam sources and gas introduction lines to a high vacuum flip-chip bonder. A water bubbling bottle was connected to the gas introduction line to humidify the oxygen gas; the gas and solvents could be replaced depending on the targeted composition of the bridging layer. Figure 10.13(b) schematically presents the bonding sequence as follows: (1) CMP-Cu film samples with a mean (R_a) roughness of 1–2 nm were introduced into this apparatus; (2) the initial adsorbate was eliminated by Ar beam bombardment; (3) oxygen gas with controlled humidity was introduced onto the clean surface at a pressure of 0.8 atm; (4) the surfaces mutually contacted; and (5) the heating temperature was applied to the samples.

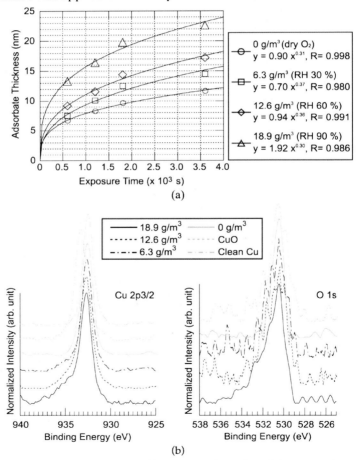

Fig. 10.14 Results of XPS depth profiling against humidity. (a) Time evolution of the adsorbate thickness on a clean Cu surface. (b) Spectra of Cu $2p3/2$ and O $1s$ at step 2 in Figure 1(a). Spectra from the bulk CuO sample are presented for comparison. The main peak intensity was normalized to emphasize differences in peak positions.

Figures 10.14(a) and (b) present the XPS depth-profiling results of the Cu surface shown in Step 3 of Figure 10.13(b); they indicate the time evolution of the adsorbate thickness and the main peaks of Cu and O against humidity, respectively. The results show that the adsorbate growth on the clean surface was apparently influenced by the humidity. For every humidity condition, increases in the thickness approximately followed the cube law, although the thicknesses for the same exposure times increased concomitantly with increased humidity. This suggests that two limiting factors simultaneously affect adsorbate growth: (a) the amount of Cu volume diffusion from the inside of the films to the adsorbate layer (i.e., diffusion factor); and (b) the number of water molecules that collide with the surface for a unit period of time (i.e., chemisorption factor). As described in step 5 of Section 2.3, excellent bond quality was ensured until an oxide layer that was approximately 2 nm thick was generated on the clean CMP-Cu surface. By comparing the ratio of the self-diffusion lengths for Cu, which can be derived from the one-dimensional solution to Fick's law, the diffusion length increases to around 15 nm at 150°C. Therefore, an exposure time of 600 s should ensure sufficient bondability for all humidity conditions at 150°C. However, the chemical structure of the bridging layer differed depending on the humidity. The synthetic peaks of Cu $2p_{3/2}$ show that $Cu(OH)_2$ was detected in humid conditions, whereas the peak of Cu_2O was dominant in the dry oxygen condition. The hydroxide component was considered hydrated because the peaks attributable to water molecules were observed in the O $1s$ spectra.

These differences influenced the interfacial structure after bonding. Figures 10.15(a) and (b) represent the TEM images and electron energy-loss spectroscopy (EELS) spectra, respectively, of the bond interface created by heating at 150°C for 600 s, where a film breakage was observed in the die-shear testing. The TEM images show that a nominally void-free interconnection was obtained through a polycrystalline CuO layer with a thickness of 10–15 nm; this was attributed to Cu_2O oxidation. This CuO layer was partially broken mainly because of the deformation of surface asperities at the moment of contact. Direct Cu-Cu interconnections occurred at those positions. The interfacial electrical resistivity, which was measured using 10 × 10 μm microfabricated electrodes, was in the same range as that of conventional SAB Cu-Cu interfaces regardless of the CuO interfacial layer; the increment ratio was less than 10% after high-temperature storage testing was carried out at 150°C for 1000 h.[26]

TEM and EELS analyses showed that the Cu surfaces could also be bonded in humid conditions through amorphous interfacial layers consisting of Cu hydroxide-hydrate and physically adsorbed hydrocarbonaceous compounds. A considerably high shear strength was obtained at temperatures higher than 100°C, which was lower than the dry oxygen condition. In these amorphous layers, precipitation of CuO grains was observed upon additional heating at temperatures higher than 100°C after bonding; the number and size of CuO grains increased concomitantly with humidity. Although detailed investigations on the influence of water molecules on the bondability are still ongoing, the electrostatic adhesion of water molecules and their dissociation can be inferred to contribute to lowering the bonding temperature. The thickness of these amorphous layers increased dramatically with humidity but can be modified. Therefore, the interfacial structure should change to a thin CuO layer similar to Figure 10.15(a) by controlling the humidity, and high bonding strength and good electrical conductivity should result even in humid conditions.[27]

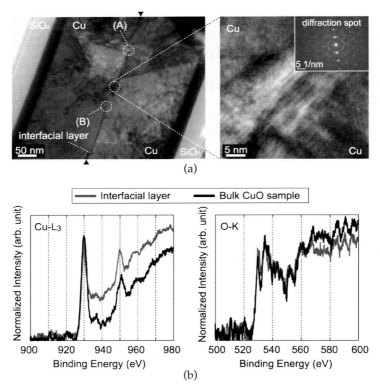

Fig. 10.15 Interfacial analysis results for the sample bonded with heating at 150°C for 600 s in dry O_2 conditions. (a) TEM cross-sectional images. (b) EELS spectra of Cu-L_3 and O-K. In (a), no readily visible interfacial layer was found at the positions marked as (A) and (B). In (b), the spectral intensity was normalized with that of the peak edges to emphasize the differences in peak positions.

10.4 CONCLUSIONS

In this chapter, the surface modification bonding method induced by the beam bombardment process is described as a promising assembly technology for future 3D interconnections of mixed discrete components;

the characteristics, principles, important parameters, applications, and future challenges are explained. Surface modification by physical beam bombardment, which utilizes the attractive force that acts between atomically clean surfaces in high vacuum, demonstrates excellent bond qualities of metals and covalent bond materials at room temperature. For ionic bond materials, the nano-adhesion layer method is used. In these bonding processes, beam irradiation generally induces simultaneous changes in the chemical binding condition and surface topography; therefore, it is crucial to clarify the trade-off between these factors to ensure good bonding quality. Hence, using the CMP-Cu interconnection as an example, a general optimization procedure for bond parameters is described here. However, considering the future possibility of integrating non-heat-resistant components and the resulting increase in process complexity, a bonding atmosphere at atmospheric pressure is required as well as low process temperature (e.g., lower than 150°C). Consequently, the chemical structure and growth behavior of the adsorbate, which is created on a clean surface at atmospheric pressure, have to be quantified to provide not only the binding energy but also the electrical function to the adsorbate. As a fundamental example, Cu-Cu interconnections at temperatures below 150°C in ambient air-like conditions were introduced. Although more detailed studies need to be conducted on bond mechanisms, a breakthrough in the application-oriented design process is anticipated for surface modification bonding technologies.

References

1. The source was extracted from the illustration entitled *Conversing Technologies*, found in: http://www.imec.be/wwwinter/buisiness/nanotechnology.pdf

2. H. Takagi et al., *Surface Activated Bonding of Silicon Wafers at Room temperature*, Appl. Phys. Lett. 68 (1996) 2222.

3. T. Suga et al., *Combined Process for Wafer Direct Bonding by Means of the Surface Activation Method*, Proc. IEEE Electronic Components & Technology Conf., 54th (2004) 484.

4. K. L. Johnson, *Contact Mechanics, Cambridge University Press*, 1985.

5. K. Takahashi and T. Onzawa, *The influence of surface roughness on the super-precise low energy bonding without heating and contact pressure*, J. of High Pressure Institute of Japan 35 (1997) 159. (in Japanese). One other related paper is: L. Lei et al., Proc. 8th Int'l Welding Symp. (2008) 325.

6. T. R. Chung et al., *Wafer Direct Bonding of Compound Semiconductors and Silicon at Room Temperature by the Surface Activated Bonding Method*, Appl. Surf. Sci. 117–118 (1997) 808.

7. H. Okada et al., *The Influence of Surface Profiles on Leakage in Room Temperature Seal-Bonding*, Sensors and Actuators A 144 (2008) 124.

8. A. Shigetou et al., *Bumpless Interconnect of 6-mm-Pitch Cu Electrodes at Room*

Temperature, IEEE Trans. on Advanced Packaging 31 (2008) 473.

9.   T. Suga et al., *Direct Bonding of Ceramics and Metals by Means of a Surface Activation Method in Ultrahigh Vacuum*, Proc., Material Research Society Int'l Meeting on Advanced Materials 8 (1989) 257.

10. T. Suga, *Room Temperature Bonding of Diamond and Metal*, J. New Diamond 16 (2000) 18. (in Japanese).

11. T. Suga et al., *Surface Activated Bonding and Its Application on Micro-Bonding at Room Temperature*, Proc. 9th European Hybrid Microelectronics Conf. (1993) 314.

12. S. Ozawa et al., *Low Distortion Cold Rolled Clad Sheets for PWB*, Proc. 6th IEEE CPMT Conf. on High Density Microsystem Design and Packaging and Component Failure Analysis (2004) 91.

13. Y. Wang et al., *Study on Sn-Ag Oxidation and Feasiblity of Room Temperature Bonding of Sn-Ag-Cu Solder*, Materials Trans. 46 (2005) 2431.

14. S. Nishi et al., *Real-Time Observation of Hydrogen Plasma Reflow Process with Lead-Free Solder Pasts*, Proc. Int'l Conf. on Solid State Devices and Materials (2006) 320.

15. M. M. R. Howlader et al., *Room Temperature Wafer Level Glass-Glass Bonding*, Sensors and Actuators A 127 (2006) 31.

16. M. M. R. Howlader et al., *Activation Process and Bonding Mechanism of Si-LiNbO$_3$ and LiNbO$_3$-LiNbO$_3$ at Room Temperature*, Proc. 8th Int'l Symp. on Semiconductor Wafer Bonding 6 (2005) 319.

17. A. Kaneko et al., *Surface Activated Bonding for GaN/Al and Electrical Characterization of Bonded Interface*, Proc. Int'l Conf. on Electronics Packaging, (2008) 26.

18. R. Takigawa et al, *Low-Temperature Au-to-Au Bonding for LiNbO$_3$ / Si Structure Achieved in Ambient Air*, IEICE Trans. Electronics E90-C (2007) 145.

19. Y. Zikuhara et al., *Sequential Activation Process of Oxygen RIE and Nitrogen Radical for LiTaO$_3$ and Si Wafer Bonding*, Proc. 9th Int'l Symp. on Semiconductor Wafer Bonding (2006) 191.

20. T. Suga, *Feasibility of Surface Activated Bonding for Ultra-Fine Pitch Interconnection-A New Concept of Bump-Less Direct Bonding for System Level Packaging*, Proc. 50th Electronic Components and Tech. Conf. (2000) 702.

21. H. Braunisch et al., *Electrical Performance of Bumpless Build-Up Layer Packaging*, Proc. 52nd Electronic Components and Tech. Conf. (2002) 353.

22. P. M. Enquist et al., *Ziptronix Zibond^{TM} and DBI^{TM} Technologies for 3D IC Applications*, Proc. Association of Super-Advanced Electronics Technogies (ASET) 3D System Intercration Conf. (2007) 5–3.

23. A. Shigetou et al., *Bumpless Interconnect of 6-mm-Pitch Cu Electrodes at Room*

Temperature, IEEE Trans. on Advanced Packaging 31 (2008) 473.

24. Z. Xu and T. Suga, *Room/Low Temperature Interconnection Technique on Micro-bump/Film for COC and COF System*, Proc. 7th Int'l Conf. on Electronic Packaging Technology (2006) 1.

25. A. Shgetou and T. Suga, *Modified Diffusion Bonding of Chemical Mechanical Polishing Cu at 150°C at Ambient Pressure*, Appl. Phys. Express 2 (2009) 056501.

26. A. Shigetou and T. Suga, *Modified Diffusion Bond Process for Chemical Mechanical Polishing (CMP)-Cu at 150°C in Ambient Air*, Proc. 59th Electronic Components and Tech. Conf. (2009) 365.

27. A. Shigetou and T. Suga, *Direct Interconnection of Chemical Mechanical Polishing (CMP)-Cu Thin Films at 150°C in Ambient Air*, Proc. European Microelectronics & Packaging Conf. (2009) 17–S1–12.

Chapter 11

THROUGH SILICON VIA IMPLEMENTATION IN CMOS IMAGE SENSOR PRODUCT

Xavier Gagnard and Nicolas Hotellier
STMicroelectronics

INTRODUCTION TO TSV TECHNOLOGY APPLIED TO CAMERA MODULE

The camera module business started early 2001 with the first integration of image capture capability inside a mobile phone. Over the last 10 years this market has been booming thanks to the very fast evolution of CMOS sensor performances combined with and endlessly demand for higher resolution camera and sophisticated image processing capability.

In parallel to the fast evolution of CMOS sensor silicon process the packaging technologies have followed the same trend not only on the performance side but also on the physical dimensions. Figure 1 illustrates the evolution of VGA camera module physical dimensions from 2002 to 2008.

Over the last two to three years, it was becoming very clear that camera module packaging technology will have to introduce breakthrough solutions to meet the customer expectations in term of physical dimensions. Through Silicon Via has been the answer to this demand and it is now implemented with some variants by most of the players in this field of applications.

Through Silicon Via offers multiple benefits for image sensor and camera module integrators and exceed by far the simple physical size reduction. It can be mentioned but not exhaustively; image sensor array protection to dust particles during the camera module assembly; compatibility with wafer lens technology; etc.

In addition, Through Silicon Via technology is one of the first example of process combining at the same manufacturing location front-end and back end semiconductor process and equipments showing if necessary that this represents one of the challenge for 3D integration

3D Integration for VLSI Systems
Edited by Chuan Seng Tan, Kuan-Neng Chen and Steven J. Koester
Copyright © 2012 by Pan Stanford Publishing Pte. Ltd.
www.panstanford.com

Fig. 11.1 VGA camera module evolution.

The first generation of products based on Through Silicon Via process have reached the market in 2009 but further evolutions of the technology toward full 3D integration combining silicon and non-silicon based elements will happen shortly on the market.

With the continuous development of the semiconductor technologies node, nowadays the development of the 22 nm node, each technology will get its limitations in term of density and power consumption and need to be succeeded by the new node. To overcome the limitations of each technology, the Moore law is replaced by the more than Moore vision (Fig. 11.2).

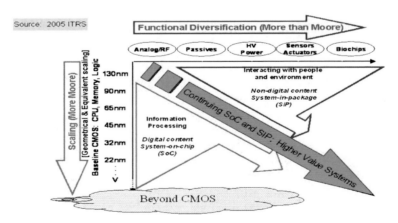

Fig. 11.2 Roadmap for the Moore law and the More than Moore concept.

The interest of the more than Moore is to integrate more functionalities on the devices,[7] creating new hybrid or autonomous device, but also to reduce the cost of the final product by optimising the package and by improving the performance (like the interconnect delay). The now called 3D integration (3Di) technology is now answering these needs, in the meantime where the spin-tronics or molecular electronic will emerge to continue the Moore law.

Many institutes in the world, like LETI in France, Fraunhofer in Germany, IMEC in Belgium or Georgia Institute in USA, work on the 3Di and propose different approaches to be able to put more functions on silicon and to get a functional product at wafer level![13] But the real 3Di program is more in the companies than in the consortiums, each company selecting their toolsets to develop and to control many of the 3Di steps, like Through Silicon Via (TSV) implementation.

The device packaging engineer is the main requester to develop the more than Moore scenario, in order to get higher functionality in a package, and because the package size approaches the chip size. To request the reduction of the thickness and the size of the die (for a footprint limitation), a miniaturization of the package itself has been already realised with package thinner than 0.5 mm and finer ball pitch. First 3Di package arrived with stack die, Package on Package (PoP), System in Package (SiP) or embedded die substrate approaches. Chip to chip stacking has been accomplished with wire bonding, sometimes till to eight stacked die, but with limitation in terms of I/O connections and electrical performances. Recently, the possibility to integrate the TSV, whatever the wafer size, allows defining very complex device integration with associated package. To perform the full die stacking, the TSV are coupled with micro-bump for the die interconnections,[15] using SnAg alloy. From 2008 to 2012, Yole development scheduled a trend from 0% to 12% of TSV using instead of wire-bonding or flip-chipping in the package integration[14] (Fig. 11.3).

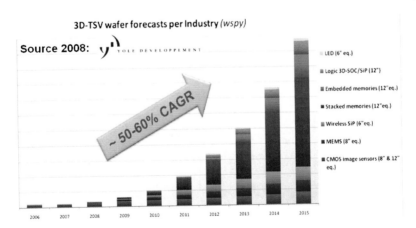

Fig. 11.3 3D-TSV market for the 1st application by Yole development June 2008 [14]. First used for image sensors, 3Di packaging technology will enter into, and grow in, various semiconductor segments (source http://www.eetasia.com).

The same analysis shows the CMOS Image Sensor (CIS) as the first application with a constant growing over the years, and schedules a strong TSV market share for the memories.

11.1 INTRODUCTION

More and more, a wide range of requests coming from customer appears to evaluate and demonstrate the feasibility of the Through Silicon Via (TSV) for a large range of Via size and Via AR either for process point of view or for performances point of view. The request is now known with the name of TSV but also with the name of 3D integration, solution using TSV option.

The main application today in the market is the CMOS Image Sensor (CIS) with the integration of Via at Aspect Ratio 1 (AR1), using a Via last approach. The Via last approach defines the way to implement the TSV when the Back End Of Line (BEOL means metallisation) of the silicon product is finished. In the contrary, the Via first or Via middle, defines the way to integrate the Via during the FEOL or BEOL process.

Now based on this first wafer level package of CIS, the integration on the z axe will continue by the wafer lens integration for a continuous form factor and low cost module[31] to enter the real 3Di for the CIS application. For the 3Di, first applications with TSV is entering the market with the Via-last approach, more simply to be developed in semiconductor manufacturing in order to secure the 3Di technologies and to promote the 3Di to customers (Fig. 11.4). Then specific design and electrical models will be developed and optimized, allowing a fast and prosperous development of the Via-first and Via-middle approaches. A challenge in the modelisation of the TSV is the understanding of the mechanical impact of the Via structure and the metal filling on the behaviour of the CMOS components and the reliability. These types of researches are progressing in various institutes and are essential for an increasing integration of TSV.

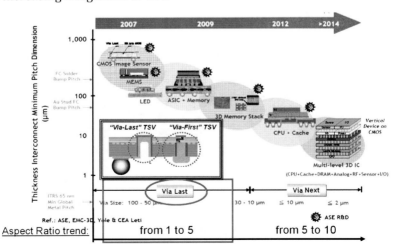

Fig. 11.4 Trend of the Via aspect ratio in function of the need for each targeted application (from ASE, Technology Venture Forum Fall 2008).

Because today, the technology continues to drive the 3D roadmap, the mechanical and thermal modelisation and 3D design toolset need to be faster developed in order to optimize the 3D module. Then the electrical testing will be a real challenge to be able to distinguish drift induced by TSV or not, to be able to isolate a Via within more than 1000 Via in a module. The electrical testing will be strictly linked to mechanical and electrical failure analysis to get feed-back in technology, actual draw-back of the 3D development. The cost of the TSV integration can be more and more important in function of the requirements of its usage. This TSV cost should be part of the economical equation of the module (the final package) but does not be act as the primary driver. The functionalities should increase faster than cost after huge work on the cost process optimisation. Indeed, throughput and materials used have a direct impact on the final price like for the packaging. Continuous perspectives of TSV integration are progressing in order to optimise actual applications or to develop new integration.

For CIS, TSV introduction is also a wafer level package, reducing the package footprint and mainly optimise the capping of the device. By integrating TSV, the capping must not only protect the device against external and clean room environment, but also protect the sensor against the humidity by a degradation of the hermiticity performance. Based on these targets, the first definition can be defined by the technology, the designer, the reliability and the device teams.

11.2 FIRST DEFINITION FOR TSV AND FAN-IN BALLING IMPLEMENTATION

The TSV option scheme for a silicon product is divided in five main parts as illustrated in the Fig. 11.5:

- The bonding of the different elements: direct or temporary bonding, wafer to wafer or die to wafer, alignment accuracy,
- The thinning of the wafer or of the die to get the thinnest thickness possible: 20 µm possible with the lowest stress achievable,[2-3]
- The Via patterning, its isolation and its metallization,
- The ReDistribution Layer (RDL) to design the module fan-in option: allow the connection to the board through the outside connection,
- The outside electrical connection for die to substrate: Cu pillar, micro-bumping or balls placement in function of diameter and pitch scenario of the connections.

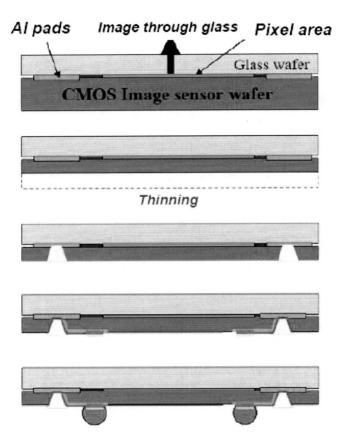

Fig. 11.5: illustration of the TSV option processing for a CIS product: the bonding, the back grinding, the Via processing, the RDL and the balling.

Based on these main bricks, others jobs need to be addressed to get a continuous improvement of the TSV integration in simple product or in 3Di product: the physical or electrical failure analysis[1] and the thermal modeling of complex circuitry.[8]

Since two years, many integration possibilities were demonstrated, but only one product is today commercialized with some of the 3Di process blocks developments: the CMOS image Sensor. The CMOS imager Sensor has received a main evolution for its package: indeed, for the consumer application, either mobile phone or computer, the request is to reduce the size of the package (thickness and footprint) but also his cost to work continuously on the form factor of the final object like the mobile phone, as seen in the following Fig. 11.6.

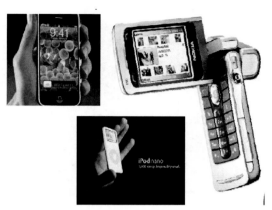

Fig. 11.6 the form factor of the mobile phone generated new challenge in the semiconductor industry, like the integration of die with the 3D integration or with the CIS-TSV more compact than standard CIS.

The evolution in the CIS application is to replace the wire bonding connection in the package by the presence of balls on the backside of the die for a direct mounting on the board. The connection of the metal layer of the CMOS with the back-side balls and RDL is realized with the Through Silicon Via TSV like seen in Fig. 11.7.

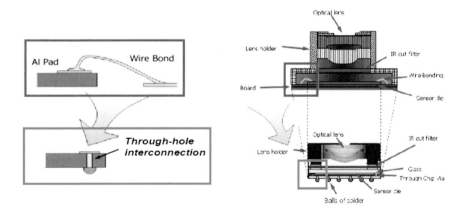

Fig. 11.7 Explanation of the advantage of the TSV vs wire bonding for the CIS package.

For the standard camera module assembly (top assembly of the Fig. 11.7), the yield can be impacted by:
- Dust
- Organic contaminant (see Fig. 11.8)
- Metal residues

In the same time, the pixel size reduction increases the constraint on particles' size and the mega pixel race asks for higher image quality, so the contaminant needs to be removed as much as possible.

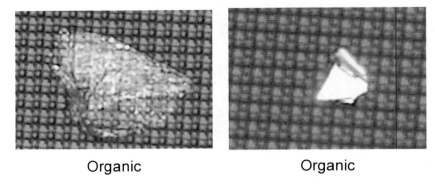

Organic   Organic

Fig. 11.8 Example of organic contamination over the colour matrix of the image sensor

TSV package improves image quality and assembly yield by protecting image sensor array at wafer level, meaning that the package is realised in a class 1 environment (semiconductor clean-room) instead of class 1000 for the packaging plant.

To summarize, the full advantage of the TSV comparing to the wire bonding is:

- Allow even smaller package outline
- No pad extension needed
- Lower sensitivity to foreign material at Camera assembly
- Wire bonding compatible layout
- Reflow process compatible
- Better interconnect routing capability

But, the TSV technology is a more complex technology using glass substrate, silicon substrate with a back-end process in a front end environment because integration at wafer level. Finally the cost can be a draw-back if the technology is not enough mature and not well known during high volume demand.

The fundamental needs of a compact camera are defined, and now the development of the CIS product with TSV option can start with the design, followed by the process development.

11.3 DESIGNS RULES FOR TSV AND FAN-IN BALLING

To initiate a design of a CMOS Image Sensor, some rules for the TSV integration need to be defined for the designer, to be able to link the Metal 1 (M1) of the CIS circuitry with the RDL lines from the back-side of the CIS through a Via, as shown in Fig. 11.9:

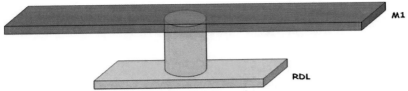

Fig. 11.9 Drawing of the Via connecting a Metal 1 of a CIS circuitry with a RDL line designed to address the fan-in balling for the board mounting.

Indeed, with the TSV last option, an important flexibility is obtained in the design by the fact that a CIS product can be either wire bonded or TSV connected in an optical package. So, the TSV design rules are directly linked to CIS design rules applied for the manufacturing of the silicon. Only some small changes in poly-silicon level of the CIS product are necessary.

Fig. 11.10 View of a CMOS Image Sensor produced by STMicroelectronics. *Centre*: COB package, 2 Mp RawBayer sensor, 3 um pixel size. *Right*: WLP TSV package, 2 Mp RawBayer sensor, 1.75 um pixel size.

Figure 11.10 shows the interest and the challenge for the TSV implementation: a compact optical module with an electrical connection on the back side.

Based on the technology used and the final module applied, the design rules are also defined taking into account the process feasibility in the manufacturing plant.

These design rules are essential to start the process development. These rules will define the dimensions and the spaces of all the structures realized on the back-side circuitry.

The first point is the definition of the Via itself (Fig. 11.11): the position with the Metal 1 connection (Critical Dimension, Overlapping), the position of a Via with others Via (Space, pitch). But the design of the Via need to take into account the design of the balls (outside connection) which are either free or defined by customer standard.

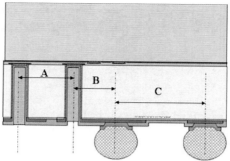

Fig. 11.11 Design Rule example for the position of the vVia and the outside connections.

Then to perform the connection between the Vias and the balls, other rules are applied to draw the RDL layer as shown in Fig. 11.12:

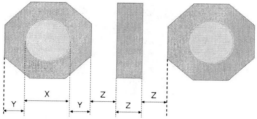

Fig. 11.12 Design rules for the RDL layer in function of Via position.

In function of the circuitry density, the line dimension, the pitch and overlap of the Via metallisation and the lines pitch, the process will be defined to fit these expectations.

For each TSV technology, the design rules is defined to take into account the TSV requests: the Via dimension, the Via aspect ratio, the line dimension and the ball positions.

With these well-defined design rules, the development of the process is easier, and an important thinking is realized to define the tools to produce the TSV. Based on these rules coupled with TSV specifications for materials properties, the process development can be started.

11.4 PROCESS DEVELOPMENT

For CIS application, the technological complexity is minimized compared to new 3D integration products, to allow high process yield and to make the process economically Viable: the Via is defined with an aspect ratio (AR) of 1. Also, because the Via is realized after the CMOS fabrication, TSV last option, the temperature of the TSV process is limited to 200°C to avoid any degradation of the BEOL of the circuitry inducing reliability failures or drift of the CMOS devices.

To achieve the first constraint of AR 1, the silicon is thinned below 100 μm, and a permanent optical carrier is used for handling issues and process-ability of 300 mm wafers. This carrier, for an image sensor application, is obligatory a glass substrate which follows severe optical and physical specifications. After the glass substrate polymer bonding, the glass carrier is the back side of the stack once the flip is realised, starting point of the TSV process with the back thinning of the silicon. The use of a glass substrate instead of a silicon substrate for the handler substrate is a more complex situation for the handling point of view of the stack due to the CTE (Coefficient of Thermal Expansion) mismatch between glass and silicon (of CIS product), mismatch which is as low as possible. Each time temperature is applied, either to bond the glass substrate to silicon or to process the Glass/Silicon stack, the CTE mismatch can induce stress or bow to the stack, generating difficulties to handle or to process the wafer. To finish, the use of the glass substrate as the back side means that the handling need to be soft as much as possible in order to avoid any degradation of the optical part, the glass substrate, when the assembly is performed.

Then the main steps are (Fig. 11.13): the deep RIE of the silicon to reach the CMOS metal layer, the oxide isolation of the Via, the barrier and seed PVD deposition and finally the Cu ECD for RDL. One drawback of the glass substrate compared to standard silicon substrate is their dielectric and thermal properties which can affect the performances of the step when RF or temperature is applied.

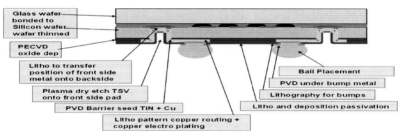

Fig. 11.13 A cross-section of the CIS developed with TSV (Source J.Michailos/STMicroelectronics – MINATEC Crossroads'08 – Grenoble – June 23-27 – Minalogic Workshop).

For all these steps, the handling of the wafers from cassette to chuck and the wafer warping on the chuck with the temperature are very important. The wafer bow evolution and the material used to process the TSV process block have a direct impact on these two points. First of all, the material for the polymer bonding and the cure bonding impact drastically the bow and leads to handling limitations. Then, following the polymer properties and the thickness of the materials deposited, the bow will be a direct function of the temperature. The reflection moiré method[9] is a good technique to follow and understand the evolution of the stress at wafer level, either at 200mm or at 300mm, and to work on the optimization of the stress within the silicon and at the bonding interface.

The four mains steps for the TSV implantation are fully described hereafter, starting with the wafer bonding of glass substrate to silicon to create the Wafer Level Package of the CMOS Image Sensor.

11.4.1 Wafer bonding

The Wafer bonding is the first step realised for the TSV implementation. This is the bonding of glass substrate with a CMOS Image Sensor which assumes the role of a packaging. By this bonding, the Wafer Level package realised in a Class 1 clean room assures a very low level of particles closed on the sensor, increasing the optical yield at module level.

To bond a substrate to another, different techniques exist for the wafer bonding [KLA02], and are:

- Direct bonding or thermal bonding, which is a bonding without assistance of pressure, electrical field or intermediate layer. It requests very smooth and flat surfaces to achieve reliable and high yields bond so involves wafer surface preparation and cleaning. After the bonding itself, an annealing step is applied to increase and to stabilise the bond strength. Typical bond is Si to Si or Si to SiO2, and annealing between 600°C to 1200°C to get strong bonds like in SOI products.

- Anodic bonding or field assisted bonding, to join an electron conductive material with a material with ion conductivity through heating (18 500°C) and voltage (200 – 1500 V). Electric field allows the fusing of both interfaces. This technique is more tolerant to surface roughness than direct bonding and allows strong and hermetic bonds.

- Adhesive bonding uses organic or inorganic intermediate layers. Layers are deposited (spin coating, laminating, and spraying) either on 1 or both wafers and adhesive material deforms and flows under pressure and heat. This process is strongly dependant of the material used to link both wafers. This technique is CMOS compatible and mainly tolerates particles and topography over the wafers surfaces.

- Plasma activated bonding,[42] alternative to standard wafer bonding processes to obtain maximum bond strength with low temperature thermal annealing. The surface is activated (nitrogen or oxygen plasma) prior to bonding by exposing surfaces to plasma (either one or both wafers), allows a controlled change in surface chemistry to obtain the highest bond strength for low temperature annealing (maximum at 400°C) and short annealing times.

- Glass frit bonding, Solder bonding, eutectic bonding, thermo-compression bonding are commonly others techniques used for the wafer level packaging but when the package is the last step and when no further process steps at wafer level is necessary in a semiconductor clean room environment. Some of these techniques are now used for the die to wafer or wafer to wafer bonding when electrical connections between both wafers are mandatory.

When these techniques are compared for the performances point of view, for the implementation issue and for the cost impact, the adhesive bonding is the most suitable technique for the CIS constraints.

For CIS wafer level packaging, the adhesive needs to be patternable in order to be removed in the optical path, to restrict the transmission loose of the light to the glass support.

Specifications for the adhesive bondings are commonly known and related to various publications like.[39]

In the literature, the common materials used to bond two wafers by the adhesive bonding are the BCB from Dow chemicals or SU8 from MicroChem. Since many years, the BCB[19-22] or SU8[20, 23-24] are studied in laboratory or in R&D centre for their performances, but the industrialisation of the BCB is very complex due to the difficulties to get a stabilised material over time and to control it.

Also, many researches explain how to improve the adhesive bonding with BCB[25-26] or compare the performances of BCB or SU8 with new materials coming on the market in 2004,[27] because it is very difficult to reproduce the same performances run to run. Since 2004, new materials are proposed from American or Japanese companies, now in 2009, in good competition with the BCB. Based on materials evaluations, a non BCB product is used for the bonding of the glass over the CIS. The adhesive glue has been selected based on physical properties in accordance with the state of the stack after the wafers bonding but also with the response or the evolution of the stack during the TSV process.

The following parameters allow to select the right glue material:
- Thickness ability
- Young modulus
- Elongation
- Curing temperature
- Coefficient of Thermal expansion
- Stress after curing
- Water adsorption

The last point is important to get a sufficient sealing of the wafer level package when assembled in an optical module. For the moment, studies on permeation of sealed package with adhesive technique report only the BCB material.[28-30] Based on the performances of the new materials, there moisture diffusions will be soon published.

But, before bonding a glass substrate to a CIS wafer, the right glass substrate need to be identified. By defining Coefficient of Thermal Expansion, the thickness, the thickness variation over the substrate, the warp, the optical transmission, glass materials from Corning and Schott are used.

To perform the bonding, four main parameters are studied to defined the best recipe which allows the stack fitting the objectives for a safe processability for TSV implementation. The parameters are:

- The bonding temperature
- The bonding force
- The vacuum in the chamber
- The duration.

A possible recipe is illustrated in Fig. 11.14, showing the different sequences with the 2-step temperatures, the piston force step, the vacuum evolution.

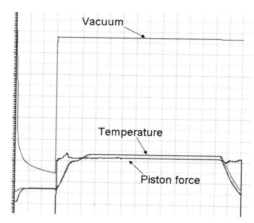

Fig. 11.14 Example of a glass to wafer adhesive bonding recipe.

The piston force is applied in order to get a uniform contact between the two layers, and the temperature activate the glue material from a liquid phase to a polymerised phase (solid phase), the cross linking of the material. The glue cross linking ratio before the bonding step has a big impact on the final bonding quality (after full polymerisation step). Following the ratio of polymerisation, voids inside the glue layer or defect or non-uniformity of thickness[40] can be detected.[41] explain that it is better to have a material defined as partially cured (>40%) than just soft cured (<40%) which improves the material stability during the bonding process (low reflow) giving better performances for the mis-alignment after bonding (less than 1 μm added) and thickness uniformity after bonding (from ~15% after soft cure to ~1% for partially cured).

Once the 100% polymerisation state is obtained at the end of the bonding step, the adhesion of two wafers is characterised by different techniques like the shear test to know bond strength, the optical control to define the defectivity or the Scanning Acoustic Microscope (SAM) to check the presence of unbonded area like shown in Fig. 11.15.

Once the bonding pressure sequence coupled to temperature sequence is defined to get good polymerised material, the CIS stacked with the glass substrate is ready for the next step of the silicon back-grinding.

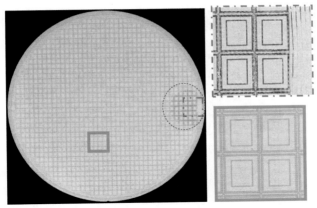

Fig. 11.15 SAM analysis of a patterned adhesive glue. SAM allows to control either Silicon to Silicon bonding or Glass to Silicon bonding. Bad and good area can be compared with the grey scale.

11.4.2 BACK-GRINDING OF THE SILICON

The second main point of the TSV process development is the silicon thickness reduction in order to get an aspect ratio of 1 for the Via used to connect the metal 1 from the silicon back-side.

The back-grinding is a strategic point of the TSV implementation but it is also the most basic process, very well known in standard semiconductor manufacturing.

Indeed, to perform the thinning of the silicon after the wafer bonding, standard back-grinder tool is used, without any trouble to work glass substrate because the same tools are already used to back-grind glass substrate for optical application.

The main specification which is defined for the recipe development of the tool is the thickness variation across the wafer. Because less than 100 μm of silicon is reached after the back-grinding, very low thickness variation need to be achieved and need to match the performance of the equipment.

Once the back-grinding is performed, the defectivity of the back-side is controlled before to move the wafers to the next process steps. This operation is duplicated for each further step because any defectivity, scratch or other foreign particles can impact the final assembly yield with a degradation of the optical performances. Back-side protection can be used to protect the back-side and to avoid any glass defectivity

For the bonding, a patternable glue is used to avoid any new material over the optical sensor. This pattern has an impact on the back-grinding, because it creates a cavity between the silicon and the glass substrate in the sensor area. In function of the targeted silicon thickness, silicon deformation can be seen due to the presence of the cavity, like shown in Fig. 11.16.

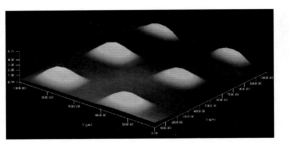

Fig. 11.16 Silicon warpage for individual sensor induced by the silicon thickness.

When the silicon is too thin, the warpage of the local sensor is too important, without impact on the wafer warpage. This sensor warpage needs to be reduced as much as possible to avoid any fragile silicon during the heating process. The silicon membrane cracking can appear when a temperature is applied due to the CTE mismatch between the silicon, the glue and the glass.

To conclude this section, the final silicon needs to be defined to reach the aspect ratio for the Via realization but thickness also not too thin to avoid silicon defectivity.

11.4.3 The Via realization with its isolation and the metallisation

Once the silicon thickness is in accordance with the designs rules, the Via processing can be started. This part is a combination of lithography, resist stripping, deep silicon etching, CVD oxide deposition for the passivation, PVD metallization for the barrier/Seed and ECD Cu growing for the Cu lines.

The heterogeneous stack, silicon bonded on glass, will have a direct impact on plasma tools. For example, the RF applied in CVD tool will change and can limit the performances of the isolation oxide deposition and the heat evacuation impacts the silicon deep trench. It means that the recipe and the hardware need to be developed in function of the handle substrate used.

For the deep trench, the goal is to get well controlled wall tilt from edge to center (with using a ESC chuck), controlled scalloping generation on the wall, uniform etch ratio wafer to wafer (uniformity of wafer temperature within the wafer), Via opening without notching at the bottom and the over-etch step due to ending on CVD oxide layer. The deep trench using Inductively Coupled Plasma (ICP) achieve anisotropic etching through the called Bosch process. This process alternates pure isotropic etching and polymerisation steps. The polymer is ionic bombarded only on the horizontal surface before to start again on the previous alternatives steps, each cycle generating unfortunately a scallop (Fig. 11.17). The Via must have extremely smooth side, because wall roughness in the Via can generate trap residue before the dielectric coating or poor step coverage, leading to electrical or reliability problems. Changes in Via or scallop profile have a direct impact on coverage of the passivation, impacting the integration of barrier/seed and thus the associated reliability. All these requirements must lead to achieve etch times short enough for production equipment whatever the Via depth.

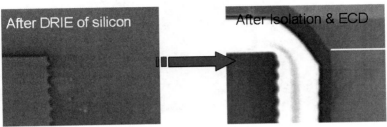

Fig. 11.17 Cross-section of top edge of Via after deep silicon etching (left) and after oxide deposition (right).

For the low temperature dielectric deposition, the thickness and step coverage control are very important for the capacitance performances of the Via in the RLC model. The Via leakage performance is a mix of silicon scalloping, silicon over-etch (bottom Via profile) and oxide step coverage. To continue the integration of the TSV Via, Via isolation needs to smooth as much as possible the wall with the lowest temperature reachable to avoid any sensor degradation (Fig. 11.16).

As requested, criteria's in barrier/seed metallization for Image sensors were centered on a Via AR 1 and low cost processes, conventional PVD was chosen as the most known and cheapest method to deposit the barrier/ adhesion layer and seed layer in order to promote electroplated copper liner.

Considering materials, Copper diffusion was targeted as a potential failure mode during sensor elaboration and lifetime, leading to the use of Ti, Ta or W based barrier layer. For deposition cost reason as well as adhesion properties, Titanium is widely used as a pure material or as Titanium nitride. Seed layer is made of pure copper with a sufficient thickness required to eliminate parasitic substrate resistance and ensure uniform ECD Cu plating.

The choice of the electrolyte is also depending on the expected mechanisms of plating. In Image sensor application, the choice of only depositing a liner of copper and not fill the Via allowed the use of cheaper additives electrolyte. This liner option was mainly chosen in order to decrease cost of metallization and to reduce thermo-mechanical stressing. As a matter of fact, filling such a thick Via would lead to long deposition times as well as subsequent process steps to remove Cu overburden and is not required for this Via sizes for which polymer filling can be added to eliminate void in the Via.

The critical parameters to be assessed for metallization step is the control of the conformality of PVD deposited layers to ensure diffusion barrier efficiency for barrier material as well as electrical continuity from copper seed, and this especially at the bottom of the Vias where scallops are more pronounced, generating local overhang (Fig. 11.18).

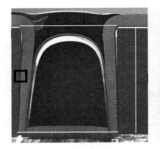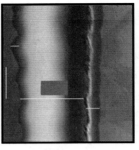

	Step coverage (%)	
	Barrier	**Cu seed**
top	100	100
mid	18	18
botttom	15	16

Fig. 11.18 Cross-section of a AR 1 TSV after DRIE of Si, Barrier/Seed PVD deposition and Cu ECD. Scallops can be seen on the magnification view on the right.

By controlling, the deep trench silicon, the barrier/seed metallization and the ECD growing, but also for the Via patterning and the Via metallization, the Vias perform the connection of the metal 1 of the CIS (at the bottom of the Via) to the metallization on the field (at the top of the Via) which are the redistribution layer which defines the fan-in approach of the CIS module.

11.4.4 The redistribution layer

The redistribution layer (RDL) is realized to route the Via to the external connections scheduled to be connected with the board. This layer is not only the field metallization with DRM constraint, it is also the metallization of the Via.

The development of this layer is done with the Via metallization, with the same techniques of Seed layer followed by Electro-chemical deposition of Cu through a resist mask.

The important step of the RDL brick is the etching of the barrier layer and seed layer once the Cu growing is done. Indeed, following the line density, any short needs to be avoid.

For a cost point of view, wet etching is the chosen technique with dedicated chemical solution to etch the barrier material and the seed material used.

Over etch is the main parameter to be controlled, because it must be long enough to avoid short-cut between the line, but short enough to limit the under-cutting under the Cu line. Indeed, in function of the line width DRM, the under-cutting of the line can be dramatic by delaminating the line from the substrate.

Wet technique coupled with optimized chemical solution and with optimized recipe allows to obtain nice RDL definition.

Then, the wafer is coated by a photosensitive layer to passivate the product, to avoid oxidation of the Cu. Because the passivation material is photosensitive, the Cu area scheduled to receive the outside connection (balls for example), is opened after exposition and development of the passivation.

This passivation is performed by the coating of a polymer or a polyimide material answering the following specifications:
- Covering the Cu of the Via,
- To planarize as much as possible the product topography,
- To get a minimum thickness on top of RDL,
- To protect the product against moisture,
- On open area, to get smooth angle.

The passivation coating can be realized by two different techniques: conventional spin coated material or vacuum laminated dry film. But it must also answer the following specifications to answer new request:
- high k capability for RF performances
- double RDL capability

Once the passivation process is done, coating-exposure-development-annealing, the wafer is ready to complete the process with the realization of the outside connection

11.4.5 The outside connection

The ballings or the outside connection to link the product to board is realised to finish the TSV implementation.

In conventional product, optical or not, the standard die to package or die to board connection is realized with wire bond like seen in Fig. 11.19.

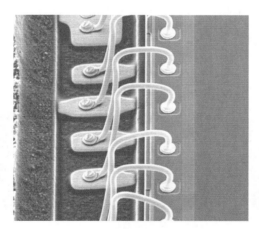

Fig. 11.19 Example of wire bonding for the die to package connection.

For the CIS with TSV option, the wire is naturally replaced by flip-chip like connects for the board integration (Fig. 11.20 right). The advantage of this approach is to get the most aggressive package rules, with an important gain in height and in footprint

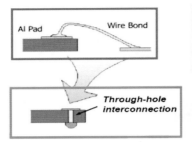
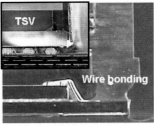

Form factor impact

Fig. 11.20 Area and cost saving by using TSV instead of wire bond, top right picture shows the compact package obtained with TSV+balls .

As seen in Fig. 11.20, the implementation of TSV with solder connection allows reducing the foot print of the CIS, and reduces also the full size of the module. A more compact module is obtained, like shown in Fig. 11.21.

His compactness and the fact that the external connection is a solder, this CIS module with TSV option has the advantage to be mounted directly on the board as a reflowable module, a big advantage for the form factor of the final application using CIS, but also for a cost point of view. Indeed, many wire bonding steps not used when TSV is applied, are identified as cost saved.

From a process point of view, the outside electrical connection can be realised with different techniques. The Cu pillar process is the most robust technique which is based on conventional Cu and SnAg obtained by Electro Chemical Deposition (see Fig. 11.22 on the right). This is also the techniques allowing the finest pitch and smallest pillar diameter, so the highest density of external connections. The second attractive technique is the solder paste printing to create micro-bumping. Following the type of the solder paste used (type 6 for 15-5 μm particles size or type 7 for 12-2 μm particles size), pitch between 120 μm to 50 μm and diameter between 100 μm to 40 μm can be reached. The main interest of this technique is the cost, cheaper than Cu pillar.[10-11]

Reflowable module on board

Fig. 11.21 On the left, example of CIS TSV compact package compared to standard wire bond package. On the right, CIS TSV is a reflowable module directly on board.

The last technique and the cheapest one is the solder balls placement. This is the placement of calibrated balls over a stencil designed to allow dispense the calibrated balls and aligned with the wafer in order to position the balls with a precision of 10µm. Before placing the balls, the aperture patterned in the passivation, opening the Cu pads, are coated with an Under Bump Metallization (UBM). Then a flux (desoxidation product) is deposited on the UBM area before the balls are placed through a stencil. Once the balls are maintained on the wafer with the flux (capillary forces), the reflow of the solder balls is realized in a furnace, allowing an auto alignment of the balls with the UBM, removing the flux and reflowing the solder. An inter-metallic alloy is created between the UBM and the solder balls given the rigidity of the balls. A result of balls placement is shown in Fig. 11.22, left.

The performances of all the techniques are characterized by the same way:

- The lateral shear test to evaluate the weight resistance to the shear
- The lateral drop test
- The vertical drop test

The customer defined the specifications for each test to guaranty a minimum lifetime against any chock of the final product.

The next section will explain the electrical test performed when the process is completed.

11.5 ELECTRICAL AND RELIABILITY RESULTS

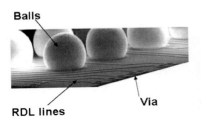

Fig. 11.22 A possible external connections, with on the left, ballings finishing and on the right, Cu pillar finishing.

Once the process has been developed with the help of cross section, the electrical and the reliability characterization can be started.

These characterizations are essential to understand the behavior of the Via but also and mainly to qualify the TSV implementation for each technology. Two steps are realized:

- The electrical characterizations of basic measurements around the Via
- The reliability characterization for the stress response of the Via

The objective of these evaluations is to know the performance of the Cu Via embedded in the silicon substrate, with the Metal 1 (M1) layer of the device BEOL and with the Cu RDL layer (Fig. 11.9).

Any short cut with the silicon needs to be eliminated to avoid parasitic signal on the circuitry functionality.

11.5.1 Electrical results

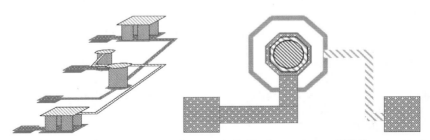

Fig. 11.23 Top: resistance structure; bottom: individual capacitance (STMicroelectronics thesis, Maxime Rousseau, 2009).

For the electrical measurements, standard tests have been defined like the capacitance measurement, the resistance measurement and the leakage measurement for the Via and the RDL.

Figure 11.23 describes two structures that can be realised to measure the Via resistance and the capacitance of an individual Via.

Once the resistance of the Via, the resistance of the RDL and the capacitance of the Via fit the product specifications, the next evaluation step can be started: the wafer level reliability.

11.5.2 Reliability

In the continuity of the electrical characterizations, the wafer level reliability is performed. The first aim is to assess the potential influence of TSV process on product reliability performing HCI, NBTI, ionic contamination and other standard characterizations. The second part of the assessment is more focused on the Via passivation stressing. The test consists in breakdown voltage stress of a TSV capacitance (LRVS).

The copper redistribution layer electro-migration is not considered as a sensitive parameter according to the line dimensions (huge compared to BEOL Cu lines).

Once the results are within specifications, it is important to pursue the reliability evaluation at module level. Indeed, some packaging steps can have a direct impact on TSV structure. For example, the mechanical stress induced by the sawing process can interact directly with the TSV structure. Typical tests at this level consist in humidity and/or temperature and/or electrical bias storage.

Finally, the TSV structure can affect directly the mechanical reliability of the final package (including lens system and can). For this item the commonly used assessment is drop testing with the module soldered on a print circuit board.

11.6 CONCLUSION AND PERSPECTIVE

11.6.1 Process point of view

The CMOS imager sensor with TSV is now produced either in 200 mm or 300 mm in a production mode (Figs. 11.3 and 11.18), which indicates that 3Di equipments are ready for the Via at AR 1.[33, 34] But what is the status for higher aspect ratio, finer pitch, smaller Via diameter? Which product needs it? Many institutes work on two different Via approach in function of the Via diameter or Via aspect ratio (AR): the Via first approach where the Cu Via are realized after the CMOS PMD, allowing Via till 2 μm diameter OR the Via last approach, like CIS, allowing actually Via till 15 μm with "packaging" like technology or till 5 μm with "Front-End" technology.

Since the CIS with TSV is announced by the most important semiconductor manufacturers, other products claiming 3Di emerge, promoting actively the 3Di, increasing the interest in the 3Di and the development of equipments to address the 3Di roadmaps. Major evolutions appear from CIS TSV at AR 1 TO 3Di product (CIS roadmap) with AR till 5 with Via last approach or till 15 with Via first or Via last approaches. The ability to scale down the TSV will be linked to the ability to achieve thinner silicon in order to cut as possible the aspect ratio of the Via, facilitating the process. In other words, if we can thin better the silicon or the die, the Via processing will be easier and the development of a TSV line can be limited to silicon thinning!

For CIS application, the presence of the silicon membrane (Fig. 11.16) will limit the minimum silicon thickness processable with temperature. Another specificity of the CIS is the mechanical robustness of the optical module that is partially linked to the silicon thickness.

An important evolution compared to the actual CIS application is the carrier which can be, in this case, a temporary carrier because in 3Di schemes, the carrier is just mandatory to carry the thin to very thin silicon substrate during some process steps. The use of a silicon substrate instead of a glass substrate can also be used in case of no optical aspect is requested and the lithography steps can be realized with Infra Red camera.

For the Via last approach and linked to CIS, the 3D Wafer Level packaging (WLP) is a good example to increase the density of the connection with a limitation in the silicon foot-print. For this cost effective option, TSV and RDL are fabricated before the final pillar connections. The Via last TSV option is independent of FEOL or BEOL processes, which means that no dedicated product is needed and the TSV outsourcing can be realized either in back-end or front-end foundries.

Moving forward in the 3Di roadmap, Via size is expected to decrease down 15 microns with aspect ratios increasing from AR1 to AR5 in the medium density integrations and even to few microns with aspect ratios higher than AR10 in the field of high density 3Di. For these trends, metallization appears to become one of the limiting process steps. As shown in Fig. 11.24, the actual (and low cost) barrier/seed process, designed for AR1 Vias and low cost applications will be limited to AR3 and alternative solution have to be found to overcome these limitations.

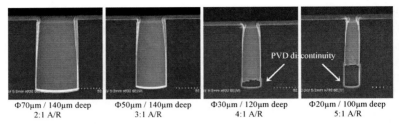

Φ70μm / 140μm deep	Φ50μm / 140μm deep	Φ30μm / 120μm deep	Φ20μm / 100μm deep
2:1 A/R	3:1 A/R	4:1 A/R	5:1 A/R

Fig. 11.24 Cross-section of various aspect ratios Via after Barrier/seed deposition and ECD Cu plating.

The chosen alternatives will have to take into account several parameters considering cost, complexity and temperature. Two approaches can be evaluated: Via first and Via last approaches.

In the case of Via last approach, which will be the primary choice for medium density integration, Temperature will be the key parameter as well as cost that have to be kept low. This approach is today targeted for aspect ratio up to 5 and advanced PVD technologies (high density plasmas, resputering solutions) are reported to be efficient and allow bottom Via coverage higher than 10% for AR5. This would be sufficient for our applications but higher deposition power will lead to an increase of wafer temperature that needs to be faced using cooled chucks. Cost of these solutions will have to be kept low which remains a challenge regarding deposited thickness compared to deposition rate as well as maintainability of the equipments.

As barrier materials generally give better conformality than copper seed in most of PVD processes, barrier appears to be pushable forward and can enable new seeding processes based on wet approaches. Seed repair process, already reported for damascene technology which consists in repairing PVD seed discontinuity by allowing electrochemical deposition of copper on barrier material coupled to seed passivation, is an attractive solution but at the expense of process complexity. Going further, many works have been conducted on direct electrochemical deposition on barrier and can be applied to 3D geometries. In these approaches, the surface preparation of barrier needs to be carefully done to ensure good nucleation of electrochemical seed as well as good adhesion. A third process was developed following the same considerations, based on electro-grafted copper. All these processes were initially developed for damascene application but their reduced costs compared to high vacuum processes make them attractive for 3D integration.

The key process, in this case will remain the barrier. Alternative barrier deposition solutions like CVD or ALD are generally developed at higher temperature not compatible with this approach. However, some Titanium precursors could allow low temperature CVD.

Then, fully filled Via or liner metallization by Cu ECD can be realized with a chemistry allowing bottom-up filling without void.

For the Via middle approach, already discussed for technology point of view above, many advantages are reported because the Via is realized before the BEOL, allowing the use of more conventional equipments: the allowed temperature can be relaxed to use ALD or PECVD techniques for Via isolation or barrier/seed deposition. The Via filling is realized more commonly with Cu even if tungsten can be used as contact plug in advanced CMOS. Poly-silicon material is the third material usable but with a different integration scheme, Via first approach: at initial step before FEOL due to the thermal budget needed. For the silicon back grinding, new challenges appear with the selective thinning until the bottom of the Vias first are exposed [4]. The Via recess will be used to connect the die to the bottom die during the die stacking with thermo-bonding techniques.

The main differences between the lower AR (from 1 to 5 with Via last) and higher AR (from 5 to 10 with Via first or Via last) is the integration usage: higher AR is commonly used for die to die integration and lower AR is developed for die to substrate connection. The goal of both is to act as a form factor and cost drivers for the module package.

11.6.2 Application point of view

A wide range of request coming from customer appears to demonstrate the feasibility of the TSV for a large range of Via size and Via AR either for process point of view or performances point of view. The main application in the market is the CMOS image sensor with the integration of Via at AR1. Based on this first success, development roadmap is defined to increase the functionality of the 1st CIS TSV product, like explained in Fig. 11.25.

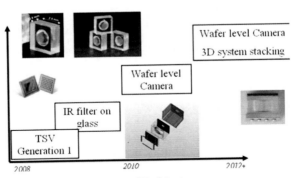

Fig. 11.25 CIS TSV roadmap to move to 3D object.

The Wafer level Camera is developed by the integration on the z axe with the wafer lens integration[18] for a continuous form factor and low cost module (Fig. 11.26). But new challenges will appear in function of the material used for the Z integration.

The temperature stability of this stack during the process steps and at the module level is one of the important questions for a safe integration.

The final challenge is the yield of the stack, multiplication of each layer yield, wafer lens, spacer ... The yield of each strata needs to be very high to get a competitive product.

First 3Di applications with TSV is entering the market with the Via-last approach, more simply to be developed in semiconductor manufacturing in order to secure the 3Di technologies and to promote the 3Di to customers (Fig. 11.28).[35, 36, 37] Then specific design and electrical models will be developed and optimized allowing a fast and prosperous development of the Via-first approach.

A challenge in the modelisation of the TSV is the understanding of the mechanical impact of the Via patterning and the metal filling on the behavior of the CMOS components and the reliability. These types of research are progressing in various institutes[12] and essential for an increasing integration of TSV.

Source: Suss

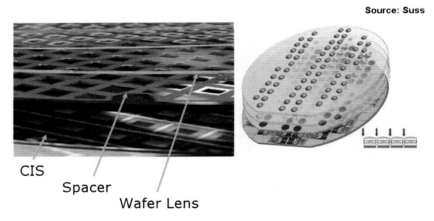

CIS
Spacer
Wafer Lens

Fig. 11.26 Illustration of the wafer lens integration from Suss.[18]

Because today, the technology continues to drive the 3D roadmap, the mechanical and thermal modelisation and 3D design toolset need to be more activated, to be developed faster in order to optimize the 3D module [38]. For example in CIS application, the bump (balls on the back-side for the connection to the board) needs to be analysed deeply to understand the mechanical stability of each bump in function of the silicon thickness of the product, in function of the board used, in function of the bump density, but also in function of the RDL density or the proximity of the Via. First experiment are already done as shown in Fig. 11.27.

Fig. 11.27 Drawing of the Via to bump (left) for the mechanical modelisation of the bumps used for the CIS application (right).

Then the electrical testing will be a real challenge to be able to distinguish drift in the right strata, to be able to isolate a Via within than more than 10000 Via in a module. The electrical testing will be strictly linked to mechanical and electrical failure analysis to get feed-back in technology, actual drawback of the 3D development.

The cost of the 3Di and the TSV integration is more and more important and look as a primary driver even if the functionalities increase faster than cost! Some steps have been already identified to be the more costly steps: bonding & Via filling. Indeed, throughput and material used have a direct impact on the final price.

Continuous perspectives of TSV integration are progressing in order to optimise actual applications or to develop new integration. First challenging integration is the interposers with 3D interconnection allowing devices mounting on both side, like passive device integration or building of micro-cooling channels.[38].

In the wafer level package, TSV is now introduced to reduce the package footprint and mainly simplify the capping of device,[6] like for the MEMS.[5] Indeed by integrating TSV, the capping must not only protect the device against external environment, but also to perform electrical path through the cap, without degrading the hermiticity performance.

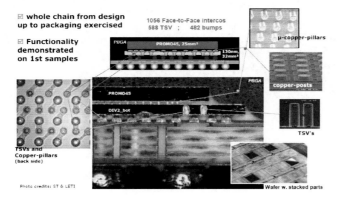

Fig. 11.28 First 3Di demonstrator at STMicroelectronics with 45 nm & 120 nm node for top & bottom die respectively.

To finish, the sentence of Yann Guillou[32] is the right situation: "The (3D) roadmaps need to be based on application requirements and not driven by technology ONLY. 3D Integration with TSV is not a scaling based concept Does it make sense today to think about submicron Via diameter or dice thinner than 30 lm for example?" Applications need to take a risk by using 3D TSV technology!

Acknowledgment

The following pictures show the success of the TSV development for the CIS application. This success has been based on a joint R&D program with the CEA-LETI, followed by the transfer to the STMicroelectronics manufacturing line.

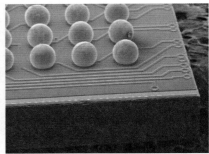

The authors would like to acknowledge all the STMicroelectronics and CEA-LETI teams working on this project.

References

1. Y. Yung, H. Bender, K. Arstila, B. Swinnen, B. Verlinden, I. De Wolf, Detection of failure sites by focused ion beam and nano-probing in the interconnect of three-dimensional stacked circuit structures, Microelectronics Reliability 48 (2008) 1517-1520

2. W. Sun, W.H. Zhu, F.X. Che, C.K. Wang, A.Y.S. Sun, H.B. Tan, Ultra thin die characterization for stack die packaging, Electroncis Components and Technology Conference IEEE, 2007, pp 1390-1396

3. V.P. Ganesh, C. Lee, Overview and emerging challenges in wafer thinning process for handheld applications, International Electronic Manufacturing Technology, IEMT 2006 proc., pp 20-26

4. P. Heinze, T. Chabert, Thin, Strong, Cheap Improving 3D chip quality by remote cold dry etching, Advanced Packaging (www.apmag.com), August/September 2008, pp 14-18

5. J. Chae, J.M. Giachino, K. Najafi, Fabrication and characterization of a wafer level MEMS vacuum package with vertical feedthroughs, journal of Microelectromechanical Systems, IEEE 2008, Vol 17, N°1, February 2008, pp 193-200

6. C.W. Lin, C.P. Hsu, H.A. Yang, W.C. Wang, W. Fang, Implementation of silicon on glass MEMS devices with embedded through wafer silicon vias using the glass reflow wafer silicon vias using the glass reflow process for wafer level packaging and 3D chip integration, J. Micromech. Microeng. 18 (2008), 025018, 1-6.

7. A. Braun, 3D Integration lacking in design and test support, Semiconductor International (www.semiconductor.net), article ID CA6615469.

8. J. Zhang, M.O. Bloomfield, J.Q. Lu, R.J. Gutmann, T.S. Cale, Modeling thermal stresses in 3D IC interwafer interconnects, IEEE Transactions on Semiconductor Manufacturing, IEEE 2006, Vol. 19, N°4, November 2006, pp 437-448

9. 9 C.S. Ng, K.Y. Chua, M.T. Ong, Y.C. Goh, A.K. Asundi, Warpage of thin wafers using computer aided reflection moiré method, Advanced Characterisation techniques for optics, Semiconductors and Nanotechnologies III, Proc. Of SPIE Vol. 6672, 667203 (2007), pp 1-11

10. R.W. Kay, E. de Gourcuff, M.P.Y. Desmulliez, Stencil printing technology for wafer level bumping at Sub-100 micron pitch using Pb-free alloys, Electronic Components and Technology Conference IEEE, 2005, pp 848-854

11. R. Lathrop, Solder paste printing and stencil design considerations for wafer bumping, SEMI Int'l Electronics Manufacturing Technology Symposium, IEEE 2004, 9p

12. S. Aravamudhan, D. Santos, G. Pham-Van-Diep and F. Andres, A study of solder paste release from small stencil apertures of different geometries with constant volumes, Semicon West 2002 IEMT Symposium (IEEE 2002), pp 159-165

13. J. Jozwiak, R.G. Southwick, V.N. Johnson, W.B. Knowlton, A.J. Moll, Integrating through wafer interconnects with active devices and circuits, IEEE Transactions on Advanced Packaging, Vol 31, N°1, February 2008, pp 4-13

14. RTI conference, 3D architectures for Semiconductor Integration and Packaging, 17-19 Nov 2008, San Francisco

15. STMicroelectronics Internal workshop, January 2009.

16. Y. Kurita, S. Matsui, N. Takahashi, K. Soejima, M. Komuro, M. Itou, C. Kakegawa, M. Kawano, Y. Egawa, Y. Saeki, H. Kikuchi, O. Kato, A. Yanagisawa, T. Mitsuhashi, M. Ishino, K. Shibata, S. Uchiyama, J. Yamada, H. Ikeda, A 3D stacked memory integrated on a logic device using SMAFTI technology, 2007 Electronic Component and Technology Conference, IEEE 2007, 821-829.

17. Semiconductor International December 2008, Interposers paly a key role in 3-D ICs, www.semiconductor.net, ref:CA6605618

18. Yann Guillou, Advanced Packaging, January 2009, 3D Jargon – Getting it straight

19. R. Voekel and R. Zoberbier, Inside Wafer Level Cameras, Semiconductor Semiconductor International, Feb 2009, pp 28-32

20. S. Seok, N. Rolland and P.-A. Rolland, Zero level packaging using BCB adhesive bonding and glass wet-etching for W-band applications, Electronics Letters, 22nd June 2006, Vol. 42 N°13, 2p

21. M. Wiemer, C. Jia, M. Toepper and K. Hauck, Wafer bonding with BCB and SU8 for MEMS packaging, Elec. Syst. Tech. Conf., Dresden Germany, IEEE 2006, pp 1401-1405

22. A. Polyakov, M. Bartek and J. N. Burghartz, Area selective adhesive bonding using photosensitive BCB for WL CSP applications, Journal of Elec. Pack., March 2005, Vol. 127, pp 7-11

23. S. Seok, N. Rolland and P.-A. Rolland, Packaging methodology for RF devices using a BCB membrane transfer technique, J. Micromech. Microeng. 16 (2006) pp 2384-2388

24. D. Reuter, A. Bertz, G. Schwenzer and T. Gessner, Selective adhesive bonding with SU-8 for zero level packaging, Proc. Of SPIE vol. 5650 (Bellingham, WA, 2005), Micro and Nanotechnology, pp 163-171

25. C. Iliescu, F. E. H. Tay, J. Miao and M. Avram, Wafer level packaging of pressure sensor using SU8 photoresist, Proc. of SPIE Vol. 5649 (Bellingnham, WA, 2005), Smart structures, Devices and Systems II, pp 297-305

26. J. Oberhammer and G. Stemme, BCB contact printing for patterned adhesive full wafer bonded 0 level packages, J. Microelec. Syst., Vol. 14, N°2, April 2005, pp 419-425

27. D. N. Pascual, Effects of softbake parameters on a Benzocyclobutene (BCB) adhesive wafer bond, Mat. Res. Symp. Proc. Vol. 782, 2004 Materials Research Society, pp 187-192

28. C.T. Pan, P.J. Cheng, M.F. Chen and C.K. Yen, Intermediate wafer level bonding and interface behaviour, Microelectronics Reliability 45 (2005) pp 657-663

29. D. Veyrie, M. Budinger, J.L. Roux, F. Pressecq, A. Tetelin and C. Pellet, Modeling of water vapour permeation inside BCB-sealed packages for Microsystems, DTIP Conf., Montreux (Switzerland), 01-03 June 2005, pp 244-249

30. A. Tetelin, C. Pellet, J.Y. Deletage, B. Carbonne and Y. Danto, Moisture diffusion in BCB resins used for MEMS packaging, Microelectronics Reliability 43 (2003) pp 1939-1944

31. A. Telelin, A. Achen, V. Pouget, C. Pellet, M. Topper and J.L. Lachaud, Water solubility and diffusivity in BCB resins used in microelectronics packaging and sensor applications, IMTC 2005 Instrumentations and Measurement Conf., Ottawa, Canada, 17-19 May 2005, pp 792-796

32. → 18 R. Voekel and R. Zoberbier, Inside Wafer Level Cameras, Semiconductor International, 2/1/2009, www.semiconductor.net, 6p, article ID: CA6634202

33. Y. Guillou and E. Saugier, 3D Jargon – Getting it straight, www.xxx

34. D. Henry, F. Jacquet, M. Neyret, X. Baillin, T. Enot, V. Lapras, C. Brunet-Manquat, J. Charbonnier, B. Aventurier, and N. Sillon, Through Silicon Vias Technology for CMOS Image Sensors Packaging, Proc. 58th Elec. Comp. and Tech. Conf. (IEEE ECTC 2008), 2008, pp. 556-562.

35. D. Henry, J. Charbonnier, P. Chausse, F. Jacquet, B. Aventurier, C. Brunet-Manquat, V. Lapras, R. Anciant, N. Sillon; B. Dunne, N. Hotellier, J. Michailos, Through Silicon Vias Technology for CMOS Image Sensors Packaging: Presentation of Technology and Electrical Results, Proc. Of 10th EPTC 2008 (IEEE EPTC 2008), pp 35-44.

36. D. Henry, S. Cheramy, J. Charbonnier, P. Chausse, M. Neyret, C. Brunet-Manquat, S. Verrun, N. Sillon, L. Bonnot, X. Gagnard, E. Saugier, 3D integration technology for set-top box application, International Conference on 3D System Integration 2009, IEEE proc., to be published, 7pp

37. S. Cheramy, D. Henry, A. Astier, J. Charbonnier, P. Chausse, M. Neyret, C. Brunet-Manquat, S. Verrun, N. Sillon, L. Bonnot, X. Gagnard, J. Vittu, 3D integration process flow for set-top box application: description of technology and electrical results, EMPC 2009, to be published.

38. X. Gagnard and T. Mourier, Through silicon via: From the CMOS imager sensor wafer level package to the 3D integration, 2009 Elsevier, On line 8 June 2009, http://dx.doi.org/10.1016/j.mee.2009.05.035, to be published

39. G. Poupon, packaging avancé sur silicium : état de l'art et nouvelles tendances, Chapitre 6. Management thermique -X. Gagnard, pp, Hermes Science Publications, ISBN : 978-2-7462-1950-2

40. CT Pan, H. Yang, S.C. Shen, M.C. Chou, H.P. Chou, A low temperature wafer bonding technique using patternable materials, Journal of Micromechanics and microengineering, 12(2002) 611-615.

41. F. Niklaus, R.J. Kumar, J.J. McMahon, J. Yu, J.Q. Lu, S; Calet, R.J. Gutmann, Adhesive wafer bonding using partially cured benzocyclobutene for three-dimensional integration, Journal of the Electrochemical Society, vol. 153 (n° 4) G291-G295 (2006).

42. Frank Niklaus, Adhesive Wafer bonding for microelectronics and microelectromechanical systems, PhD repport, Microsystem Technology department of signals, sensors & systems, Royal Institute of Technology, Stockholm, ISSN 0281-2878, 2002.

43. V. Dragoi, S. Farrens, P. Lindner, Plasma activated wafer bonding for MEMS, SPIE – Microtechnologies for New Millenium, Symp. Smart Sensors, Actuators & MEMS, May 9-11, 2005 Sevilla, Spain, Vol. 5836, pp179-187.

Chapter 12

A 300-MM WAFER-LEVEL THREE-DIMENSIONAL INTEGRATION SCHEME USING TUNGSTEN THROUGH-SILICON VIA AND HYBRID CU-ADHESIVE BONDING

F. Liu, R. R. Yu, A. M. Young, L. Shi, K. A. Jenkins, X. Gu, N. R. Klymko, S. Purushothaman, S. J. Koester and W. Haensch

IBM Research

A 300-mm wafer-level three-dimensional integration (3DI) process using tungsten (W) through-silicon vias (TSVs) and hybrid Cu/adhesive wafer bonding is demonstrated. The W TSVs have fine pitch (5 μm), small critical dimension (1.5 μm), and high aspect ratio (17:1). A hybrid Cu/adhesive bonding approach, also called transfer-join (TJ) method, is used to interconnect the TSVs to a Cu BEOL in a bottom wafer. The process also features thinning of the top wafer to 20 μm and a Cu backside BEOL on the thinned top wafer. The electrical and physical properties of the TSVs and bonded interconnect are charaterized. The results show RLC values that satisfy both the power delivery and high-speed signaling requirements for high-performance 3D systems. A reliability evaluation of a 300-mm-compatible 3DI process is then presented. The interface bonding strength, deep thermal cycles test, temperature and humidity test, and ambient permeation oxidation all show favorable results, indicating the suitability of this technology for VLSI applications.

3D Integration for VLSI Systems
Edited by Chuan Seng Tan, Kuan-Neng Chen and Steven J. Koester
Copyright © 2012 by Pan Stanford Publishing Pte. Ltd.
www.panstanford.com

12.1 INTRODUCTION

3DI is a promising technology to further improve the performance of computational systems.[1-5] 3DI enables increased functions per physical die area, and can dramatically increase the available cache memory in multi-core architectures. While many approaches exist, the ones with wider integration flexibility and higher reliability can better adapt to various application needs. Previously, W TSVs for silicon carrier applications have been demonstrated with both high yield and reliability.[6] In addition, a hybrid metal/adhesive TJ process has been shown to be a highly reliable inter-chip connection process in multi-chip modules.[7] However, neither of these technologies has been demonstrated in a full integrated process on a 300-mm platform. Here, we utilize the combination of W TSVs and transfer-join assembly on 300-mm wafers, and demonstrate the functionality and robustness of this process.

All 3DI technologies require some form of through-silicon vias (TSVs) for signal delivery and power delivery through various component silicon layers. Hence, it is critical to understand TSV electrical properties over wide operation frequencies. Up to date, there have been relatively few studies on TSV inductance and capacitance characterizations. Particularly, it is difficult to extract TSV frequency dependent properties with high sensitivity due to the relatively small value of inductance and capacitance. The second part of the chapter, we will present three techniques to characterize TSV electrical properties.[8] (1) TSV low frequency capacitance; (2) TSV inductance; (3) frequency dependent TSV capacitance extraction based on transmission line method. For each technique, we will show specific test structures, discuss measurement metrologies, and compare measurement to simulation. Besides, we will evaluate TSV electrical properties from circuit and system perspective.

In the third part of chapter, further step is taken to evaluate the reliability of this technology. To date, most published reliability results have been reported based on chip-level assembly technique.[9] Here, we will study the reliability of the hybrid Cu-adhesive wafer bonding structure on a 300-mm platform from the following aspects: structural, thermal-mechanical, and chemical stability.

12.2 PROCESS DESCRIPTION

Figure 12.1 presents the integration flow of our 3D wafer-level process. The TSVs are formed by continuous duty reactive ion etching (RIE) and have 25 μm depth, 1.5 μm width, 6 μm length, and 17:1 aspect ratio. Figure 12.2 summarizes the TSV etch depth as a function of TSV width at the center and edge of a wafer. At fixed via width, a cross-wafer depth variation of less than 0.8 μm is achieved. The TSVs use thermal SiO_2 and/or high density plasma (HDP) oxide as the insulating liner, and chemical vapor deposition (CVD) W metal as metallic connection. After W TSV formation, insulation and Cu studs are formed on the top wafer, followed by a planarized deposition of a polymer adhesive. The bottom wafer utilizes Cu pads with recessed dielectric openings. The wafers are then aligned using infrared system, and bonded in vacuum using transfer-join (TJ) adhesive/metal hybrid bonding method. The TJ bonding technique involves the use of mechanical lock-and-key structures with metal/adhesive hybrid bonding to create a 3D interconnect. After bonding, the backside of the top wafer is ground and chem-mechanical polished (CMP) to a thickness of approximately 40 μm. The topside of the bonded pair is further RIE trimmed down to 20 μm to expose the TSVs. Then, a mechanical backside knock-off is introduced to approximately level the TSVs with the backside wafer surface in order to minimize the TSV protrusion. In the end, backside Cu metallization is patterned.

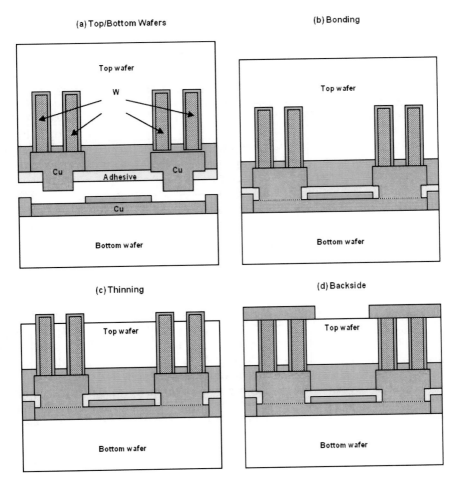

Fig. 12.1 Schematic diagram of 3D process flow.

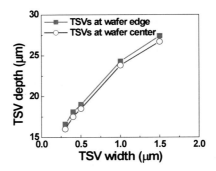

Fig.12.2 Plot of TSV etch depth vs. TSV width for center (squares) and edge (circles) dies on a 300-mm wafer. For a fixed width, the center to edge variation is less than 0.8 μm.

12.2.1 Integration Scheme

Compared to chip-to-chip and chip-to-wafer process, wafer-to-wafer 3D process allows high TSV density, lower manufacture cost, and the capability of creating integrated passive components. However, wafer-to-wafer 3D process demands more challenges in terms of process controllability and robustness. To successfully demonstrate 300-mm wafer-to-wafer 3D integration, face-to-face via-first integration scheme is utilized. Face-to-face integration is favored because of its process simplicity, and process controllability. No handling wafer is involved in the integration scheme, and majority of the processes are done on regular Si substrates. These avoid the complicity associated with glass wafer handling and wafer distortions in the strata. The avoiding lamination and handling wafer release steps also reduce process cost. Via first scheme gives more flexibility in terms of TSV insulating liner and metallic material selections, allowing processing TSV to meet different design requirements.

12.2.2 W TSV vs. Cu TSV

W TSV is superior in two major aspects than Cu TSV. First, W has good fill characteristics. The aspect ratio of W TSV can be 2~3 times higher than Cu TSV. Figure 12.3 shows chemical vapor deposition (CVD) W TSV with 17:1 aspect ratio. The W fills in TSV nicely from top to bottom of TSV. Second, W TSV is expected to have good reliability characteristics. Since the coefficient of thermal expansion (CTE) of W $(4.5*10^{-6}/K)$ is close to the CTE of Si $(2.5*10^{-6}/K)$, comparing to the CTE of Cu $((17.6*10^{-6}/K))$, W TSV survives through consequent 3DI process, and shows good reliability in thermal cycle related stress tests (shown later). Besides, W TSVs show almost zero strain. W TSV may build up strain due to W residue strain. For example, CVD W used in the process has large tensile strain. One the other hand, due to the close CTE between W and Si, thermal processes will play little role in W TSV strain. Figure 12.4 illustrates lateral strain profile of Si regions near W TSVs. As grown W TSV array of 5-µm-pitch is used for micro-Raman spectroscopy characterization. The wavelength of the UV light is 325 nm, which corresponds to the penetration depth of 10 nm in Si. For the sample using oxide and HDP as insulating liner, regions within ~ 1 µm of TSVs show 0.1% tensile strain, while compressive strain is observed in field regions. The keep away distance for devices is about 1-2 µm. For the sample using only oxide as insulating liner, strain is almost zero for both between TSVs regions and field regions, indicating negligible strain coming from W.

Fig.12.3 (a) SEM image of of high aspect ratio TSVs (AR: 17:1) utilizing oxide liner and CVD tungsten.

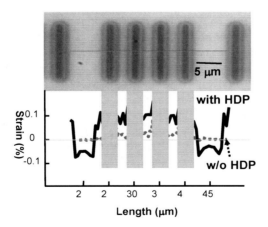

Fig.12.4 Lateral strain profile of Si regions near W TSVs as determined by UV (λ=325 nm) micro-Raman spectroscopy. For the sample using oxide and HDP as insulating liner, regions within ~ 1 µm of TSVs show 0.1% tensile strain, while compressive strain is observed in field regions. (black solid curve). For the sample using only oxide as insulating liner, strain is almost zero for both between TSVs regions and field regions (red dotted curve).

Of course, W has disadvantage in its relatively large resistivity. Also W may exhibit large residue strain (Fig. 12.5). More than 300-µm wafer bow change for tensile W film after W deposition. W CMP process, however, reduces the wafer bow back to that of before W deposition. Therefore, rectangular bar shapes is preferred to minimize the required W deposition thickness.

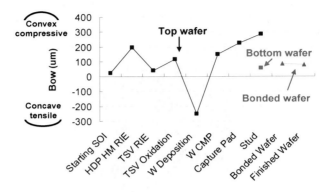

Fig.12.5 Wafer bow at each process step for top wafer, bottom wafer, and bonded wafer pair.

12.2.3 TJ Adhesive/Metal Hybrid Bonding

A variation of the TJ adhesive/metal hybrid bonding method[7] has been utilized to bond the wafers and interconnect the TSVs. Metal bonding has the advantage of forming the electrical connections during bonding process. As summarized in Table 1, adhesive bonding has strong bonding strength. Adhesive material has bonding energy tens to several hundreds J/m^2; Cu-Cu thermocompression bonding has bonding energy 3 J/m^2; Oxide-oxide bonding has bonding energy less than 1 J/m^2. Adhesive bonding has the advantage in conformality to compensate surface topology by means of reflow during bonding. However, compared to oxide-oxide and Cu-Cu bonding, the potential issues for PI as a bonding adhesive are thermal stress due to higher coefficient of thermal expansion (CTE) mismatch with copper and silicon, moisture absorption, and small but finite weight loss at BEOL processing temperatures (Fig. 12.6). In selecting the adhesive material, the properties of the desired joining material in terms of bonding strength, CTE, thermal and chemical stability are contrasted with available options in Table 12.1. Polyimide (PI) has multiple roles in the 3DI structure: conformal adhesion, environmental insulation, and stress buffer. Relative to other typical adhesives (e.g. benzocyclobutene (BCB)), PI is clearly more suitable. It is shown that the combination of adhesive and Cu-Cu bonding achieves high bonding strength and electrical connection between strata during the bonding. Cu-Cu/PI hybrid bonding utilizes the benefits and satisfies the mechanical, thermal and chemical requirements.

Lock-and-key structures[1, 10, 11] can safeguard the wafer pair from lateral shifts during bonding. In designing the lock-and-key structure, appropriate selection of adhesive thickness is critical to compensate the topology across the wafer, and to achieve very low thermal stress for bonding thermal reliability. The height of stud ought to be larger than the oxide recess thickness plus adhesive thickness, allowing the stud and the recess to form bonding connection with the expected stud height reduction due to Cu-Cu deformation.

	Oxide / Cu	**Adhesive**
Bond type	planar	cohesive
Surface	smooth	conformal
Impact	localized	diffused
Crack	easy	difficult

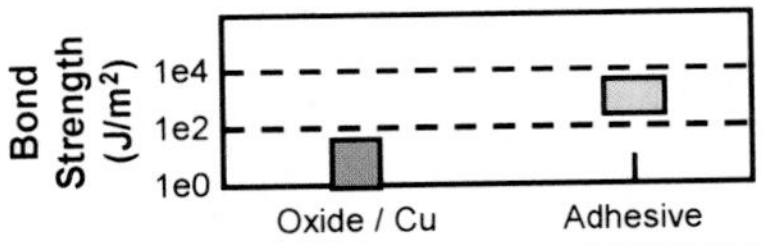

Fig.12.6 Structural comparison of oxide-oxide bonding, Cu-Cu bonding and adhesive bonding.

Table 12.1 Desired material properties for a reliable 3DI connection (structural, thermal-mechanical, thermal-BEOL, and chemical) contrasted with actual properties of candidate adhesives/bonding options. Deficiencies are highlighted in bold.

	Dielectric Material	Ideal	BCB	PI	Cu-Cu	Oxi-oxi	TJ/PI
Structural	Compliance (%)	> 50	20	45	**~ 5**	**~ 0**	45
	Modulus (GPa)	> 5	1-3	2-4	**120**	**60**	5
Thermal Mechanical	Peel strength (g/mm)	> 100	**~ 30**	> 100	**Medium**	**Low**	High
	TCE (ppm/°C)	~ 17	60	45	~ 17	3	17-50
Thermal BEOL	Weight Loss % (400°C)	0	**> 8**	0.25	0	0	0.25
	Weight Loss % (450°C)	< 0.2	**> 25**	0.44	0	0	0.44
Chemical	Moisture absorption %	< 0.1	< 0.2	**< 2**	0	0	**~ 2**

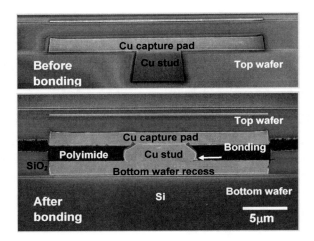

Fig.12.7 SEM cross-section of a TJ structure before and after bonding. A lock-and-key structure is formed using a matched stud (top wafer) and a recess (bottom wafer). Polyimide material fills in regions between the wafers. The thermo-compression effect on the bonding structure is apparent.

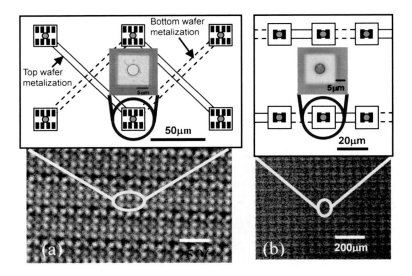

Fig.12.8 High resolution scanning acoustic images of a portion of an 80-μm-pitch 3D via chain (a) and of a 28.8-μm-pitch 3D via chain (b). The uniformly distributed bright spots (Cu joining) and dark areas (adhesive bonding) indicate a good TJ joining interface with no macroscopic defects detectable.

In the bonding process, wafer cleaning is essential to remove oxide from Cu surface. The wafer bonding is performed in vacuum under isostatic pressure. The minimum pressure needs to form Cu-Cu deformation is related with bonding structures and pattern densities. The bonding temperature is 320-340°C to form Cu-Cu bonding and adhesive reflow. Figure 12.7 shows a cross-sectional SEM of a typical Cu stud and capture pad lock-and-key structure before and after bonding. The structural effect of the Cu-Cu thermocompression process is clearly observed. Non-destructive scanning acoustic microscopy is an effective wafer to inspect bonding quality and macroscopic bonding defects. Figure 12.8 shows scanning acoustic microscope image for orthogonal zigzag 80-µm-pitch via chains and linear 28.8-µm-pitch via chains. The uniform periodic patterns in the regions indicate good bonding free of defects. Infrared microscopy is used to inspect the alignment accuracy. Post bonding alignment accuracy is observed to be within 2 µm across the entire 300-mm wafer.

12.2.4 Wafer Thinning and Backside Process

After bonding, the backside of the top wafer is ground and chem-mechanical polished (CMP) to a thickness of approximately 40 µm as shown in Fig. 12.9. Five-point thickness measurements of the post-CMP bonded pair indicated a wafer thickness standard deviation to be 3 µm, and no pronounced global dishing as shown in Table 12.2. To achieve tighter thickness variation, special substrate with etch stop layer, such as oxide or Si with different doping density, may be used to fulfill the requirements.

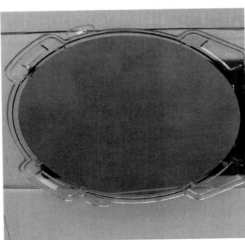

Fig.12.9 Photograph of a TJ bonded wafer pair after grinding and polishing of the top wafer to 40 µm thickness.

Table 12.2 Five-point thickness measurements of the post-CMP bonded pair indicated a wafer thickness standard deviation to be 3 μm, and no pronounced global dishing.

Top wafer after CMP	Flat	Right	Center	Left	Top
t (μm)	40	33	40	40	35

The topside of the bonded pair is further RIE trimmed down to 20 μm to expose the TSVs. An AFM scan of the trimmed surface is shown in Fig. 12.10, which indicates an RMS roughness of 25 nm. As shown in Fig. 12.11, the TSVs have a notably different appearance on the backside of the wafer as compared to the front. To minimize the TSV protrusion, a mechanical knock-off was utilized to approximately level the vias with the backside wafer surface. A photograph of a completed 300-mm 3DI wafer after depositing and patterning of the backside Cu BEOL metallization is shown in Fig. 12.12.

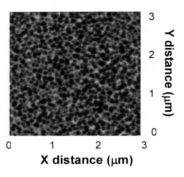

Fig.12.10 AFM scan of a region of a 300-mm wafer after grinding, polishing and RIE trimming down to 20 μm. The RMS roughness is 25 nm.

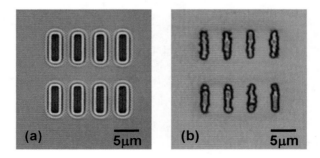

Fig.12.11 Top-down images of a bank of eight W TSVs from (a) the front side of the top wafer, and (b) from the back side of the top wafer after polishing and RIE trimming.

Fig.12.12 Photograph of a completed 300-mm 3DI wafer.

12.2.5 Completed Wafer Evaluation and Robustness

Wafer bow is recorded at key process steps (Fig. 12.5) to make sure that 3DI wafer is within tool handling requirements. Except the wafer bow change due to W deposition and CMP, it is interesting to note that post bonding wafer bow is an average of the top and bottom wafer bows before the wafer bonding, which gives guideline of controlling post bonding wafer bow. The wafer bow results are reproducible across wafers, reflecting the reproducibility and robustness of our bonding process. The curvature of the completed wafer is further characterized by XRD rocking curves X-ray technique at different location of the completed wafer. The result is summarized in Table 12.2. Regions of non-uniform curvature are observed in the region -25 mm to 15 mm away from wafer center. The average wafer curvature is 140 μm, consistent with the completed wafer bow (81 μm) obtained from topology measurement using the following relationship:

$$B = \frac{D^2}{8 \cdot R} ,$$ (12.1)

where B is wafer bow; D is the diameter of the wafer, and R is average wafer curvature.

Wafer edge cannot be ignored with device continuing attaining finer linewidths. Figure 12.13 shows SEM of wafer edge for the 300 mm completed wafer. There are no chips, cracks and blisters at the edge of the wafer. The apex, and back bevel of the bottom wafer are unaffected. The front top wafer bevel is also well preserved for both the top and bottom wafer with the help of adhesive material. However, to restore the standard wafer edge, some form of edge polishing may be needed to refine the front wafer bevel of the 3DI completed wafer.

Last but not least, the completed wafer is diced into pieces using a dicing saw. The bonding interface survives the dicing test. The edge of each chip is well preserved with no cracks and delaminations observed. The dicing test further confirms the strong bonding strength formed by the Cu/adhesive hybrid bonding approach.

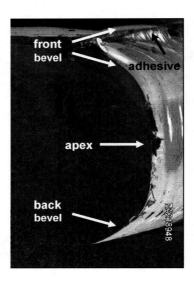

Fig.12.13 SEM of wafer edge for a 300 mm completed wafer.

12.3 ELECTRICAL RESULTS

12.3.1 TSV DC Resistance

Electrical characterization has been carried out on the completed 3D via chains. The dotted blue line in the inset of Fig. 12.10 shows a typical I-V curve recorded on an 80-μm-pitch via chain (shown in Fig. 12.8 (a)) with 1032 connecting nodes. The patterned bars in Fig. 12.14 shows a histogram of the resistance per node of the connected 80-μm-pitch via chains on an entire 300-mm wafer. The average resistance per node is extracted to be 1.0 Ω, where each node consists of banks of 8 TSVs. The resistance variation is attributed to differences in the TSV protrusion height across the wafer (Fig. 12.15). After extracting the resistance of the interconnecting metal and the resistance of the via to the backside metallization (de-embedding), the intrinsic via and bonded interface resistance is determined to be ~ 0.1 Ω (i.e., resistance per TSV to be 0.8 Ω), indicating that a considerable portion of the total node resistance is due to the excess protrusion of the via from the backside of the top wafer. For the samples with the knock-off process, the average node resistance for 80-μm-pitch via chains reduced from 1.0 to 0.8 Ω, as illustrated by both the I-V curves and the node resistance distribution histograms shown in Fig. 12.14.

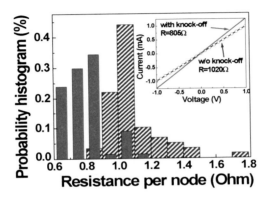

Fig.12.14 Histogram of node resistance of 80-μm-pitch 3D via chains of 1032 nodes without backside knock-off process (patterned bars) and with backside knock-off process (solid bars). Inset: I-V curve of a typical 80-μm-pitch via chain with (solid line) and without (dotted line) backside knock-off process.

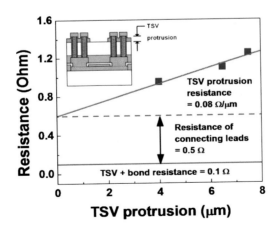

Fig.12.15 Plot showing via chain node resistance breakdown. Three major components are: (1) Resistance from thin Cu connections at TSV protrusion, (2) Resistance from Cu connecting metallization, and (3) TSV and TJ bond resistance. Inset: Schematic drawing of TSV protrusion.

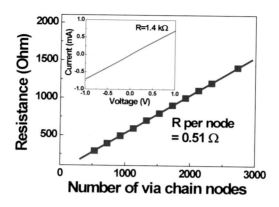

Fig.12.16 A 28.8-μm-pitch via chain resistance as a function of TSV node number. Resistance per node is extracted to be 0.51 Ω. Inset: I-V curve of a typical 28.8-μm-pitch via chain of 2724 nodes.

The inset of Fig. 12.16 is a typical I-V curve recorded on a 28.8-μm-pitch via chain (shown in Fig. 12.8 (b)) of 2724 connecting nodes with backside knock-off process. Fig. 12.16 shows a linear relationship between the resistance of the via chain and the number of connecting nodes. The resistance per node is extracted to be 0.51 Ω, where each node consists of banks of 2 TSVs. The yielding of the tighter-pitch via chains demonstrates the scalability of our W TSV, hybrid Cu-adhesive bonding process.

12.3.2 TSV Low Frequency Capacitance

The fundamental TSV test structure used for capacitance and inductance characterization consists of two W TSVs in parallel (dual-TSVs) 5 μm away from each other. One 10%10 μm² copper (Cu) capture pad is used to group the dual-TSVs as shown in Fig. 12.17.

Capacitance simulation is done using a commercial 3D electromagnetic quasi-static extraction tool (Q3D™). In the simulation, the relative permittivity and bulk conductivity of silicon are 11.9 and 100 S/m, respectively. The capacitance of the dual-TSVs is extracted with respect to the rectangular perfect electric conductor (PEC) wall wrapping around the lossy silicon. Figure 12.17 shows the charge distributions for the fundamental TSV test structure. It is shown that the coupling effect between them can be ignored due to the relative large separations (5 μm) between the two TSVs. Quasi-static simulation predicts the capacitance of the fundamental structure to be 54 fF, which includes 8 fF from capture pad and 46 fF from the dual-TSVs. Low frequency capacitance measurement is done using LCR meter at the frequency of 100 kHz. Total capacitance is measured on a top wafer before wafer bonding process. By subtracting fringing probe capacitance from total capacitance, the de-embeded capacitance of the fundamental TSV test structure is achieved to be 58 fF and the measured capacitance of the dual-TSVs is 50 fF. The measurement result is in agreement with the simulation results. Furthermore, the capacitance of the fundamental test structure shows almost no voltage dependence over the voltage range (Fig. 12.18). The lack of voltage modulation is attributed to thick oxide and possible positive charges in the Si and oxide interface.

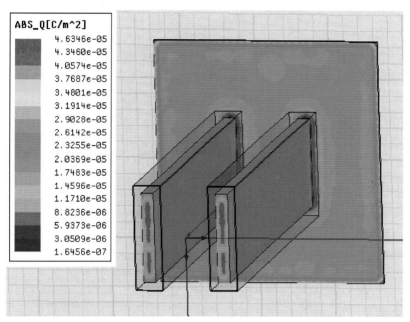

Fig.12.17 Charge distributions of the dual-TSVs structure (only charge distribution in the dual-TSVs is shown).

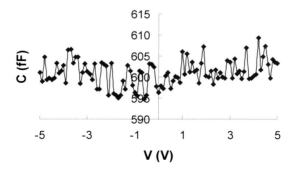

Fig.12.18 Standard C-V for the fundamental TSV test structure with voltage from -5 V to 5 V. The capacitance shows almost no voltage dependence.

12.3.3 TSV Inductance

Inductance simulation is done for one standing alone node. The simulation uses Ansoft Q3D™ for the quasi-static extraction assuming the same material properties as used in the capacitance extraction. The geometries for the simulation are shown in Fig. 12.19. The size of the silicon box is 60%60%20 μm.³ A quasi-static extraction shows the DC inductance of the dual-TSVs is 0.009 nH.

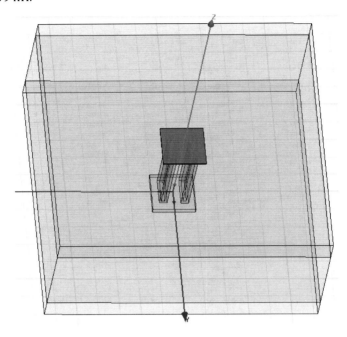

Fig.12.19 Structure and grid for dual-TSVs inductance simulation using Ansoft Q3D full wave.

Due to the relatively shallow dual-TSVs, the dual-TSVs inductance in this technology is relatively small. Hence, a 3D dual-TSVs chain is used for inductance extraction in order to achieve better signal-to-noise ratio. Figure 20 shows the layout of the TSV inductance macro. A tested dual-TSVs chain is in orthogonal zigzag shape with 40-μm-pitch. The chain has 4 (or 8) connection nodes. Each connection node consists of a fundamental TSV test structure. Another 4 (or 8) node dual-TSVs chain is inter-twisted with the tested via chain. The latter via chain serves as DC ground for the tested TSV chain. The tested dual-TSVs chain can be probed either without the ground dual-TSVs chain in Ground-Signal (GS) configuration, or with the ground dual-TSVs chain in Ground-Signal-Ground (GSG) configuration. At the same time, there is a calibration chain on the backside of the top wafer to de-embed the inductance contribution from chain connection leads.

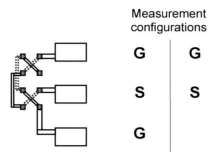

Fig.12.20 Layout of the dual-TSV inductance macro. Black solid region are pads and wirings at the backside of the top wafer; doted regions are pads and wirings at the front side of bottom wafers; rectangles are TSVs, which connect metallization between the top wafer and the bottom wafer. Right side: two kinds of probe configurations for dual-TSV S_{11} measurements: Ground-Signal-Ground (GSG) and Ground-Signal (GS).

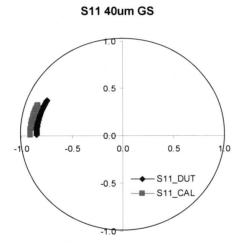

Fig.12.21 (a) S_{11} plotted in Smith chart for both the dual-TSVs chain (DUT) and the calibration chain (CAL) measured under GS configuration for a frequency up to 10 GHz.

Inductance is obtained from the frequency-dependent reflection coefficient. Reflection coefficients (S_{11}) are measured using a vector network analyzer for both the tested dual-TSV via chains (DUT) and the calibration chains. The substrate is chucked to DC ground during the measurement. The measurement technique is similar to Ref. 12. Measured reflection coefficients (S_{11}) for both the tested dual-TSVs chain and the calibration chain measured under GS configuration for frequencies up to 10 GHz is plotted in Smith chart shown in Fig. 21 (a). The measured S_{11} of the dual-TSV chain and the calibration chain are converted to impedance. Net impedance is taken to be $Z=R+j\omega L$ where $\omega=2\pi f$. The imaginary parts of Z are plotted as a function of frequency as shown in Fig. 12.21(b) for the dual-TSVs chain and the calibration chain, respectively. Inductances are obtained from the fitted slopes. The linear behavior of $Im(Z_{11})$ vs. frequency indicates near constant inductance over the frequency range. Table 12.3 summarises the inductances for dual-TSVs chains and their corresponding calibration chains for 4-node and 8-node under GS and GSG configurations. From there, de-embedded dual-TSV inductances for are obtained by subtracting inductances of the calibration chains from those of dual-TSV chains. De-embedded dual-TSV inductances of the 8 node chain are twice as much as those of the 4 node chain for both GS and GSG configurations, showing the scalability of the measurement.

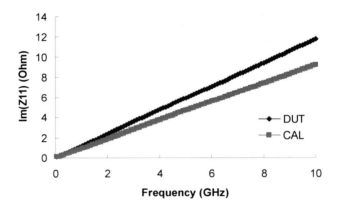

Fig.12.21 (b) Imaginary parts of Z_{11} as a function of frequency measured under GS configuration for the dual-TSVs chain (DUT) and the calibration chain (CAL), respectively. Inductances can be obtained from the fitted slopes.

Table 12.3 Inductances for dual-TSVs chains (L_DUT) and their corresponding calibration chains (L_CAL) for 4-node and 8-node cases with GS and GSG configurations. Total dual-TSVs inductances (L_TSV) for each case are obtained by subtracting inductances of calibration chains from those of dual-TSV chains.

Configuration	Layout # of nodes	Inductance (nH) L_DUT	L_CAL	L_TSV L_DUL-L_CAL
GS	4	0.188	0.146	0.041
GS	8	0.374	0.291	0.083
GSG	4	0.296	0.228	0.068
GSG	8	0.510	0.395	0.115

It is worth noting that extracted TSV node inductances are 0.010 nH under GS configuration and 0.015 nH under GSG configuration, even though the same dual-TSVs chain is used for the measurements. As described previously, the simulation shows the TSV node inductance of 0.009 nH. The slight discrepancy is due to the capacitive coupling to Si substrate as described in the following. Since the measured chains are shorter than microwave length, a lump model (shown in Fig. 12.22) is sufficient for equivalent circuit. In the model, each chain consists of a resistor in series with an inductor and shunted with a capacitor. The capacitor represents the coupling of electric fields from the chain to DC ground through the oxide and silicon. For the sake of simplicity, the conductance element which represents the silicon loss [8] is neglected. In the measurement, the real part of Z_{11} of the chain is much smaller than the imaginary part of Z_{11}, therefore, the imaginary part of Z_{11} can be re-written as:

$$\text{Im}(Z_{11}) \approx \frac{\omega L}{1 - \omega^2 LC} \qquad (12.2)$$

where L is dual-TSV inductance, and C is the coupling capacitance from the chain to DC ground. Due to the existing of coupling C, the imaginary part of Z_{11} overestimates the inductance dual-TSV chain. Therefore, the TSV inductance obtained from the reflection coefficient method tends to be larger than the intrinsic dual-TSV inductance. This is especially true when the dual-TSV chains are grounded using the ground dual-TSV chain under GSG configuration. Better inductance extraction accuracy can be obtained by minimizing coupling (such as using GS configuration) or by taking coupling capacitance effect into inductance extraction.

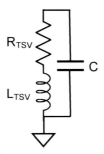

Fig.12.22 Equivalent circuit to explain the effect of capacitance coupling to dual-TSVs inductance extraction.

12.3.4 TSV Frequency Dependent Capacitance

Resistance, inductance, and capacitance of TSV are, in general, functions of frequency. The frequency dependence is especially pronounced for TSV capacitance. As the frequency increases above characteristic frequency, the substrate capacitance acts in series with the oxide capacitance for a net capacitance reduction. The reduction amount depends on the ratio of oxide capacitance to substrate capacitance. The substrate capacitance depends not only on geometry, but also on the distance of the return path. To characterize this behavior, we introduce S-parameter based transmission line characterization method. Similar approach was recently applied to TSV characterization for capacitance extraction.[13, 14] However, the capacitance extracted from measurement in Ref.[14] shows large oscillations with many non-physical negative values at certain frequencies. In our experiment, we design dual-TSV chains and calibration chains both with 32 connection nodes at the pitch of 40 μm to satisfy the transmission line requirement and to achieve better signal-to-noise ratio. Similar to inductance measurement, two structures are implemented, one with the dual-TSV chain grounded only through Si substrate, and the other with the dual-TSV chain grounded through both Si substrate and nearby a ground dual-TSV chain. The layouts of the two configurations are shown in Fig. 12.23(a) and (b). Calibration chains are also built to de-embed connection lead effect.

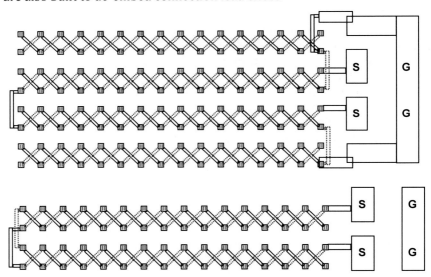

Fig.12.23 Layout of frequency dependent capacitance extraction macro. Black solid region are pads and wirings at the backside of the top wafer; doted regions are pads and wirings at the front side of bottom wafers; rectangles are TSVs, which connect metallizations between the top wafer and the bottom wafer. (a) GSG configuration with ground dual-TSVs chain, (b) GS configuration without ground dual-TSVs chain.

Two-port S-parameters (S_{11}, S_{12}, S_{21}, S_{22}) are measured as a function of frequency from DC to 30GHz using a vector network analyzer for both the dual-TSV chains and the calibration chains. During the measurement, Si substrate is chucked to DC ground. The measurement data can be analyzed based on transmission line model. Each node in the chains is modeled by

RLGC as shown in Fig. 12.24, and the frequency-dependent RLGC can be extracted from measured S-parameters:[15]

$$Z = Z_0 \sqrt{\frac{(1+S_{11})^2 - S_{21}^2}{(1-S_{11})^2 - S_{21}^2}} \qquad\qquad 12.3$$

$$\gamma = \frac{1}{d} \ln\{\frac{1-S_{11}^2 + S_{21}^2}{2S_{21}} + \sqrt{\frac{(S_{11}^2 - S_{21}^2 + 1)^2 - (2S_{11})^2}{(2S_{21})^2}}\} \qquad\qquad 12.4$$

$$R = \text{Re}\{\gamma Z\} \qquad\qquad 12.5$$

$$L = \text{Im}\{\gamma Z\}/\omega \qquad\qquad 12.6$$

$$G = \text{Re}\{\gamma/Z\} \qquad\qquad 12.7$$

$$C = \text{Im}\{\gamma/Z\}/\omega \qquad\qquad 12.8$$

where d is the number of connection node, in our case $d=32$ unit.

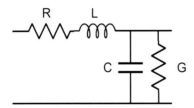

Fig.12.24 Equivalent circuit for each node used by S-parameter transmission line characterization method.

Figures 12.25(a) and (b) shows the extracted chain capacitance per node with and without ground dual-TSV chain configurations, respectively. Frequency dependent capacitance of the de-embedded dual-TSVs can be obtained by subtracting calibration chain capacitances from TSV chain capacitances. There are several interesting points worth noting. First, the de-embeded dual-TSV capacitance is obtained to be 60fF at the frequency close to DC, in consistent with low frequency capacitance measurement. Second, pronounced capacitance frequency dependences are observed for both the dual-TSV chains and calibration chains as predicted by the physical model shown in the inset of Fig. 12.25(a).

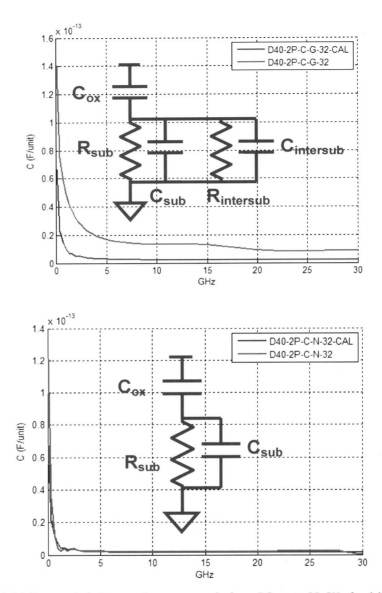

Fig.12.25 Extracted chain capacitance per node from DC up to 30 GHz for (a) GSG configuration with ground dual-TSVs chain, and (b) GS configuration without ground dual-TSVs chain. Inset: the corresponding equivalent circuits to explain frequency dependence capacitance concept and the difference between the two measurement configurations.

Moreover, the chain capacitances at high frequency (5-30 GHz) strongly depend on ground return path and the values are is only a small fraction of those at low frequency. Third, even though the capacitances of the dual-TSV chains at low frequency behave similar for the GSG and GS configurations, those at high frequency show different characteristics. For the GS configuration, the dual-TSV chain and its corresponding calibration chain have almost the same high frequency capacitance, indicating that the two structures have similar electric field coupling strength to the DC ground. For the GSG configuration, however, the dual-TSV chain exhibits a larger high frequency capacitance than its corresponding calibration chain, which is attributed to the extra capacitance of the dual-TSV to surrounding ground TSV in parallel with the capacitance of the dual-TSV to Si substrate as shown in the inset of Fig. 12.25(b). As a result, the high frequency capacitance of dual-TSV chain is higher than that of corresponding calibration chain as well as that of dual-TSV chain in GS configuration.

12.3.5 TSV Performance Prediction

The simulation is done using Spectre simulator. The input signal is pseudo-random data (PRBS15) with 50% switching activity. The rise and fall time is 10% of clock. The driver is CMOS inverter made of 6 μm NFET and 8 μm PFET using 45 nm technology with VDD=1V. Figure 12.26 shows the simulated eye diagrams of pseudo-random data transmitted over the four dual-TSVs in series with data rate of 10Gb/s (a) and 30Gb/s (b), respectively. Table 12.4 summaries vertical and horizontal eye openings, driver power consumption, and signal delay (from driver input to receiver input) for pseudo-random data transmitted over the four dual-TSVs in series at 10 GB/s, 15 GB/s, 20 GB/s and 30 GB/s. Good signal integrity (97% vertical eye opening, 63% horizontal eye opening), reasonable power consumption (1.2 mW), and short delay time (1.8 ps) can be achieved with data rate up to 30 GB/s.

The simulation also confirms that the frequency dependent capacitance is a critical design factor for TSV high frequency applications. TSV capacitance at high frequency determines the signal response at the switches, while TSV capacitance at low frequency determines the signal response later on. As a result, signal delay is mainly determined by TSV capacitance at high frequency, and eye openings and power consumption are mainly determined by TSV capacitance at low frequency.

Shown in Table 12.5, the measured and simulated RLC parameters for the dual-TSVs obtained from the work have been compared with the system-level requirements. These values meet the power delivery and interlayer signaling requirements anticipated for high-performance 3D systems.

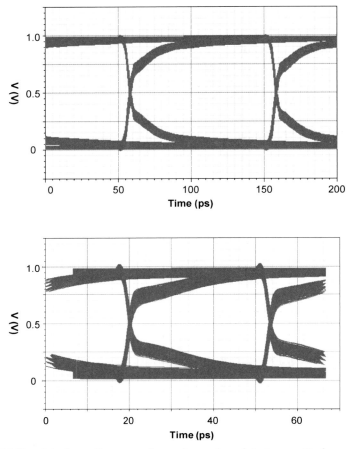

Fig.12.26 Simulated eye diagrams of pseudo-random data transmitted over the four dual-TSVs in series with data rate of 10 GB/s (a) and 30 GB/s (b), respectively.

Table 12.4 Performance prediction for pseudo-random data transmitted over the four dual-TSVs in series.

	Vertical eye opening (%)	Horizontal eye opening (%)	Driver power consumption(mW)	Signal delay (driver + TSV) (ps)
10 GB/s	80%	98%	0.8	3.3
15 GB/s	76%	98%	0.9	2.5
20 GB/s	73%	98%	1	2.1
30 GB/s	63%	97%	1.2	1.8

Table 5 A comparison between system requirements and RLC simulation and experiment values for the dual-TSVs.

	TSV Parameters	System Requirement	Simulation	Experiment
Power Delivery	Resistance	$> 1.5*10^{-3}$ $W^{-1}mm^{-2}$	$2.5*10^{-2}$ $W^{-1}mm^{-2}$ *	$2.5*10^{-2}$ $W^{-1}mm^{-2}$ *
Signal Delivery	Capacitance at low frequency	< 100 fF at low frequency	46 fF	50 fF
		< 100 fF at high frequency	2-5 fF **	2-5fF **
	Inductance	< 200 pH for digital < 26 pH for 60 GHz RF	9 pH ***	10 pH ***

12.4 RELIABILITY

12.4.1 Deep Thermal Cycle Test

The impact of thermal stress on the Cu/adhesive bond structure is modeled as shown in Figure 12.27. Figure 12.28 shows that very low thermal stresses can be achieved by appropriate selection of adhesive CTE and thickness. Our 3DI bonding structure has been carefully designed based upon these results to minimize the impact from PI CTE. To confirm modeling results, 3DI via chains have been subjected to -25 to 125°C deep thermal cycle (DTC) up to 1000 cycles. Both 80-μm-pitch via chains with 1032 connecting nodes and 28.8-μm-pitch chains with 2724 connecting nodes have been utilized.

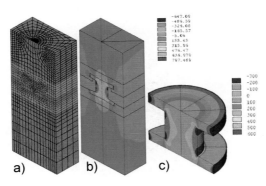

Fig.12.27 Stress map of the thermal mechanical modeling of Cu/adhesive structure: (a) modeling grid, (b) stress in 3D Si stack, (c) stress in Cu/adhesive bonding.

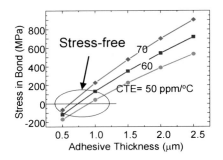

Fig.12.28 Stress modeling of Cu/adhesive structure as a function of adhesive thickness for different CTE's. Low stress zone is indicated by the black circle.

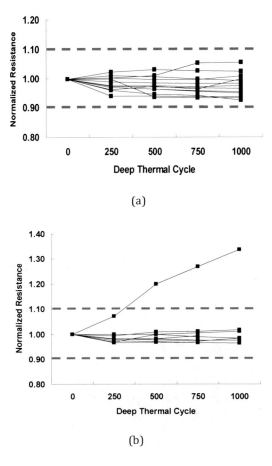

Fig.12.29 Via-chain resistance as a function of number of deep thermal cycles for (a) 80-μm-pitch chains, and (b) 28.8-μm-pitch chains.

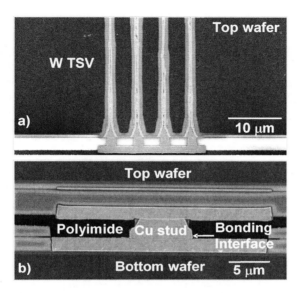

Fig.12.30 SEMs of (a) TSV and (b) bonding interface after 1000 deep thermal cycles. No stress-induced defects are observed.

The resistance change for those via chains are less than 10% except for one 28.8-μm-pitch chain as shown in Fig. 12.29. The failure via chain is inspected by I-V measurement of segmented portions of the chain and optical microscope inspection indicate the failure nodes are caused by the water condensation on the bare Cu lead connection. Furthermore, SEMs of TSVs and bonding interface of an 80-μm-pitch chain after 1000 DTC are shown in Figs. 12.30(a) and (b), respectively. No stress-induced defects are observed. The experiment suggests the technology can pass DTC requirements.

12.4.2 Temperature and Humidity

Another key reliability issue is the resistance of the bonded structure to ambient or humidity-induced oxidation. To evaluate the moisture absorption related issues, the via-chains have been subjected to a temperature and humidity (T&H) stress test up to 1000 hours at 85°C and 85% relative humidity. The top BEOL Cu connections are protected with a 1 μm thick BCB coating. After 200 hours, almost no change in via chain resistance is observed. However, after 1000 hours, resistance degradation becomes obvious as shown in Fig. 12.31. The degradation is likely due to insufficient top-side BCB protection, since slight surface oxidation is observable for the top Cu connection shown in Fig. 12.32(a). However, no abnormality is seen in the TSVs and bonding interface shown by SEM in Fig. 12.32(b).

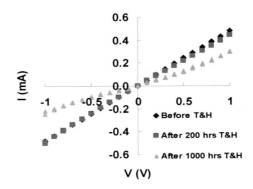

Fig.12.31 I-V characteristics of a 28.8-μm-pitch via chain for various T&H times.

Fig.12.32 (a) Photo image of top surface Cu BEOL with BCB protection, and (b) SEM of a 28.8-μm-pitch chain after 1000-hour T&H.

To understand the impact of T&H to adhesive material, FTIR spectra of blank PI are collected. Figure 12.33 shows transmission measurements through the bulk of the film for 0 hours stress sample, 500 hours T&H stress sample, and 1000 hours T&H stress sample (a), and attenuated total reflectance (ATR) measurements with sampling depth of 600 nm for 0 hours stress sample, 500 hours T&H stress sample, and 1000 hours T&H stress sample (b), respectively. The FTIR spectra show no significant differences among the three samples from both measurements.

To further understand ambient air oxidation of Cu, the oxidation rates of Cu for various structures are compared with blanket unpassivated Cu at elevated temperatures as indicated in Fig. 12.33. For Cu-Cu bonding without seal, the rate is only slightly lower than blanket Cu. This is due to gaps between the interfaces which allow the ambient O_2 to permeate the structure. For the hybrid Cu/PI adhesive approach, the oxidation rate is reduced by nearly two orders of magnitude and no difference in Cu oxidation is observed between a PI sealed bond and an oxide-oxide sealed bond. These results offer promise that, with further process optimization, the molecular sealing provided by the adhesive bonding could sufficiently slow the rate of oxygen ingress to the Cu bonds to meet stringent manufacturing T&H stress criteria.

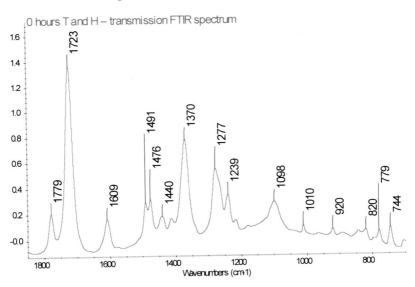

peak position, cm-1	Assignment
1779	imide C=O symmetric stretch
1723	imide C=O asymmetric stretch
1609	aromatic skeletal ring breathing vibrations
1491	"
1476	"
1440	"
1370	cyclic imide C-N stretch
1277	ether -C-O-
1239	"
1098	in-plane C-H bending on aromatic rings
1010	"
920	out-of-plane C-H bending on aromatic rings
820	"
779	"
744	imide C=O bending

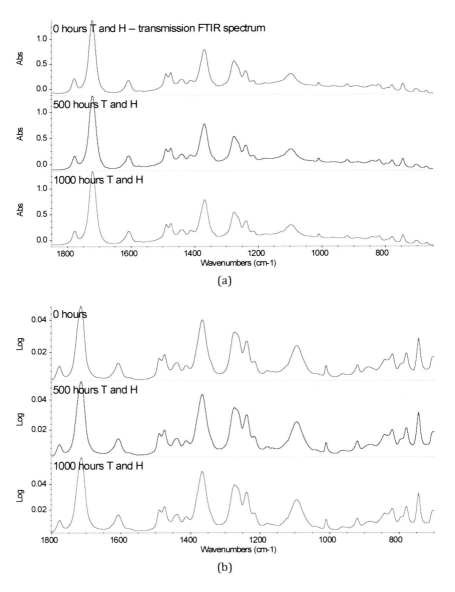

Fig.12.33 (a) Transmission measurements through the bulk of the film for 0 hours stress sample, 500 hours T&H stress sample, and 1000 hours T&H stress sample. (b) Attenuated total reflectance (ATR) measurements with sampling depth of 600nm for 0 hours stress sample, 500 hours T&H stress sample, and 1000 hours T&H stress sample, respectively. The FTIR spectra show no significant differences among the three samples from both transmission measurement and ATR measurement.

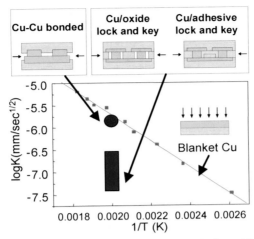

Fig.12.34 Arrhenius plot of oxidation rate for various Cu-based bonding structures compared to blanket Cu.

12.5 OUTLOOK

A particular integration scheme may be more beneficial for certain applications than the others. However, CMOS-compatibility, manufacturability, reliability, and cost are critical considerations. By exploring new processes and tooling, further refining of the key process steps (i.e., TSV formation and metallization, bonding technology, wafer thinning and finishing process) are expected. Inline non-destructive metrologies are needed to realize controllable and repeatable 3D offers. Reliability of TSV, bonding interface, and their impacts to active circuits demands further study. At the same time, the understanding the system impacts of TSV and precisely extracting its small signal model may require innovations on the 3DI testing structures and strategies.

12.6 SUMMARY

A general-purpose CMOS-compatible 300-mm wafer-level 3D integration process is successfully demonstrated by combining high aspect ratio tungsten TSVs at tight pitches and a transfer-join adhesive/metal hybrid connection method to form a robust bonding interface. Resistance, inductance and capacitance of the tungsten TSVs can meet interlayer power and signaling requirements for high-performance 3DI systems. Moreover, the tungsten TSV and hybrid Cu-adhesive bonding is a promising approach to meet the wide variety of reliability specifications required for a 3D integration technology, including high mechanical strength, and robustness against thermal-induced stress and oxidation.

Acknowledgements

This work was partially supported by DARPA under contract # N66001-04-C-8032.

References

1. F. Liu, R. R. Yu, A. M. Young, J. P. Doyle, X. Wang, L. Shi, K.-N. Chen, X. Li, D. A. Dipaola, D. Brown, C. T. Ryan, J. A. Hagan, K. H. Wong, M. Lu, X. Gu, N. R. Klymko, E. D. Perfecto, A. G. Merryman, K. A. Kelly, S. Purushothaman, S. J. Koester, R. Wisnieff, and W. Haensch, International Electron Devices Meeting (IEDM), 2008.

2. R. R. Yu, F. Liu, R. J. Polastre, K.-N. Chen, X. H. Liu, L. Shi, E. D. Perfecto, N. R. Klymko, M. S. Chace, T. M. Shaw, D. Dimilia, E. R. Kinser, A. M. Young, S. Purushothaman, S. J. Koester, and W. Haensch, Symposium on VLSI Technology, 2009.

3. P. Morrow, M. J. Kobrinsky, S. Ramanathan, C.-M. Partk, M. Harmes, V. Ramachandrarao, H.-M. Park, G. Kloster, S. List, and S. Kim, *Proceedings of the UC Berkeley Extension Advanced Metallization Conference (AMC)*, 2004.

4. B. Swinnen, W. Ruythooren, P. De Moor, L. Bogaerts, L. Carbonell, K. De Munck, B. Eyckens, S. Stoukatch, D. Sabuncuoglu Tezcan, Z. Tőkei, J. Vaes, J. Van Aelst, E. Beyne, *International Electron Devices Meeting (IEDM), 2006.*

5. H. J. Tu, W. J. Wu, J. C. Hu, K. F. Yang, H. B. Chang, W. C. Chiou and C. H. Yu, *International Interconnect Technology Conference(IITC), 2008.*

6. C. K. Tsang, P. S. Andry, E. J. Sprogis, C. S. Patel, B. C. Webb, D. G. Manzer, J. U. Knickerbocker, *Material Research Society, (MRS)*, Fall 2006.

7. R. Yu, *VLSI/ULSI Multilevel Interconnection Conference, (VMIC)*, 2007.

8. F. Liu, X. Gu, K. A. Jenkins, E. A. Cartier, Y. Liu, P. Song, and S. J. Koester, Electronic Components and Technology Conference (ECTC), 2010.

9. B. Swinnen, W. Ruythooren, P. De Moor, L. Bogaerts, L. Carbonell, K. De Munck, B. Eyckens, S. Stoukatch, D. Sabuncuoglu Tezcan, Z. Tőkei, J. Vaes, J. Van Aelst, E. Beyne, International Electron Devices Meeting (IEDM), 2006.

10. B. Vandevelde, C. Okoro, M. Gonzalez, B. Swinnen, E. Beyne, International Conference on Thermal, Mechanical and Multi-Physics Simulation and Experiments in Microelectronics and Micro-Systems, (EuroSimE), 2008.

11. R. Yu, The International Conference on Electronic Packaging Technology & High Density Packaging (ICEPT-HDP), 2008.

12. Keith A. Jenkins, and Chirag S. Patel, International Interconnect Technology Conference (IITC), 2005.

13. Chunghyun Ryu, Jiwang Lee, Hyein Lee, Kwangyong Lee, Taesung Oh, and Joungho Kim, Proc. Electronics System integration Technology Conference, Dresden, Germany, 2006

14. C. Bermond, L. Cadix, A. Farcy, T. Lacrevaz, P. Leduc, B. Fléchet, Proc. Electronic Components and Technology Conference (ECTC), 2008.

15. William R. Eisenstadt, and Yungseon Eo, IEEE Transactions on Component, Hybrids, and Manufacturing Technology, Vol 15, No 4, Page 483-490, 1992.

Chapter 13

POWER DELIVERY IN 3D IC TECHNOLOGY WITH A STRATUM HAVING AN ARRAY OF MONOLITHIC DC-DC POINT-OF-LOAD (POL) CONVERTER CELLS

Ronald J. Gutmann and Jian Sun

Rensselaer Polytechnic Institute

Wafer-level three-dimensional (3D) integration offers the potential of enhanced performance and increased functionality, combined with the low manufacturing cost inherit from monolithic IC processing. This chapter addresses the problem of power delivery in microprocessors, application-specific ICs (ASICs) and system-on-a-chip (SoC) implementations, based upon a wafer-level 3D technology platform with arrays of monolithic DC-DC converter cells in one stratum providing power locally to the signal electronics strata. The chapter consists of five main sections: (1) a technical background of the increasing difficulty of delivering power to advanced ICs and the limitations of current technologies, (2) a brief review of 3D platforms that enable implementation of this power delivery architecture, (3) a description of the design of a prototype completely-monolithic, two-phase, point-of-load (PoL) buck converter cell, (4) a summary of the fabrication (using a 180 nm SiGe BiCMOS foundry process) and performance of the converter cell, and (5) a discussion of technologies and designs offering the promise of improved monolithic power converter performance. This power delivery architecture provides extreme power delivery flexibility, including dynamic control, at the expense of an additional stratum in a 3D chip stack. In addition, the power quality is controlled within the 3D chip stack, and the number of I/O power and ground pins is reduced.

3D Integration for VLSI Systems
Edited by Chuan Seng Tan, Kuan-Neng Chen and Steven J. Koester
Copyright © 2012 by Pan Stanford Publishing Pte. Ltd.
www.panstanford.com

13.1 TECHNICAL BACKGROUND

Power delivery has become an important concern in the design of ICs, both as the power density of high performance microprocessors, ASICs and SoCs has increased with minimum feature size below 100 nm, and as the number of voltage levels increased to both minimize the total power dissipation of digital electronics and to optimize performance of mixed-signal functions. Conventional implementations incorporate DC-DC converters and/or voltage regulator modules (VRMs) near the high performance IC chips, with power delivery to the IC using multiple power and ground pins as depicted in Fig. 13.1. Power integrity is achieved by distributing decoupling capacitors in both the package and IC, resulting in a complex L-C-R network between the PoL converter (or VRM) and the signal electronics.

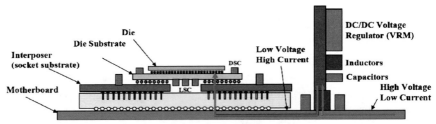

Fig. 13.1 Conventional 2D power delivery architecture using a VRM mounted on the motherboard

A simplified power delivery network for Intel Pentium4 processors at the 90 nm technology node is shown in Fig. 13.2.[1] The parasitic inductance of the long interconnect line between the VRM and processor generates large di/dt noise, forcing the use of a large number of decoupling capacitors at various locations along the power delivery path. The improvement in both power devices and passive components has enabled the development of more compact and efficient VRMs that can be mounted in close proximity to the processor core to reduce interconnect parasitic inductance, thereby reducing the amount of required decoupling capacitance. However, the decoupling capacitors remain a critical design constraint. Beside cost implications, there will also not be enough space around the processor core for these capacitors in the future as the supply voltage continues to decrease and the current slew rate keeps increasing. Another problem with the existing 2D power delivery platform is the large number of power and ground pins required, which increases the complexity of the first-level package.

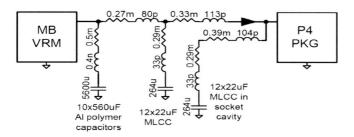

Fig. 13.2 Simplified power delivery network for Intel Pentium4 processor on 90 nm process showing the use of large number of decoupling capacitors in order to maintain voltage regulation at the processor.[1]

The recent trend toward multi-core processors has further increased the demands on microprocessor power delivery. In order to enhance processor speed without increasing IC power dissipation, multi-core processors require fine-grain power control (FGPC). The FGPC needs to be dynamically controllable (i.e. temporally) and variable from core-to-core (i.e. spatially). The rapid temporal control requires a high frequency network between the PoL converter and the microprocessor core, which is limited in conventional approaches such as depicted in Figure 1 by the inductance in the 1-10 cm-length conductors. In comparison, the conductor length in the proposed 3D power delivery architecture is approximately half the core length, or ~ 1 mm. In addition, the variation in core-to-core power is best achieved with separate converters (i.e. converter cells) for each core to enable the most design flexibility. An array of DC-DC PoL converter cells in one stratum of a 3D stack is an ideal architecture for prime power delivery for multi-core processors or ASICs requiring FGPC.

13.2 WAFER-LEVEL 3D TECHNOLOGY PLATFORMS AND RESULTING POWER DELIVERY ADVANTAGES

Wafer-level three-dimensional technology platforms require four main process steps: wafer-to-wafer alignment, wafer bonding, wafer thinning, and micron-sized inter-wafer interconnections. The principal differentiator is the type of wafer bonding, either metal-to-metal (usually Cu-Cu) bonding, oxide bonding or adhesive bonding. Important further differentiators are (1) via-first or via-last process flows and (2) either face-to-back or face-to-face bonding with or without a handling wafer. These and other generic platform considerations are discussed elsewhere.[2-3]

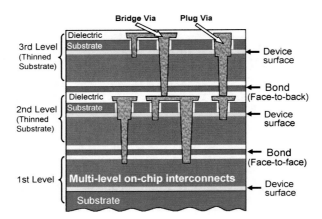

Fig. 13.3 Three-wafer stack Cu/BCB 3D technology platform with via-last with blanket BCB bonding and interwafer interconnects after top-wafer thinning.[4]

Two platforms established by Rensselaer, depicted in Figs. 13.3 and 13.4 for three-wafer stacks, are based on adhesive wafer bonding using partially-cured benzocyclobutene (BCB) with face-to-face bonding, Cu inter-wafer interconnects and without a handling wafer. A via-last process flow with blanket BCB bonding, shown in Fig. 13.3, requires processing of inter-wafer interconnects after bonding and top wafer thinning. A simplified process flow that employs patterned Cu/BCB bonding or redistribution layer bonding (RLB) and results in even a higher density of through-silicon vias (TSVs), but with additional constraints on pre-bonding planarity, is shown in Fig. 13.4. Results to date indicate that planarity conditions can be achieved with conventional CMP-based planarization. The potential of the latter platform is attractive if either metal-metal bonding strength is not sufficient for package processes and system applications or for prototyping of 3D designs. Both platforms have been summarized elsewhere,[4] and various applications enabled by wafer-level 3D platforms pursued by our group summarized elsewhere as well.[5]

The potential for the via-first platform is considered particularly attractive in a variety of applications. In the near term, the patterned Cu/BCB bonding (or RLB) provides a transition between the larger via dimensions and lower pitch in wafer-level chip-scale 3D packages. Thus, the approach provides an orderly transition from the tens of microns in wafer-level packaging to the micron-scale in wafer-level ICs. In the long term, the patterned Cu/BCB bonding is a manufacturing alternative if the lower bonding strength of Cu-Cu bonding proves inadequate for low-cost robust IC manufacturing; if such capability is not needed, the patterned Cu/BCB bonding platform could be useful as a development platform due to the inherent redistribution layer provided between strata.

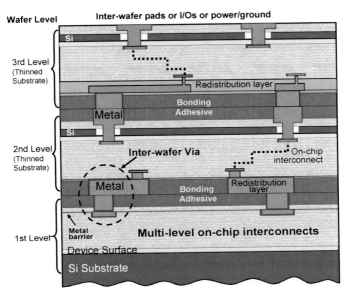

Fig. 13.4 Three-wafer stack Cu/BCB 3D technology platform with via-first Cu/BCB redistribution layer bonding.[4]

Note that a typical microprocessor architecture with three strata would have the processor core and first-level cache in the lower stratum nearby the heat sink (not shown), with second-level cache in the middle stratum and an array of completely monolithic PoL converters in the top stratum. As discussed later in the manuscript, an option with significant performance advantages would include an additional stratum containing only high-performance passives on an insulating substrate for a two-strata PoL converter. Such an implementation is analogous to the "power supply on chip" developed commercially, with monolithic implementation of the power devices and control electronics but with externally mounted passives. Here the separately fabricated passives are also processed with monolithic IC processing, but with an insulating substrate and high processing temperature compatibility.

Compared to existing 2D power delivery methods, the 3D architecture offers several important advantages:

1. minimum interconnect parasitics between the power supply and the processor;
2. easy generation and distribution of multiple, individually regulated supply voltages;
3. wide control bandwidth of high-frequency monolithic converters enabling dynamically controllable supply voltages;

4. significantly reduced number of power and ground pins in the stacked die package;
5. minimize the effect of the first-level package power distribution network on the power delivery to the signal electronics in the 3D stack, due to the PoL conversion and decoupling provided in the power conversion stratum;
6. low cost in high volume production and high reliability, due to the monolithic IC-type interconnectivity.

Advantages 3 and 4 are especially important for multi-core processors, where fine-grain (both spatially and temporally) power delivery is important to minimize power consumption.

Appreciable completely monolithic converter design and technology need to be developed to fully explore this power delivery platform. In our research, a prototype, monolithic DC-DC converter was designed, fabricated and evaluated, with the main purposes of establishing an experimental baseline and delineating future needs. The design is presented in Section 13.3, the prototype fabrication and electrical performance presented in Section 13.4, and future technology needs and performance expectations discussed in Section 13.5.

13.3 PROTOTYPE MONOLITHIC DC-DC CONVERTER DESIGN

A monolithic DC-DC converter cell was designed using a Si CMOS-based foundry so that the MOSFETs can be operated at much higher switching frequencies than typical converters and VRMs. While design simulations indicated that the desired 90% conversion efficiency could not be obtained in any monolithic approach with the power density desired for multi-core processors, this baseline prototype cell would allow technology and design limitations to be quantified. Key features of the design are described here, with additional information published previously.[6-7]

A two-cell interleaved buck converter with linear feedback control was designed and fabricated using a 180 nm SiGe BiCMOS process. The converter operates at a switching frequency near 200 MHz with a control bandwidth near 10 MHz. The BiCMOS process has enhanced passive components compared to lower-cost Si CMOS processes available to us. The power stage was implemented with CMOS devices, with bipolars only used to simplify design of the control loop.

Power Stage Design

A simplified circuit diagram of the buck converter cell emphasizing the power flow, including the gate drivers, the control switch, the synchronous rectifier, the output filter inductor (L) and capacitor (C) and an internal active load to allow converter testing, is shown in Fig. 13.5. Each such cell is designed to convert an input voltage of 1.8 V to an output voltage of 0.9 V at a nominal current of 500 mA.

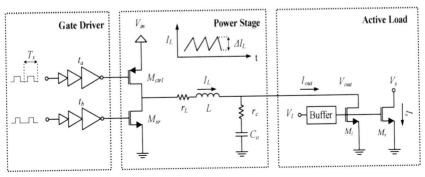

Fig. 13.5 Circuit diagram of a buck converter cell.

A switching frequency of 200 MHz was targeted to limit the size of the on-chip inductors and capacitors. The PMOS control switch was designed with an equivalent width of 16.6 mm and an on-resistance of 152 mohms, while the NMOS synchronous rectifier has an equivalent width of 11.0 mm and an on-resistance of 62 mohms. Large equivalent-width devices were needed to handle the current levels and provide low on-resistance, with an extensive number of interlevel vias needed to minimize interconnect resistance. The output 8.22 nF capacitor (with 1 mohm effective series resistance (ESR)) was implemented with MOS structures because of the high capacitance density.

An air spiral inductor with two shunted metal layers was selected for the on-chip inductor L, with maximum width of the spiral tracks (25 um) and minimum track spacing (5 um) to minimize the parasitic resistance. A 3.5 turn spiral with an outer diameter of 290 um provides an inductance of 2.14 nH with a DC resistance of 200 mohms. A patterned metal ground plane under the inductor reduces capacitive substrate losses and boosts inductor Q by approximately 30%. Table 13.1 summarizes the parameters of the major power stage components.

Table 13.1 Parameters of key power stage components.

Components	Design Parameters	Equivalent Resistance
Control Switch	Width = 16.6 mm	$R_{DS(ON)}$ = 152 mΩ
Syn. Rectifier	Width = 11 mm	$R_{DS(ON)}$ = 62 mΩ
Inductor	L = 2.14 nH	R_{DC} = 201 mΩ
Capacitor	C = 8.22 nF	R_{ESR} = 1 mΩ

Since the primary objective of the design was to provide a prototype for performance evaluation, an on-chip active load was incorporated to all fast transient load currents to be evaluated, thereby emulating actual loading by a microprocessor or high-performance ASIC. The active load is a large MOSFET operating in the saturation region and acting as a current sink. A current buffer is used to drive the MOSFET such that fast load transients (10-100 A/us) can be generated.

Compensator and Control Loop Designs

Control designs for on-chip dc-dc converters based on hysteretic [8] and peak-current control[9] have been reported in the literature. Both methods are nonlinear and avoid the need for a wide bandwidth compensator as well as a pulse-width modulator (PWM), making these methods attractive for high-frequency converters. However, hysteretic control is very difficult to synchronize, hence is not suitable for applications where the operation of multiple converter cells needs to be coordinated (e.g. interleaved as in our design). Peak-current control uses a clock signal to initiate the turn-on of the switch, making it possible to coordinate the operation of multiple converter cells. Peak-current control also facilitates current-sharing control among parallel converters, a feature which is important for conventional VRM design. This, however, is not critical for monolithic converters where the design of parallel cells can be matched to ensure equal current sharing without active control. In terms of control performance, dynamic responses of peak-current control to load current changes are slower than that under voltage-mode control.[10] Considering these, a voltage-mode control was selected.

To maximize control bandwidth, a voltage feedback loop is typically closed at a frequency above the power stage resonant frequency. However, since our power stage design uses very small inductor and capacitor, placing the loop crossover frequency above the resonant frequency makes the loop very difficult to stabilize. Considering this, we placed the loop crossover frequency at 10 MHz, with a phase margin of 95°. This is a relatively conservative design for a 200 MHz converter, but still provides sufficient bandwidth for control of the output voltage, as presented in Section 13.4.

A compensator is traditionally implemented using an operational amplifier (op amp) in an inverting configuration, with two complex impedances realized using capacitors and resistors. This approach works well at low frequencies where the op amp can be treated as an ideal component, but is not suitable for high-speed compensators where the inherent poles and zeros of the op amp are close to the required external poles and zeros. An integral design approach is employed in our design to realize the required compensator transfer function.

Fig. 13.6 shows a simplified diagram of the implemented voltage compensator. The locations of the poles and zeroes (in rad/s) of this circuit are defined by the following expressions where C_1 is the internal capacitance at the output of the first stage of the op amp (differential pair):

$$\omega_z = \frac{1}{C_c(g_{M_s}^{-1} - R_z)},$$

$$\omega_{p1} \approx \frac{-1}{C_c R_1 (1 - g_{M_s} R_2)},$$

$$\omega_{p2} \approx \frac{-g_{M_s} C_c}{C_1 C_2 + C_c(C_1 + C_2)}$$

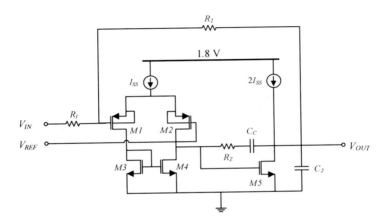

Fig. 13.6 A simplified diagram of the compensator design.

In order to realize the designed compensator transfer function, a compensation capacitor (C_c) of 66 pF, a feed-through resistor (R_z) of 84 Ω, and a load capacitor (C_2) of 8 pF were chosen. A gain of 36 dB is set by resistors R_1 and R_2, yielding a steady-state error of 12 mV. Since the output impedance of the op amp is 1.4 kΩ, R_2 must be larger to prevent loading; R_1 and R_2 were chosen to be 1 kΩ and 100 kΩ, respectively.

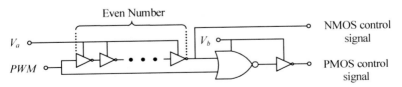

Fig. 13.7 Circuit for controlling the dead time between the control switch and the synchronous rectifier.

The control circuitry also includes a high-speed pulse-width modulator and an active dead-time controller. The modulator employs a high-gain (52 dB) comparator designed by cascading numerous high-speed low gain stages. The dead time between the control switch and the synchronous rectifier is controlled by a series of inverters and NOR gates, shown in Fig. 13.7. The dead time can be adjusted by changing voltages V_a and V_b.

13.4 FABRICATED PROTOTYPE AND PERFORMANCE

The two-phase interleaved buck converter prototype using the designs outlined above and implemented in a 180 nm SiGe BiCMOS process is shown in Fig. 13.8. The power stage was duplicated to give the two phases, while a common control loop (including pulse-width modulator and dead-time controller) is used by both phases. The gate control signals for the two phases are shifted by half a switching cycle to achieve maximum ripple cancellation.

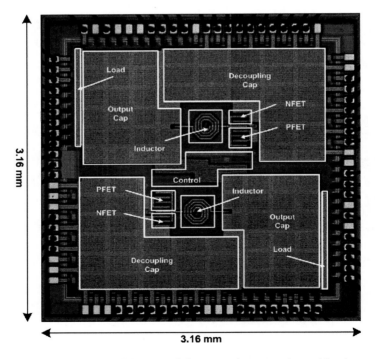

Fig. 13.8 Micrograph of the fully monolithic two-phase interleaved buck converter test chip.

The total chip area is about 10 mm^2, including input/output pads, test vias and the on-chip active load for dynamic response testing. The area is dominated by capacitors (~58%), almost half being the output capacitors; all other converter components combined (including power MOSFETs, on-chip inductor, control circuitry and active load) consume ~11%, with the bond pads and electrostatic discharge (ESD) protection circuitry ~31%.

The fabricated converter chips were measured to evaluate both steady-state and dynamic performance. Since the second phase contains probe pads which significantly degraded electrical performance, only one-phase performance (effective chip area of 5 mm^2) has been fully characterized and presented here.

The measured efficiency of the converter under different switching frequencies and load conditions is presented in Fig. 13.9. The efficiency is relatively independent of switching frequency from 160 MHz to 220 MHz with a modest decrease with increasing output current; to first order, the input and output voltages were held constant at 1.8 V and 0.9 V, respectively. Maximum efficiency of 64% is achieved at 200 MHz with an output current of 500 mA for the one-phase circuit (or 200 mA/mm^2).

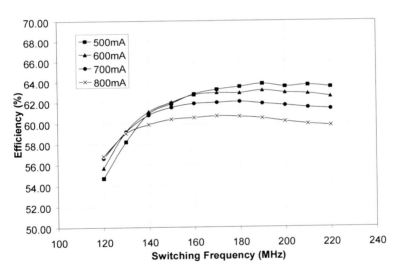

Fig. 13.9 One-phase efficiency measured under variable load and switching frequency.

The output voltage transient response to a step-up in the load current is shown in Fig. 13.10, with a current step of 225 mA and slew rate of 10 A/us. The voltage shows an overshoot of approximately 88 mV and returns to the desired level in 86 ns. The output voltage decreases approximately 6 mV at the higher current due to the relatively small DC gain of the compensator (34 dB). Response to a step-decrease of the load current is similar. Overall, the one-phase baseline converter responds to a current transient with a slew rate of 10 A/us while keeping the output voltage within a window of approximately 225 mV.

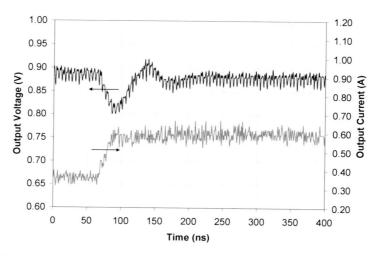

Fig. 13.10 One-phase dynamic response of output voltage measured with a load current step of 225 mA at a slew rate of 10 A/us.

13.5 FUTURE DIRECTIONS FOR PERFORMANCE ENHANCEMENT

The prototype converter results demonstrate that a monolithic converter has the potential to meet the dynamic regulation requirements of microprocessors, ASICs and SoCs, due to the low device and interconnect impedance possible with monolithic IC processing and the fast control loop enabled by high-frequency switching. However, the static performance of the baseline converter cell needs to be improved significantly as discussed in this section.

The output current density of 200 mA/mm², corresponding to 20 A/cm², of the prototype converter cell is within an order of magnitude of future needs. For example, with a microprocessor array of 64 cores (8×8 array), the prototype two-phase converter cell could provide 64 A of current with a converter stratum area of 440 mm² (excluding the bond wire area of the prototype). A reasonable goal would be 80-100 mm², or ~1 A/mm² current density for multi-core microprocessors, although future technology nodes have not been considered in depth. However, the area of the prototype is dominated by bond pads and ESD circuitry (31%) and capacitor requirements (58%). Clear paths to increasing the output current density to ~1 A/mm² include multi-phase converter architectures and higher density capacitors (as well as the obvious reduction in bond pad area and ESD circuitry for each converter cell). For example, it was estimated in Ref. 6 that, by taking advantage of ripple cancellation under interleaved PWM control, the required total input filtering capacitance of ten 500 mA converter cells can be smaller than that required by a single converter cell. Fig. 13.11 compares the input current ripple of a single converter switched at 100 MHz and providing 500 mA output current with that of 10 such converter cells operated in parallel and under interleaved PWM control. The total peak-peak ripple current of 10 interleaved converter cells is smaller than that of a single converter while its frequency is increased by 10 times (to 1 GHz). Output current density, therefore, is not considered to be a major factor limiting implementation of the proposed PoL converter array.

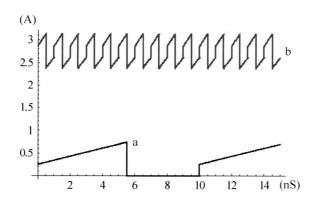

Fig. 13.11 Comparison of theoretical input current waveforms of a 2 V input, 1 V/500 mA output converter cell (a) with that of 10 such converter cells operated in parallel and under interleaved PWM control (b). A 55% duty ratio is assumed in the analysis.

However, the 64% conversion efficiency for the 2:1 buck converter is significantly lower than a conventional converter/VRM ($\sim$ 95%), and significant improvement is needed. Three promising options are to (1) include a second passive stratum with only passive components (i.e. inductors and capacitors), (2) incorporate MOSFETs with device structure and parameters optimized for power device performance, and (3) consider alternative converter topologies. With a passive component stratum, the monolithic converter would be implemented in two strata as illustrated in Fig. 13.12: (1) a silicon substrate with active devices, both the MOSFETs in the power train and semiconductor devices and passives for the control and sensor circuitry (DC-DC wafer) and (2) a dielectric substrate (e.g. glass with a temperature coefficient of expansion (TCE) matched to silicon for bond integrity) with inductors and capacitors for the power train (passive wafer). With such an implementation, thicker and wider conductors could be incorporated in the passive layer for reduced inductor loss, and higher dielectric constant films implemented to reduce capacitor area. We project an efficiency of as high as 75-80% is possible for the 2:1 buck converter, with an output current density of 400 mA/mm^2. Incorporating power device design principles in the PoL active stratum, an efficiency of 80-85% is projected, although additional design work is needed to better quantify the obtainable conversion efficiency.

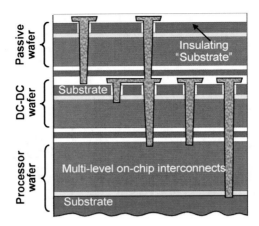

Fig. 13.12 Two-strata implementation of the monolithic converter with inductors and capacitors on a passive stratum with insulating substrate.

Use of new converter topologies may also significantly increase the cell conversion efficiency, hopefully with an increase in output current density as well. While the use of coupled inductors[11] and a two-stage conversion circuitry are promising pursuits, a breakthrough in conversion efficiency is needed to enable the proposed monolithic converter cell topology to be compatible with future multi-core processor requirements.

13.6 SUMMARY AND DISCUSSIONS

Wafer-level 3D cellular power delivery architectures overcome the limitations of conventional 2D approaches by eliminating most of the interconnect parasitics between the PoL converter and the IC. Fully monolithic DC-DC converters compatible with a 3D platform appear to be capable of meeting the needs of future microprocessors, ASICs and SoCs with dynamically scalable, sub-1 V supply voltages. While the 3D platform might accommodate less efficient PoL converters and voltage regulators, increased conversion efficiency is needed in second-generation prototypes. A passive stratum containing high permeability inductors and high permittivity capacitors on an insulating substrate offers the potential for increased efficiency and higher output current density, although a new power conversion circuit topology may be needed to achieve acceptable conversion efficiencies. The cellular architecture enables the full benefits of interleaving to reduce filtering requirements and is ideally suited for 3D integration and fine-grain power control required by future multi-core processors.

Acknowledgments

This work was supported by the Interconnect Focus Center (IFC) sponsored by MARCO, DARPA and NYSTAR, the Center for Power Electronics Systems (CPES) sponsored by NSF and the RPI Broadband Center sponsored by IBM.

References

1. G. Schrom, P. Hazucha, J. Hahn, V. Kursun, D. Gardner, S. Narenda, T. Karnick, and V. De, "Feasibility of Monolithic and 3D-Stacked DC-DC Converters for Microprocessor in 90 nm Technology Generation," in Proc. ISLPED, 2004, pp. 263-268.

2. C.S. Tan, R.J. Gutmann and L.R. Reif, "Overview of Wafer-Level 3-D ICs", Chapter 1 in C.S. Tan, R.J. Gutmann and L.R. Reif, editors, "*Wafer-Level Three-Dimensional (3-D) IC Process Technology*", Springer, 2008.

3. P. Garrou, C. Bower and P. Ramm, "*Handbook of 3D Integration: Technology and Applications of 3D Integrated Circuits*", Wiley, 2008.

4. J.Q. Lu, T.S. Cale and R.J. Gutmann, "Adhesive Wafer Bonding Three-Dimensional (3D) Technology Platforms", Chapter 9 in C.S. Tan, R.J. Gutmann and L.R. Reif, editors, "*Wafer-Level Three-Dimensional (3-D) IC Process Technology*", Springer, 2008.

5. R.J. Gutmann and J.Q. Lu, "Wafer-Level Three-Dimensional Integration for Advanced CMOS ICs", Chapter 2.6 in K. Iniewski, editor, "*VLSI Circuits for NanoEra: Communications, Imaging and Sensing*", CRC Press, 2008.

6. J. Sun, J.-Q. Lu, D. Giuliano, T.P. Chow and R.J. Gutmann, "3D Power Delivery for Microprocessors and High-Performance ASICs", in *Proceeding of IEEE 2007 Applied Power Electronics Conference* (APEC), pp. 127-133, 2007.

7. J. Sun, D. Giuliano, S. Devarajan, J.-Q. Lu, T.P. Chow and R.J. Gutmann, "Fully Monolithic Cellular Buck Converter for 3D Power Delivery", *IEEE Transactions on VLSI Systems*, Vol. 7, No. 3, March 2009, pp. 447-451.

8. G. Schrom, P. Hazucha, J. Hahn, D. Gardner, B. Bloechel, G. Dermer, S.Narendra, T. Karnik and V. De. "A 480-MHz, Multi-Phase Interleaved Buck DC-DC Converter with Hysteretic Control," in *Records of 2004 IEEE Power Electronic Specialists Conference*, pp 4702-4707.

9. C. F. Lee and P. K. T Mok, "A Monolithic Current-Mode CMOS DC-DC Converter with On-Chip Current-Sensing Technique," *IEEE Jour. Solid State Circuits*, Vol. 39, No. 1, January 2004, pp. 3-4.

10. J. Sun, "Control Design Considerations for Voltage Regulator Modules," in *Proceedings of IEEE INTELEC'03*, 2003, pp. 84-91.

11. J. Wibbin and R. Harjani, "A High-Efficiency DC-DC Converter using 2 nH Integrated Inductors," IEEE Jour. Solid State Circuits, Vol. 43, No. 4, April 2008, pp. 845-855.

Chapter 14

THERMAL-AWARE 3D IC DESIGNS

Xiaoxia Wu, Yuan Xie and Vijaykirshnan Narayanan

Penn State University

Power and thermal issues have become the primary concerns in traditional 2D IC design. Although emerging 3D technology offers several benefits over 2D, the stacking of multiple active layers in 3D design leads to higher power densities than its 2D counterpart, exacerbating the thermal issue. Therefore, it is essential to conduct thermal-aware 3D IC designs. This chapter presents an overview of thermal modeling for 3D IC and outlines solution schemes to overcome the thermal challenges at Electrical Design Automation (EDA) and architectural levels.

14.1 INTRODUCTION

3D technology offers several benefits compared to traditional 2D technology. Such benefits include: (1) The reduction in interconnect wire length, which results in improved performance and reduced power consumption; (2) Improved memory bandwidth, by stacking memory on microprocessor cores with TSV connections between the memory layer and the core layer; (3) The support for realization of heterogeneous integration, which could result in novel architecture designs; (4) Smaller form factor, which results in higher packing density and smaller footprint due to the addition of a third dimension to the conventional two dimensional layout, and potentially results in a lower cost design. However, one of the major concerns in the adoption of 3D technology is the increased power densities that can result from placing one power hungry block over another in the multi-layered 3D stack. Since the

3D Integration for VLSI Systems
Edited by Chuan Seng Tan, Kuan-Neng Chen and Steven J. Koester
Copyright © 2012 by Pan Stanford Publishing Pte. Ltd.
www.panstanford.com

increasing power density and the resulting thermal impact are already major concerns in 2D ICs, the move to 3D ICs could accentuate the thermal problem due to increased power density, resulting in higher on-chip temperatures. High temperature has adverse impacts on circuit performance. The interconnect delay becomes slower while the driving strength of a transistor decreases with increasing temperature. Leakage power has an exponential dependence on the temperature and increasing on-chip temperature can even result in thermal runaways. In addition, at sufficiently high temperatures, many failure mechanisms, including electromigration (EM), stress migration (SM), time-dependent dielectric (gate oxide) breakdown (TDDB), and thermal cycling (TC), are significantly accelerated, which leads to an overall decrease in reliability. Consequently, it is very critical to model the thermal behaviors in 3D ICs and investigate possible solutions to mitigate thermal problems in order to fully take advantage of the benefits that 3D technologies offer.

This chapter reviews recent efforts on thermal modeling for 3D ICs and outlines solution schemes to overcome the thermal challenges. We first focus on thermal modeling in Section 14.1.2. Then, in Section 14.1.3, we present several thermal-aware EDA solutions such as thermal-aware floorplanning, placement, and routing. Section 14.1.4 reviews several architecture solutions that help mitigate thermal impacts in 3D ICs.

14.2 THERMAL MODELING FOR 3D ICs

This section surveys several thermal modeling approaches for 3D ICs. A detailed 3D modeling method is first introduced, and then compact thermal models are introduced. An overview of a widely used architecture-level thermal tool, called HotSpot, and its 3D extension are described. Then a detailed 3D modeling is presented.

14.2.1 A detailed 3D thermal model

Sapatnekar *et al.* proposed a detailed 3D thermal model.[5] The heat equation (1), which is a parabolic partial differential equation (PDE) defines on-chip thermal behavior at the macroscale.

$$\rho C_p \frac{\partial T(r,t)}{\partial t} = k_t \nabla^2 T(r,t) + g(r,t) \tag{14.1}$$

where p represents the density of the material (in kg/m3), c_p is the heat capacity of the chip material (in J/(kg K)), T is the temperature (in K), r is the spatial coordinate of the point where the temperature is determined, t is time (in sec), k_t is the thermal conductivity of the material (in W/(m K)), and g is the power density per unit volume (in W/m3).

The solution of Eq. (14.1) is the transient thermal response. Since all derivatives with respect to time go to zeroes in the steady state, steady-state analysis is needed to solve the PDE, shown in Eq. (14.2), which is the well-known Poisson's equation.

$$\nabla^2 T(r) = -\frac{g(r)}{k_t} \qquad (14.2)$$

A set of boundary conditions must be added in order to get a well-defined solution to Eq. (14.1). It typically involves building a package macro model and assuming a constant ambient temperature is interacted with the model. There are two methods to discretize the chip and form a system of linear equations representing the temperature distribution and power density distribution: finite difference method (FDM) and finite element method (FEM). The difference between them is that FDM discretizes the differential operator while the FEM discretizes the temperature field. Both of them can handle complicated material structures such as non-uniform interconnects distributions in a chip.

In FDM, a heat transfer theory is adopted, which builds an equivalent thermal circuit through the thermal-electrical analogy. The steady-state equation represents the network with thermal resistors connected between nodes and with thermal current sources mapping to power sources. The voltage and the temperature at the nodes can be computed by solving the circuit. The ground node is considered as a constant temperature node, typically the ambient temperature.

The FEM has different perspective to solve Poisson's equation. The design space is discretized into elements, with shapes such as tetrahedral or hexahedra. A reasonable discretization divides the chip into 8-node rectangular hexahedral elements. The temperature within an element then is calculated using an interpolation function.

A simple 3D thermal model is given in Fig. 14.1. A distributed power source feeds a distributed resistive network, connected to a thermal resistance, which models the heat sink. The voltage represents the temperature on the chip by using the thermal-electrical analogy.

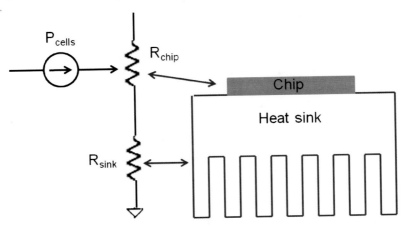

Fig. 14.1 A simple thermal model for a 3D chip.

14.2.2 HotSpot tool

There are numerous thermal models for traditional 2D design. One widely used architecture-level thermal tool, called HotSpot,[1] is based on an equivalent circuit of thermal resistances and capacitances that correspond to microarchitecture components and thermal packages. In a well-known duality between heat transfer and electrical phenomena, heat flow can be described as a "current" while temperature difference is analogous to a "voltage". The temperature difference is caused by heat flow passing through a thermal resistance. It is also necessary to include thermal capacitances for modeling transient behavior to capture the delay before the temperature reaching steady state due to a change in power consumption. Like electrical RC constants, the thermal RC time constants characterize the rise and fall times led by the thermal resistances and capacitances. In the rationale, the current and heat flow are described by exactly the same different equations for a potential difference. These equivalent circuits are called compact models or dynamic compact models if thermal capacitors are included. For a microarchitecture unit, the dominant mechanism to determine the temperature is the heat conduction to the thermal package and to neighboring units.

In HotSpot, the temperature is tracked at the granularity of individual microarchitectural units and the equivalent RC circuits have at least one node for each unit. The thermal model component values do not depend on initial temperature or the particular configurations. HotSpot uses a simple library that generates the equivalent RC circuit automatically and computes temperature in the center of each block with power dissipation over any chosen time step. Figure 14.2 shows a typical example of current package. The die is placed against a spreader plate, often made of aluminum, copper, or some other highly conductive material, which is placed against a heat sink of aluminum or copper that is cooled by a fan. Figure 14.3 illustrates the equivalent circuits for the chip and typical thermal package. The die layer is divided into blocks based on microarchitectural blocks and their floorplan. For simplicity, only three blocks are included in the die layer, while 10-20 blocks may be included in a realistic model. The RC model consists of a vertical model and a lateral model for the die, spreader, and sink layers. In the vertical model, the heat flow passes through from one layer to the next, moving from the die through the package and then into the air. The lateral model captures heat diffusion between adjacent blocks within a layer. In the dynamic simulation, the power dissipated in each unit of the die is modeled as a current source (not shown) at the node in the center of that block at each time step.

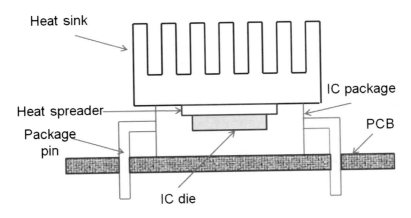

Fig. 14.2 Side view of a typical package.

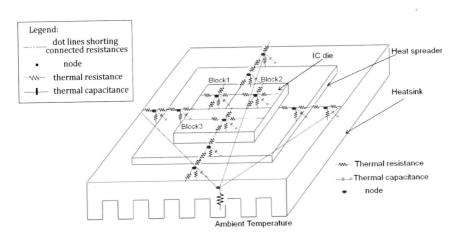

Fig. 14.3 Example HotSpot RC model for a floorplan with three architectural units, a heat spreader, and a heat sink.

14.2.3 HS3D

Here we briefly introduce a 3D thermal estimation tool named HS3D,[2] which is an extension of HotSpot. HotSpot provides a simple interface with the temperature profiling of chips. This library has only two inputs, a device floorplan at the function unit level and the power consumption of each unit. It is easily integrated with other tools, such as Wattch power modeling tool.[3] Although this library is powerful, it has some limitations. First, the restrictive input format makes the library useless in automated tools, in which location or power consumption of the entire device is not necessarily completed. Second, there are unreasonably high performance penalties when HotSpot evaluates several floorplans since it is intended for single chip evaluation. Third, the HotSpot library does not support 3D modeling. To address these limitations, HS3D is proposed by extending HotSpot library and optimizing the performance.

HS3D allows 3D thermal evaluation despite of largely unchanged computational model and methods in HotSpot. The inter-layer thermal vias can be approximated by changing the vertical thermal resistance of the materials. HS3d library allows incompletely specified floorplans as input and ensures accurate thermal modeling of large floorplan blocks. Many routines have been recreated for optimizations of loop accesses, cache locality, and memory paging. These improvements offer reduced memory usage and runtime reduction by over 3 orders when simulating a large number of floorplans. To guarantee the correctness and efficiency of HS3D library, the comparison between this new library and a commercial FEM software is performed.[4] First a 2D sample device and package is used for the verification. The difference of the average chip temperatures from HS3D and FEM software is only 0.02°C. Multi-layer (3D) device modeling verification is provided using 10 micron thick silicon layers and 2 micron thick interlayer material. The test case includes two layers with a sample processor in each layer. The experiment results show that the average temperature misestimation 3°C. In addition, the thermal analysis using FEM software costs 7 minutes while only costs one second using HS3D. It indicates that HS3D provides not only high accuracy but also high performance (low run time). The extension in HS3D was integrated into HotSpot in the later versions to support 3D thermal modeling.

14.3 THERMAL-AWARE EDA SOLUTIONS

Physical design tools play an important role in 3D design automation since one essential difference between 2D and 3D tools is from physical design. It is crucial to develop new placement and routing tools to adopt 3D technologies. As indicated in Section 14.1, one major concern in 3D ICs is the increased power density due to placing one thermal critical block over another in the 3D stack. Consequently, thermal-aware physical design tools should be developed to address both thermal and physical tool issues. Recently, thermal-aware design tools have been investigated.[6-8] In this section, we first present a thermal-aware floorplanner for 3D microprocessors.[6] Then we review thermal-aware 3D placement and routing tools.[7-8]

14.3.1 Thermal-aware floorplanning

Floorplanning is a critical step in the process of physical since it significantly influences the performance of the final design. 3D floorplanning is even more important because the multiple layers dramatically enlarge the solution space and the increased power density accentuates the thermal problem. A 3D thermal-aware floorplanner was introduced by Hung *et al.*[6] In contrast to other work, this floorplanner considers the interconnect power consumption in exploring a thermal-aware floorplan. The model used in this 3D thermal-aware floorplanner is based on the B*-tree representation proposed by Chang *et al.*[9] The B*-tree is an ordered binary tree which can represent a nonslicing admissible floorplan. Figure 14.4 shows a B-tree structure and its corresponding floorplan. The bottom left corner node is the root of a B*-tree. The admissible floorplan can be obtained by depth-first search procedure in a recursive fashion. The original B*-tree structure was used for the 2D floorplanning problem. This model is extended to explore a multilayer 3D floorplanning and modify the perturbation functions. Figure 14.5 shows the extension model for 3D floorplanning. There are six perturbation operations in this algorithm: node swap, rotation, move, resize, interlayer swap, and interlayer move. To obtain floorplanning solutions, a simulated annealing engine is used in this 3D floorplanner. This algorithm takes the area of all functional modules and interconnects among modules as inputs. During the simulated process, each module dynamically adjusts itself to fit with the adjacent modules. The 3D floorplanner uses a two-stage approach, which is different from 2D floorplanning. The first stage partitions the blocks to appropriate layers, trying to minimize the packed area difference between layers and minimize total wire length using all perturbation operations. However, since the first stage tries to balance the packed areas of different layers, the floorplan of some layers may not be packed compactly. The second stage is intended to address this issue. It focuses on adjusting the floorplan of each layer using the first four operations.

The experiment is conducted using Alpha microprocessor and four MCNC benchmarks to verify this floorplanner. The results show that in 2D architecture, the corresponding thermal-aware floorplanner can reduce the peak temperature by 7% on average while increasing wire-length by 18% and providing a similar chip area compared to the results without thermal consideration. When move to 3D architecture without thermal consideration, the peak temperature increased by 18% on average due to the increased power density. However, the wire length and chip area reduced by 32% and 50%, respectively. Our 3D thermal-aware floorplanner can lower the peak temperature by 6% with little area penalty compared 3D placement without thermal consideration.

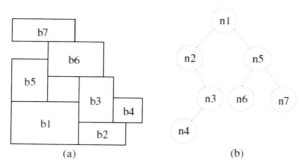

Fig. 14.4 (a)an example floorplan; (b)the corresponding B* tree.

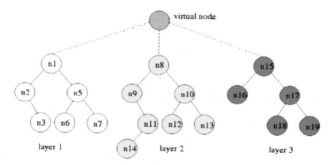

Fig. 14.5 B* tree model extension for 3D floorplanning.

14.3.2 Thermal-aware placement

Placement is an important step in the physical design flow since the quality of placement results has significant impact on the performance, power, temperature and routability. Therefore, a thermal-aware 3D placement tool is needed to fully take advantage of 3D IC technology. We briefly introduce a 3D thermal-aware placement tool proposed by Cong et al.[7]

The 3D thermal-aware placement problem is defined as: given a circuit with a set of cell instances and a set of nets, the device layer number, and the per-layer placement region, a placement of the cells in that region, find a placement for every cell so that the objective function of weighted total wirelength is minimized, under certain constraints such as overlap-free constraints, performance constraints and temperature constraints.

Since the weighted total wirelength is a widely accepted metric for placement qualities, it is considered as the objective function of the placement problem. The objective function can be obtained from a weighted sum of the wirelength of instances and the number of through-silicon vias (TS vias) over all the nets. For thermal consideration, temperatures are not directly formulated as constraints but are appended as a penalty to the wirelength objective function. There are several ways to formulate this penalty: the weighted temperature penalty with transformation to thermal-aware net weights, or the thermal distribution cost penalty, or the distance from the cell location to the heat sink during legalization.

The 3D placement framework is shown in Fig 14.6. The existing tools are marked with dashed boundary. First, a 2D placement for the target design is generated using a 2D wirelength-driven and/or thermal-driven placer. This initial placement has significant impact on the final 3D placement. Then the 2D placement is transformed into a legalized 3D placement using the given 3D technology. During the transformation step, wirelength, TS via number and temperature are considered. To further reduce the TS via number and reduce the maximum on-chip temperature, a refinement process through layer reassignment will be carried out after 3D transformation. Finally, a 2D detailed placer will further refine the placement result for each device layer.

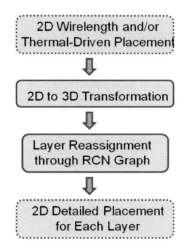

Fig. 14.6 3D placement framework.

Before the transformation from 2D placement to 3D placement, the initial optimized 2D placement is performed on a chip with N times area compared to one layer of the 3D chip, where N is the number of device layers. After the 2D placement targeting wirelength minimization, local stacking transformation can achieve even shorter wirelength for the same circuit under 3D IC technology. To perform the transformation, two folding-based transformation schemes, folding-2 and folding-4, can be used to generate 3D placement with very low TS via number. Furthermore, TS via number and wirelength tradeoffs can be achieved by the window-based stacking/folding. These transformation methods guarantee wirelength reduction of 3D placement over the initial 2D placement results.

14.3.3 Thermal-aware routing

A thermal-driven multilevel routing tool is presented by Cong *et al.*[8] In this work, an efficient 3D multilevel routing approach is proposed. This approach also includes a novel through-the-silicon via (TS-via) planning algorithm. More specifically, it features an adaptive lumped resistive thermal model and a two-step multilevel TS-via planning scheme. The routing problem is defined as given the design rule, height and the thermal conductivity of each material layer, a 3D placement or floorplan result with white space reserved for interconnect, a given maximum temperature, route the circuit so that the weighted cost of wirelength and the total number of TSVs is minimized according to the design rules.

In this thermal-driven 3D multilevel routing with TS-via planning, a weighted area sum model is used to estimate the routing resources, which are reserved for local nets at each level. In addition, the average power density of each tile is computed during the coarsening process. In the initial routing stage, each multipin net generates a 3D Steiner tree, which starts from a minimum spanning tree (MST). Then at each edge of the initial MST a point-to-path maze-searching engine is applied. Steiner edges are generated when the searching algorithm reaches the existing edges of the tree. After maze-searching algorithm, the number of dummy TS-vias is estimated through binary search in order for the insertion. The upper bound of dummy TS-vias is estimated by the area of white space between the blocks. After the estimation, a TSV-via number distribution step is applied to assign the dummy TS-vias.

The TS-vias are refined to minimize wirelength and maximum temperature during each refinement step. During the planning process, the multilevel routing system communicates with the thermal model, which provides the latest thermal profile. The thermal model returns a temperature map by reading the tile structure and the TS-vias number at each tile and assigns one temperature values to each tile.

During the TSV planning step, every device layer of the 3D design is divided into planning windows PW. TS-via planning is conducted for each planning window. The problem of TS-via planning is described as: given a planning window PW, and tiles with position and capacity, assign each TS-via to one of the tiles, so that the total number of TS-via assigned to each tile does not exceed its capacity, also with minimized wirelength and peak temperature. The TSV planning procedures are repeated at all device layers. The temperature profile is updated after finishing the planning for every device layer. Therefore, the following plannings can use a more accurate temperature map. This multi-level routing scheme is shown to be effective to reduce the dummy TS-vias and decrease the circuit temperature to a required level.

14.4 THERMAL-AWARE ARCHITECTURE

Three-dimensional integration provides many new exciting opportunities in computer architecture area. 3D offers several benefits for new architecture designs such as high bandwidth due to core-cache or core-memory stacking and heterogeneous stacking, i.e., each layer can have specific circuitry type such as RF, analog, memory, MEMS, digital, etc. or from different technology node such as 130 nm, 90 nm, 65 nm, 45 nm, 32 nm etc. However, as mentioned in Section 14.1, thermal problem is a big challenge for the adoption of 3D technology, it is necessary to take thermal into account when designing new 3D architectures. In this section, we give a survey on thermal-aware 3D architecture designs and outline possible solutions.

14.4.1 3D Network-in-Memory Architecture

We present a 3D network-in-memory architecture and the solution to mitigate the thermal challenges of 3D stacking [10]. The proposed architecture for multiprocessor systems with large shared L2 caches involves placement of CPUs on several layers of a 3D chip with the remaining space filled with L2 cache banks. Most 3D IC designs observed in the literature so far have not exceeded 5 layers, mostly due to manufacturability issues, thermal management, and cost. As previously mentioned, the most valuable attribute of 3D chips is the very small distance between the layers. A distance on the order of tens of microns is negligible compared to the distance traveled between two network on-chip routers in 2D (1500 μm on average for a 64 KB cache bank implemented in 70 nm technology). This characteristic makes traveling in the vertical (inter-layer) direction very fast as compared to the horizontal (intra-layer). One inter-layer interconnect option is to extend the NoC into three dimensions. This requires the addition of two more links (up and down) to each router. However, adding two extra links to a NoC router will increase its complexity (from 5 links to 7 links). This, in turn, will increase the blocking probability inside the router since there are more input links contending for an output link. Moreover, the NoC is, by nature, a multi-hop communication fabric, thus it would be unwise to place traditional NoC routers on the vertical path because the multi-hop delay and the delay of the router itself would overshadow the ultra fast propagation time. It is not only desirable, but also feasible, to have single hop communication amongst the layers because of the short distance between them. To that effect, the authors propose the use of dynamic Time-Division Multiple Access (dTDMA) buses as "Communication Pillars" between the wafers, as shown in Fig. 14.7. These vertical bus pillars provide single-hop communication between any two layers, and can be interfaced to a traditional NoC router for intra-layer traversal using using minimal hardware. Furthermore, hybridization of the NoC router with the bus requires only one additional link (instead of two) on the NoC router. This is the case because the bus is a single entity for communicating both up and down.

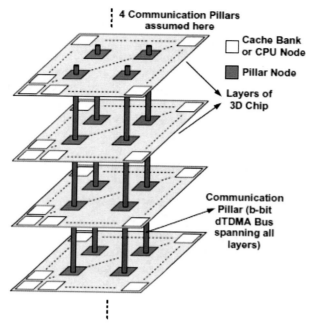

Fig. 14.7 Proposed 3D Network-in-Memory architecture.

The dTDMA pillars provide rapid communication between layers of the chip. There is a dedicated pillar associated with each processor to provide fast inter-layer access, as shown in Fig. 14.7. Such a configuration gives each processor instant access to the pillar, additionally providing them with rapid access to all cache banks that are adjacent to the pillar. By placing each processor directly on a pillar, its memory locality (the number of banks with low access latency) is increased in the vertical direction (potentially both above and below), in addition to the pre-existing locality in the 2D plane. Such an increase in the number of cache banks with low access latency can significantly improve the performance of applications.

Stacking CPUs directly on top of each other would give rise to thermal issues. Increased temperature due to layer stacking is a major challenge in 3D design, and is often a major determining factor in component placement. Since the CPUs are expected to consume the overwhelming majority of power (they are constantly active, unlike the cache banks), it would be thermally-unwise to stack any two or more processors in the same vertical plane. Furthermore, stacking processors directly on top of each other on the same pillar would affect the performance of the network as well, as it would create high congestion on the pillar. Processors are the elements which generate most of the L2 traffic; therefore, forcing them to share a single link would create excessive traffic. To avoid thermal and congestion problems, CPUs can be offset in all three dimensions (maximal offsetting), as shown in Fig. 14.8.

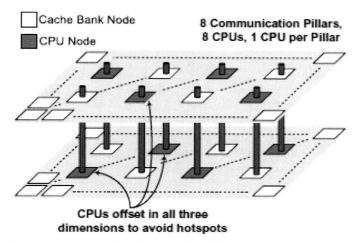

Fig. 14.8 Hotspots can be avoided by offsetting CPUs in all three dimensions.

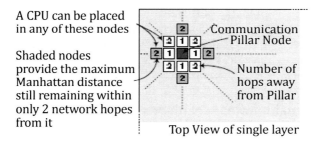

Fig. 14.9 Placement pattern of CPUs around the pillars.

If, however, the manufacturing technology provides only low via densities, the designer might be forced to use fewer pillars than the number of CPU cores to minimize the wasted device-layer area. In such cases, multiple CPUs would have to share a single pillar, warranting a careful CPU placement methodology, to ensure maximum performance with minimum thermal side-effects. The best way to achieve this is to offset the processors at various distances from the pillar, while keeping them within the direct vicinity. Moving the processors far away from the pillars would be detrimental to performance, since changing layers would inflict lengthy 2D traversal times to and from the pillar. In their approach, the CPUs are placed according to the pattern illustrated in Fig. 14.9. The processors are placed at most two hops away from a pillar to minimize the adverse effect on performance.

To perform the placement of the CPUs based on the restrictions of Fig. 14.9, a simple algorithm was developed to achieve uniform offset of all CPUs in both the vertical and horizontal directions for the configurations studied. The algorithm assumes placement of 2 or 4 CPUs per pillar per chip layer. The parameter k is the offset distance from a pillar in number of network hops. In the proposed implementation, k was chosen to be 1. If thermal issues require more mitigation, then k can be increased at the expense of performance. Parameter c is the number of CPUs assigned to each pillar on each layer. Assigning more than 4 CPUs per pillar per layer is not desirable, since it will increase bus contention dramatically, thus degrading network performance. The location of the pillars is assumed to be predetermined by the designer, and is given as a constant to the algorithm. The pillars need to be placed as far apart from each other as possible within the layer to avoid the creation of network congested areas. On the other hand, the pillars should not be placed on the edges because such placement would limit the number of cache banks which are in the vicinity of the pillar.

To validate the proposed CPU placement methodology proposed above, thermal tool HS3D is used. The thermal simulations is conducted by a system with 256 64 KB L2 cache banks (in total 16 MB) and 8 CPU cores. As expected, moving from a 2D layout to a 3D layout causes the average temperature of the chip to increase. However, of critical importance is the avoidance of hotspots, i.e., places where the peak temperature is extremely high. Hotspots can substantially degrade performance and adversely affect the lifetime of the chip. Hotspots can be avoided through careful offsetting of the processor cores. Offsetting the cores in all three dimensions provides the best results, causing an increase in peak temperature of only $8°C$ when using two layers, as compared to the 2D case.

14.4.2 Thermal-herding techniques

In this section, a family of thermal herding, microarchitecture techniques for controlling HotSpots in high performance 3D processors, is presented.[11] Thermal herding techniques herd or steer the switching activity in the 3D processor to the die, which is closest to the heat sink and reduce the total power and power density while still maintaining the performance benefits. It adopts different strategies for different components in the processor such as placing the frequently switching 16-bits on the top die in the significance-partitioned datapath, proposing a 3D-aware allocation scheme for instruction scheduler, an address memorization approach for the load and store queues, a partial value encoding for the L1 data cache, and exploiting a form of frequent partial value locality in target addresses for a branch target buffer.

Thermal herding is proposed based on some observations made by prior work that only a few of the least significant bits are needed in the data used in many integer instructions,[12] i.e., many 64-bit integer values only requires 16 or fewer bits for the representation. In addition, it is highly predictable for an instruction's usage of low-width values.[13] Therefore, the datapath can be organized by assigning each 16 bits to a separate dies (4 dies) in 3D stacking. The least significant bits can be placed on the top layer which is closest to the heat sink so that the power on the other three dies may be saved using the instruction's width prediction. A prediction is made for each instruction whether to use low-width (≤ 16 bits) or full-width (≥ 16 bits) values. A simple program counter indexed two-bit saturating counter predictor is used. When the predictor indicates the data is low-width usage but it is actually a full-width usage then the result is an unsafe misprediction, which requires pipeline stalls in related stages. Some critical microarchitecture components are discussed as follows.

Register Files: each 64-bit entry in the register file is partitioned into four layers with the least significant bits closest to the heat sink. The width prediction is used to determine gating control signals for other three layers before the actual register file access. If low-width instruction is predicted then only the top layer portion of the register file is active. In this case, the power density is similar to that of a 2D register file design. A width memorization bit is also provided for each entry in the top layer, indicating whether the remaining three layers contain non-zero values. The processor compares this bit to the predicted width when reading it. If the width prediction is low and the actual width is full the processor stalls the previous stages of the register file and activate the logic on the other three layers. It also corrects the prediction to prevent further stalls.

Arithmetic Units: only 3D integer adder is presented but the concept can be extended to other arithmetic units. A tree-based adder is partitioned into four layers with the least significant bits reside on top layer. The processor uses predication information to decide if the clock gate should be activated for other three layers. Two possible unsafe width mispredictions should be handled properly. One is misprediction on an instruction's input operands. In this scenario, since the arithmetic unit is not fully active at the start of execution, one cycle stall is needed to activate the upper 48 bits. Another scenario is misprediction on the output, in which the width misprediction is not known at the beginning of the computation. The instruction needs to be executed again to guarantee the correctness, causing performance penalty. Therefore, the accuracy of the width predictor is very important.

Bypass Network: the bypass network does not need additional circuitry to handle the width prediction since the unsafe misprediction is handled by the arithmetic units. With a correct low-width prediction, only the dirvers/wires on the top layer consume dynamic power. In addition, the partitioning in 3D reduces the wire length so that the latency and power are reduced accordingly.

Instruction Scheduler: in the instruction scheduler, there is one reservation station (RS) for each instruction dispatched but not executed. When an instruction is ready to issue, the instruction broadcasts its destination identifier to notify dependent instructions. The instruction scheduler is partitioned based on the RS entries so that the length of the broadcast buses is significantly reduced, resulting in latency and power reduction. A modified allocation algorithm is also adopted to move instructions to the top layer so that the active entries are closer to the heat sink. If no available entries in the top layer, the allocator will check the layer that is next closest to the heat sink.

Load and Store Queues: the load and store queues are used to track the data and addresses of instructions. The queues are partitioned in a similar fashion with the main datapath due to its similarity with register file. Load and store addresses are normally full-width values but the upper bits of the addresses do not change frequently. To take advantage of this feature, a partial address memorization (PAM) scheme is proposed. The low-order 16 bits of a load or store's address is broadcast on the top layer. In addition, whether the remaining 48 bits are the same as the most recent store address is also indicated. To summarize, the PAM approach tries to herd the address broadcasts and comparisons to the top layer.

Data Cache: the L1 data cache is organized in a word-partitioned manner due to its similarity with register file. Memoization bits are also provided to detect unsafe width mispredictions. In addition, the definition of a "low-width" value for load and store instructions is broadened to increase the frequency of low-width values. Two bits instead of one single width memorization bit are stored to encode the upper 48 bits. Value "00" means the upper 48 bits are all zeros. Value "01" indicates all ones. Value "10" means the upper bits are identical to those of the referencing address. Value "11" means the upper bits are not encodeable.

Front End: data-centric approaches for partitioning the front end are not effective since no data values are handled. A register alias table (RAT) is implemented to place the ports of each instruction on different layers so that unnecessary die-to-tie vias are avoided. The instruction that requires the most register name comparisons is placed on the top layer so that most switching activities are moved to the top layer. For branch predictor based on two-bit saturating counters, the counters are partitioned into two separate arrays: direction bit array and hysteresis array. The more frequently used direction-bit array is place on the layer closer to the heat sink. For the branch target buffers (BTBs), since most branch targets are located relatively close to the originating branch, they are organized like the data cache. The low-order 16 bits are place on the top layer with one extra target memorization bit indicating whether the bits on the other three layers should be accessed.

3D thermal herding techniques can improve IPC by reducing the pipeline depth and L2 latency but it also can reduce IPC due to width mispredictions. By conducting the experiments on several benchmarks, the overall performance is improved since the pipeline reduction benefits outweigh the performance penalties caused by width mispredictions. The power consumption is also reduced due to two reasons: the length of wires is reduced because of 3D stacking and thermal herding reduces the switching activities because of clock gating. The thermal experiments show that thermal herding techniques successfully control power density and mitigate 3D thermal issues.

14.4.3 MIRA architecture

In this section, we review a 3D stacked Network-on-Chip (NoC) router, called MIRA.[14] Various architectural alternatives are investigated for designing a high performance, energy efficient NoC router. The design is based on the concept of dividing a traditional 2D NoC router along with the rest of the on-chip communication fabric into multiple layers, with the objective of exploiting the benefits of the 3D technology in enhancing the design of the router micro-architecture for better performance and power conservation. The multi-layer NoC design is primarily motivated by the observed communication patterns in a Non-Uniform Cache Architecture (NUCA)-style CMP.[15] The NoC in a NUCA architecture supports communication between the processing cores and the 2nd-level cache banks. The NUCA traffic consists of two kinds of packets, data and control. The data packets contain certain types of frequent patterns like all 0's and all 1's. A significant part of the network traffic also consists of short address/coherence-control packets. Thus, it is possible to selectively power down the bottom layers of a multi-layer NoC that have redundant or no data (all 0 word or all 1 word or short address flits), helping in energy conservation, and subsequently mitigating the thermal challenges in 3D designs. Furthermore, such a multilayered NoC design complements a previous study, where a processor core is partitioned into multiple (four) layers. It was shown that by switching off the bottom three layers based on the operand manipulation characteristics, it is possible to achieve significant power savings. These savings in power result in minimizing the thermal impacts compared to a standard 3D stacking.

Three 3D design options for a mesh interconnect are studied: (1) 3D baseline router (3DB): direct extension of a 2D NoC into 3D by providing two additional ports in each router for connections in the third dimension. This structure is called the 3D baseline router (3DB). (2) 3D multi-layered router (3DM): analyze the design implications of splitting the router components such as the crossbar, virtual channel allocator (VA) and buffer in the third dimension, and the consequent vertical interconnect (via) design overheads. It shows that the multi-layered design renders several advantages in terms of reduced crossbar size and wire length with respect to the 3DB design. (3) 3D multi-layered router with express paths (3DM-E): The saving in chip area in 3DM approach can be used for enhancing the router capability, and is the motivation for the third design, which is called a 3D router with express paths (3DM-E). These express paths between nonadjacent nodes reduce the average hop count, and helps in boosting the performance and power behavior. The design of major components in 3DM is described as follows.

Input Buffer: assuming that the flit width is W bits, we can place them onto L layers, with W/L bits per layer. For example, if W=128 bits and L=4 layers, then each layer has 32 bits starting with the LSB on the top layer and MSB at the bottom layer. Typically, an on-chip router buffer uses a register-file type architecture and it is easily separable on a per-bit basis. In this approach, the word-lines of the buffer span across L layers, while the bit-lines remain within a layer. Consequently, the number of inter-layer vias required for this partitioning is equal to the number of word-lines. Since this number is small in the case of on-chip buffers (e.g., 8 lines for 8 buffers), it makes it a viable partitioning strategy. The partitioning also results in reduced capacitive load on the partitioned word-lines in each layer as the number of pass transistors connected to it decreases. In turn, it reduces the driver sizing requirements for the partitioned word-lines. Since buffers contribute about 31% of the router dynamic power, exploiting the data pattern in a flit and utilizing power saving techniques can yield significant power savings. The lower layers of the router buffer can be dynamically shutdown to reduce power consumption when only the LSB portion has valid data. We define a short-flit as a flit that has redundant data in all the other layers except the top layer of the router data-path. For example, if a flit consists of 4 words and all the three lower words are zeros, such a flit is a short flit. The clock gating is based on a short-flit detection (zero-detector) circuit, one for each layer. The overhead of utilizing this technique in terms of power and area is negligible compared to the number of bit-line switching that can be avoided.

Crossbar: In the proposed 3DM design, a larger crossbar is decomposed into a number of smaller multi-bit crossbars positioned in different layers. The crossbar size and power are determined by the number of input/output ports ("P") and flit bandwidth ("W") and therefore, such a decomposition is beneficial. In the 2DB case, P=5 and the total size is $5W \times 5W$, whereas in the 3DM case with 4 layers, the size of the crossbars for each layer is $(5W/4) \times (5W/4)$. If we add up the total area for the 3DM crossbar, it is still four times smaller than the 2DB design. A matrix crossbar is used for illustrating the ideas. However, such 3D splitting method is generic and is not limited to this structure. In the design, each line has a flit-wide bus, with tri-state buffers at the cross points for enabling the connections from the input to the output. In the 3D designs, the inter-layer via area is primarily influenced by the vertical vias required for the enable control signals for the tri-state buffers (numbering $P \times P$) that are generated in the topmost layer and propagated to the bottom layers. The area occupied by these vias is quite small making the proposed granularity of structure splitting viable.

Inter-router Link: Inter-router links are a set of wires connecting two adjacent routers, and therefore, they can also be distributed as in the crossbar case above. Assuming that the link bandwidth is W and the number of layers is L, the cross-section bandwidth across L layers is $W \times L$ in the 3DB case. To maintain the same cross-section bandwidth in the proposed 3DM case for fair comparison, this total bandwidth should be distributed to multiple layers and multiple nodes. For example, if we assume 4 layers (L=4), the 3DB architecture has 4 separate nodes (A,B,C, and D), with one node per layer whereas in the 3DM case, we have only 2 separate nodes, since now the floor-plan is only half of the size of a 3DB node. Consequently, in the 3DB design, 4 nodes share $4 \times W$ wires; while in the 3DM design, 2 nodes share $4 \times W$ wires. This indicates that the available bandwidth is doubled from the perspective of a 3DM node.

Routing Computation (RC) Logic: A physical channel (PC) in a router has a set of virtual channels (VCs) and each VC is associated with a routing computation (RC) logic, which determines the output port for a message (packet); however, a RC logic can be shared among VCs in the same PC, since each PC typically takes at most one flit per cycle. Hence, the number of RC logic blocks is dependent on the number of VCs (or PCs if shared) per router. Since the RC logic checks the message header and is typically very small compared to other logics such as arbiters, it is best to put them in the same layer where the header information resides; we can avoid feeding the header information across layers, thereby eliminating the area overheads from inter-wafer vias. Here, the number of VCs per PC if fixed to be 2. The increase in RC logic area and power is evaluated. The choice of 2 VCs is based on the following design decisions: (i) low injection rate of NUCA traffic (ii) to assign one VC per control and data traffic, respectively (iii) to increase router frequency to match CPU frequency, and (iv) to minimize power consumption. However, the proposed technique is not limited to this configuration.

Virtual-Channel Allocation (VA) Logic: The virtual channel allocation (VA) logic typically performs a two-step operation. The first step (VA1) is a local procedure, where a head flit in a VC is assigned an output VC. If the RC logic determines the output VC in addition to the output PC, this step can be skipped. Here an assumption is made that the RC logic assigns only the output PC and requires P×V V:1 arbiters, where P is the number of physical channels and V is the number of virtual channels. The second step (VA2) arbitrates among the requests for the same output VC since multiple flits can contend for the same output VC. This step requires P×V PV:1 arbiters. As the size of VA2 is relatively large compared to VA1, the VA1 stage arbiters is placed entirely in one layer and distribute the P×V arbiters of the VA2 stage equally among different layers. Consequently, it requires P×V inter-layer vias to distribute the inputs to the PV:1 arbiters on different layers. It should also be observed that the VA complexity for 3DM is lower as compared to 3DB since P is smaller. Hence, 3DM requires smaller number of arbiters and the size of the arbiters is also small (14:1 vs. 10:1).

Switch Allocation (SA) Logic: since the SA logic occupies a relatively small area, it is kept completely in one layer to help balance the router area in each layer. Further, the SA logic has a high switching activity due to its per-flit operation in contrast to the VA and RC logics that operate per packet. Hence, the SA logic is placed in the layer closest to the heat sink.

The thermal result shows that 3DM obtains benefit from reduced overall temperature since lower layers will not be active for short messages, thereby reducing the overall power density. When the injection rate increases, more temperature reduction is expeted due to increased number of flit activities in the router, which triggers more activities in separable modules (buffer, crossbar, inter-router links).

14.4.4 PicoServer

In this section, PicoServer, an architecture to reduce power and energy consumption using 3D stacking technology, is presented.[16] The basic idea is to stack on-chip DRAM main memory instead of using stacked memory as a larger L2 cache. The on-chip DRAM is connected to the L1 caches of each core through shared bus architecture. It offers wide low-latency buses to the processor cores and eliminates the need for an L2 cache, whose silicon area is allocated to accommodate more cores. Increasing the number of cores can help improve the computation throughput, while each core can run at a much lower frequency, and therefore result in an energy-efficient many core design.

The PicoServer is a chip multiprocessor, consists of several single issue in-order processors. Each core runs at 500 MHz, containing an instruction cache and a data cache, which uses a MESI cache coherence protocol. The study showed that the majority of the bus traffic is caused by cache miss traffic instead of cache coherence due to the small cache for each core. This is one motivation to stack large on-chip DRAM, which is hundreds of megabytes, using 3D technology.

In PicoServer, a wide shared bus architecture is adopted to provide high memory bandwidth and to fully take advantage of 3D stacking. A design space is explored by running simulations varying the bus width on a single shared bus, which ranges from 128 bits to 2048 bits. The impact of bus width on the PicoServer is determined by measuring the network performance. The results showed that a relatively wide data bus is needed to achieve better performance and satisfy the outstanding cache miss requests. The bus traffic increase caused by narrowed bus width will result in latency increase. Wide bus widths can speedup DMA transfer since more data can be copied in one transaction. The simulation also shows that a 1024 bit bus width is a reasonable for different configurations including 4, 8, and 12 multiprocessors due to performance saturation at this point.

The stacked on-chip DRAM contains 4 layers in order to obtain a total size of 256 MB, which may be enough depending on the workload. More on-chip DRAM capacity can be obtained with aggressive die stacking. In order to fully take advantage of 3D stacking, it is necessary to modify the conventional DDR2 DRAM interface for PicoServer's 3D stacked on-chip DRAM. In the conventional DDR2 DRAMs, a small pin count is assumed. In addition, address multiplexing and burst mode transfer are used to compensate the limited number of pins. With 3D stacking, there is no need to address multiplexing so that the additional logic to latch and mux narrow address/data can be removed.

In servers with large network pipes such as PicoServer, one common problem is how to handle large amount of packets arrive at each second. Interrupt coalescing, a method to coalesce non-critical events to reduce the number of interrupts, is one solution to solve this problem. However, even with this technique, the number of interrupts received by a low frequency processor in PicoServer is huge. To address this issue, multiple network interface controllers (NICs) with their interrupt lines are routed to a different processor. One NIC is inserted for 2 processors to fully utilize each processor. Such NIC should have multiple interface IP addresses or an intelligent method to load balancing packets to multiple processors. In addition, it needs to keep track of network protocol states at the session level.

Thermal evaluation using HopSpot[1] showed that the maximum junction temperature increase is 5~10°C in 5-layer PicoServer architecture. This comes from power and energy reduction caused by core clock frequency reduction and improvement on high network bandwidth.

14.5 CONCLUSION

Increasing power density in 3D ICs can result in higher on-chip temperatures, which can have a negative impact on performance, power, reliability, and the cost of the chip. Consequently, thermal modeling and thermal-aware design techniques are very critical for future 3D IC designs. This chapter presents an overview of thermal modeling for 3D IC and outlines solution schemes to overcome the thermal challenges at Electrical Design Automation (EDA) and architectural levels.

References

1. Skadron, K., Stan, M. R., Huang, W., Velusamy, S., Sankaranarayanan, K., and Tarjan, D. (2003). Temperature-aware microarchitecture. In Proceedings of the 30th Annual international Symposium on Computer Architecture, pp. 2-13.

2. Link, G.M. and Vijaykrishnan, N.. (2006). Thermal trends in emerging technologies, International Symposium on Quality Electronic Design, pp.627-632.

3. Brooks, D., Tiwari, V., and Martonosi, M. (2000). Wattch: A framework for architectural-level power analysis and optimizations. In Proceedings of the 27th Annual international Symposium on Computer Architecture, pp. 83-94.

4. Flomerics Corp. Flotherm modeling software

5. Sapatnekar, S. S. (2009). Addressing thermal and power delivery bottlenecks in 3D circuits In Proceedings of the Asia and South Pacific Design Automation Conference, pp. 423-428.

6. Hung, W.-L., Link, G.M., Yuan Xie, Vijaykrishnan, N. and Irwin, M.J. (2006). Interconnect and thermal-aware floorplanning for 3D microprocessors, International Symposium on Quality Electronic Design, pp. 99-104.

7. J. Cong, G. Luo, J. Wei, and Y. Zhang. (2007). Thermal-Aware 3D IC Placement via Transformation, Proceedings of the 12th Asia and South Pacific Design Automation Conference, pp. 780-785.

8. Cong, J. and Yan Zhang. (2005). Thermal-driven multilevel routing for 3D ICs, Asia and South Pacific Design Automation Conference, pp. 121-126.

9. Chang Yun-Chih, Chang Yao-Wen, Wu Guang-Ming, and Wu Shu-Wei. (2000). B*-trees: a new representation for non-slicing floorplans. Design Automation Conference, pp. 458–463.

10. F. Li,C. Nicopoulos, T. Richardson, Yuan Xie, N. Vijaykrishnan and M. Kandemir. (2006). Design and Management of 3D Chip Multiprocessors using Network-in-memory. Proceedings of the Annual International Symposium on Computer Architecture (ISCA), pp. 130-141.

11. Puttaswamy, K. and Loh, G. H. (2007). Thermal Herding: Microarchitecture Techniques for Controlling Hotspots in High-Performance 3D-Integrated Processors. In Proceedings of the IEEE 13th international Symposium on High Performance Computer Architecture, pp. 193-204.

12. D. Brooks and M. Martonosi. (1999). Dynamically Exploiting Narrow Width Operands to Improve Processor Power and Performance. In Proc. of the 5th Intl. Symp. on High Perf. Comp. Arch., pp. 13-22.

13. O. Ergin, D. Balkan, K. Ghose, and D. Ponomarev. (2004). Register Packing: Exploiting Narrow-Width Operands for Reducing Register File Pressure. In Proc. of the 37th Intl. Symp. on Microarchitecture, pp. 304--315.

14. Park, D., Eachempati, S., Das, R., Mishra, A. K., Xie, Y., Vijaykrishnan, N., and Das, C. R. (2008). MIRA: A Multi-layered On-Chip Interconnect Router Architecture. In Proceedings of the 35th international Symposium on Computer Architecture, pp. 251-261.

15. K. Changkyu, B. Doug, and W. K. Stephen. (2002). An adaptive, non-uniform cache structure for wire-delay dominated on-chip caches. Proc. of the ASPLOS-X, pp. 211-222.

16. Kgil, T., D'Souza, S., Saidi, A., Binkert, N., Dreslinski, R., Mudge, T., Reinhardt, S., and Flautner, K. (2006). PicoServer: using 3D stacking technology to enable a compact energy efficient chip multiprocessor. In Proceedings of the 12th international Conference on Architectural Support For Programming Languages and Operating Systems, pp. 117-128.

Chapter 15

3D IC DESIGN AUTOMATION CONSIDERING DYNAMIC POWER AND THERMAL INTEGRITY

Hao Yu and Xiwei Huang

Nanyang Technological University, Singapore

15.1 INTRODUCTION

The high-performance VLSI integration by the technology scaling has been confronted with the dramatically increased design cost resulting from noise, power and process variation. System integration has emerged as an alternative design paradigm to improve performance by increasing the throughput. However, in today's two dimensional (2D) Systems-on-chip (SoC) integration, the memory is surrounded by logic circuits and its performance, in terms of, is limited by the length of long interconnect. Thanks to the recent advance in three dimensional (3D) integration[2, 4, 5, 9, 14, 17, 18] a 3D integration can reduce the physical distance between the memory and the logic circuits and hence, has shown a promising potential to integrate hundreds of cores with scaled performance superior to that of the 2D integration. Multiple device layers can be vertically connected by through-silicon via (TSV), resulting in not only a shorter length but also a denser density. Obviously, the system size can be dramatically reduced with diverse components compactly integrated. In addition, the use of device-stacking enables different layers to be fabricated by a number of independently optimized technologies. As a result, the components in one system from analog/RF, MEMS, or digital processing can be integrated together with a low fabrication cost and a high yield-rate. More importantly, as there is a significant boost in communication bandwidth, the 3D integration is well suited for I/O-centric system with concurrent data processing, ideally for the many-core computing system integration.

3D Integration for VLSI Systems
Edited by Chuan Seng Tan, Kuan-Neng Chen and Steven J. Koester
Copyright © 2012 by Pan Stanford Publishing Pte. Ltd.
www.panstanford.com

Since there are large numbers of devices densely packed in a number of device layers, it brings a significant burden to the supply voltage in 3D ICs. This chapter discusses an allocation algorithm of (TSV) to simultaneously consider both the and in 3D ICs. It first illustrates the need for a high-performance 3D design driven by dynamic power and thermal integrity, and then discusses how to allocate TSVs to simultaneously deliver power supply and remove heat. More importantly, to cope with the large-scale design complexity, the modern technique is applied to handle not only large numbers of dynamic inputs/working-loads but also large sizes of interconnection networks that distribute the power/heat. Experiments applying the design automation for 3D integration presented by this chapter show promising results to reduce both runtime and resource.

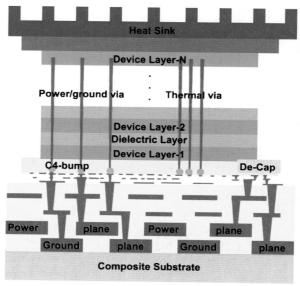

Fig. 15.1 A typical 3D stacking with through silicon vias.

15.2 DYNAMIC POWER AND THERMAL INTEGRITY

Figure 15.1. illustrates a typical 3D stacking of multiple device layers within one package. In the following, the importance of dynamic power and thermal integrity is exposed in the content of 3D integration using TSVs.

The supply voltage is delivered from the bottom power/ground planes in the package, passed through the vias and C4 bumps, and connected to the on-chip power/ground grid on active device layers. We call the through vias that deliver the supply voltage. The 3D integration, by definition, has integrated more than one layer of active devices. They draw much larger current from package power/ground planes than 2D ICs. This can obliviously result in the IR drop for the horizontal on-chip power/ground grid. The surge from the injecting current further leads to a large (SSN) for those I/O drivers at the chip package interface. Figure 15.2. shows a detailed view of how to place signal and power/ground vias through package planes. Clearly, the regions to place power vias or ground vias decide the path of the returned current and the loop area for those signal nets connecting I/Os. They form a number of different sized loop-inductance that have significant couplings with each other. We call the voltage bounce at I/Os *power integrity* in this work. Compared to the allocation of off-chip decoupling capacitors (mm^2), placing vias (um^2) have a smaller cost of area. This is important for the high density integration in 3D ICs as there may be a large number of signal nets to be delivered from the package to the chip.

On the other hand, due to increased power density and slow heat-convection at inter-layer dielectrics, heat dissipation is another primary concern in 3D ICs. The excessively high temperature can significantly degrade the reliability and performance of interconnects and devices.[3,10] We call the temperature gradient at active device layers *thermal integrity*. As shown in Figure 15.1., a heat-sink is placed at the top of device layers and it is the primary heat-removal path to the ambient air. One observation is that there are through vias delivering supply voltages or signals from the bottom package through each active device layer. Since the metal vias are good thermal conductors, the through vias can provide additional heat-removal paths passing the inter-layer dielectrics to the top heat-sink. This leads to the concept of adding *dummy thermal vias* or directly inside chips[3] to reduce effective thermal resistances. Its physical arrangement is further studied in Refs.[4,5]

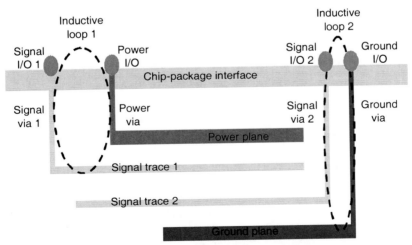

Fig. 15.2 The power delivery by vertical power/ground vias and its impact to inductive current loops.

In modern VLSI designs, such as clock-gating and uncertainty from the workload can lead to time-varying power inputs. This results in a spatially and temporally variant thermal model. The inputs are time-varying thermal power (see Fig. 15.3.)[7, 11] defined by the running-average of the cycle-accurate (often in the range of *ns*) power over several thermal time constants (often in the range of *ms*). They are injected at input ports of each layer. As such, a temporally and spatially variant temperature at output ports can be considered by defining an *integrity integral* with respect to time and space.[17] As a result, the temperature gradient can have either a sharp-transition with a large peak value, or a time-accumulated impact on the device reliability. In addition, different regions can reach their worst-case temperatures at different times.

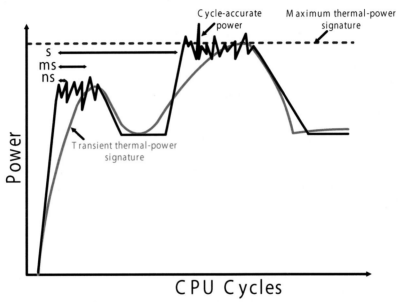

Fig. 15.3 The definitions of the cycle-accurate power, transient thermal-power, and maximum thermal-power at the different scale of time-constant.

A dynamic thermal-integrity constraint is thereby needed to accurately guide the physical level resource allocation. Since the active device layer at the bottom (See. Fig. 15.1.) has the longest path to the heat-sink at the top, in this work, a *dynamic thermal integrity* is defined as the integrated temperature fluctuation at p_o output ports on the bottom device layer. As shown in Refs.[3, 10, 14, 15, 17, 18] a dynamic thermal integrity can accurately capture not only the sharp-transition of temperature change due to dynamic power management, but also time-accumulated temperature impact that can affect device reliability. Similar to static thermal-integrity analysis, dynamic thermal integrity assumes the worst-case input from a limited number of thermal-power inputs. However, since the dynamic integrity has a more accurate transient temperature profile, it leads to a smaller allocation when compared to the static thermal-integrity based design.[4, 5]

Note that the dynamic power-integrity has already been employed in many on-chip or off-chip power integrity verifications and designs.[12, 16, 19] A similar *dynamic power integrity* in this work is defined as the time-integrated voltage bounce at power/ground I/Os, which are located on the interface between the bottom device layer and the package.

To calculate the dynamic thermal or power integrity, we represent the 3D ICs by two distributed models: a thermal-RC model for the heat-removal and an electrical-RLC model for the power-delivery. They (without power/ground vias) can be described in the state-space by $\mathcal{G}$

$$\mathcal{G}x(t) + C\frac{dx(t)}{dt} = B\mathbf{I}(t), y(t) = \mathcal{L}^T x(t) \tag{15.1}$$

or in frequency (s) domain

$$(\mathcal{G} + sC)x(s) = B\mathbf{I}(s), y(s) = \mathcal{L}^T x(s). \tag{15.2}$$

Note that B is the topology matrix to describe p_i input ports with injected input sources, and $\mathcal{L}$ is the one to describe p_o output ports for probing thermal or power integrity and adjusting via density.

15.3 TSV ALLOCATION PROBLEM

We notice that previous thermal via allocations[4, 5] assume adding dummy vias to conduct heat. They ignore the fact that power/ground vias can help remove heat as well. Therefore, the reusing of the power/ground via as the thermal via can save the routing resource for signal nets. More importantly, the allocation of power/ground vias can minimize not only dynamic power integrity, i.e., voltage bounce for those I/Os at package and chip interface, but also thermal integrity, i.e., the temperature gradient at those active device layers.

Similar to the work in Goplen and Sapatnekar,[5] this work assumes that the via allocation is after the placement and global routing but before the detailed routing of the signal nets. The power ground vias are placed at centers of tiles between two layers, and follow an aligned path from the bottom package I/Os to the top heat-sink. We call those aligned paths *vertical tracks* or *tracks*. As vias are aligned, the p_o tracks pass both p_o output ports of the electrical-RLC model and p_o output ports of the thermal-RC model. The density of power/ground vias at each track is the primary design parameter considered in this work. The density is adjusted to satisfy two requirements at output ports. The first is the integrity constraint of the temperature gradient and voltage bounce. The second is the resource constraint with provided signal net congestion.

Moreover, to capture the sharp-transition of temperature change as well as the time-accumulated temperature impact, we employ the *thermal-integral* used in Ref.[17] as the measure of dynamic thermal integrity at jth ($j=1,...,p_o$) output port:

$$f_j^T = \int_{t_o}^{t_p} max[y_j(t), Tc]\, dt = \int_{t_s}^{t_e} [y_j(t) - T_r]\, dt, \qquad (15.3)$$

with a pulse-width (t_s, t_e) in a sufficient long time-period t_p all in the scale of thermal-constant (ms). $y_j(t)$ is the transient temperature waveform at jth output port, and T_r is the reference temperature.

To further consider the spatial difference of p_o output ports at the bottom device layer, the overall thermal integrity is defined by a normalized summation:

$$f^T = \frac{\sum_{j=1}^{P_o} f_j^T}{t_p^T \cdot p^o}. \qquad (15.4)$$

Such a measure of thermal integrity takes into account both the temporal and spatial variation of temperature. Similarly, the power-integrity integral f_j^V is defined at jth power/ground I/O with reference voltages Vdd and ground, and integrated at the period $t_p{}^V$ in the scale of electrical-constant (ns). The overall power integrity f^V is defined similarly to f^T for p_o power/ground I/Os with the reference voltage V_r (0 for ground vias and Vdd for power vias).

Accordingly, we have the following problem formulation:

Given the targeted voltage bounce V_t for p_o output ports at power/ground I/Os, and the targeted temperature gradient T_t for p_o output ports at bottom device layer, the via-allocation problem is to minimize the total via number, such that the temperature gradient f^T is smaller than T_t and the voltage bounce f^V is smaller than V_t.

Such a via-allocation problem simultaneously driven by power and thermal integrity can be represented by

$$min \sum_{j=1}^{P_o} n_j$$
$$s.t.\ f^V \le V_t, f^T \le T_t$$
$$and\ n_{min} \le n_j \le n_{max} \qquad (15.5)$$

Note that n_j is the via density at the jth track and V_t and T_t are the targeted voltage bounce and temperature gradient. f^V and f^T are the metrics of power integrity and thermal integrity, respectively.

As discussed later, n_j is decided according to the power and thermal sensitivities obtained from the macromodel. As our power/ground vias are allocated after the placement and global routing of signal nets at each active device layer, the densities of those inter-layer signal nets are available to calculate a maximum density n_{max} for the power/ground vias. In addition, for the sake of the reliability concern of the large current, the via density n_j on the other hand cannot be smaller than the minimum density n_{min}. These parameters $(n_{max}, n_{min}, V_t, T_t)$ can be estimated and provided by users.

15.4 INTEGRITY OPTIMIZATION WITH MACROMODEL

To consider dynamic integrity during the design optimization of our TSV allocation problem, main difficulties to apply the above state-space equation come from three-fold. Firstly, there are so many inputs to try and so many outputs to probe. Secondly, the dimension of the distributed thermal-*RC* and electrical-*RLC* models are too large to analyze. Lastly, for the sake of design optimization, we are more interested in the than the nominal response. In the following, we will show how to compress the states, then I/Os and further how to generate sensitivity for the design automation.

15.4.1 Complexity Compression of States

As the layouts of active device layers and power/ground planes are discretized sufficiently, the size of resulting state-matrices thereby can be huge. As such, the distributed thermal-*RC* and electrical-*RLC* model are difficult to be employed for either the integrity verification or the optimization.

Grimme; and Odabasioglu *et al.*[6, 8] finds the dominant state variables and obtains compact macromodels. As shown in Refs.,[6, 8] the dominant state variables are related to the block Krylov subspace

$$\mathcal{K}(\mathcal{A}, \mathcal{R}) = \{\mathcal{A}, \mathcal{AR}, \dots \mathcal{A}^{q-1} \mathcal{R}, \dots\},$$

constructed from moment matrices

$$\mathcal{A} = (\mathcal{G} + s_0 C)^{-1} C, \mathcal{R} = (\mathcal{G} + s_0 C)^{-1} B$$

by expanding the system transfer function

$$\mathrm{H}\,(s) = \mathcal{L}^T (\mathcal{G} + sC)^{-1} B$$

at one frequency s_0.

By applying the block Arnoldi iteration,[8] a small dimensioned projection matrix Q ($N \times q \times p_i$) can be found to contain qth-order block Krylov subspace

$$\mathcal{K}(\mathcal{A}, \mathcal{R}, q) \subseteq Q.$$

Using Q the original system can be reduced by projection

$$(\hat{\mathcal{G}} = Q^T \mathcal{G} Q \quad \hat{C} = Q^T C Q \quad \hat{B} = Q^T B \quad \hat{\mathcal{L}} = Q^T \mathcal{L}.$$

Accordingly, the reduced system transfer function becomes

$$\hat{H}(s) = \hat{\mathcal{L}}^T (\hat{\mathcal{G}} + s\hat{C})^{-1} \hat{B}.$$

As proved in Refs.,[6, 8] the reduced $\hat{H}$ approximates the original system transfer function $H(s)$ by matching the first q block moments expanded at the frequency-point s_0. This procedure can be applied to generate compact macromodels for both thermal-*RC* and electrical-*RLC* circuits.

Our thermal-*RC* and electrical *RLC* models have multiple inputs and multiple outputs (MIMO). This brings challenges for the projection based model order reduction. The dimension of the reduced MIMO system

$\hat{H}$ ($\in R^{p_o \times p_i}$) depends on the input-port number p_i and output-port number p_o. In general, there are large numbers of injecting inputs. An accurate monitoring of integrity also needs large numbers of pre-designated regions to probe outputs. Therefore, an effective macromodel needs to further compress the number of ports when p_i and p_o are both large. In the following, we identify a much smaller number of principal input and output ports by studying the correlation.

15.4.2 Complexity Compression of I/Os

Generally, there can be thousands of thermal-power sources injected at each active layer or hundreds of switching-current sources injected at I/Os. The size of the macromodel increases with the number of ports, and hence the computational cost to solve the macromodel is still high. Since the electrical signals may share the same clock and operate within a similar logic function, their waveforms in the time-domain at certain input ports can show a correlation. Similarly, the thermal power may differ significantly between those regions with and without the clock gating, but can be quite similar inside the region with the same mode since inputs have similar duty-cycles over time. Based on the correlation, we can reduce the in I/Os by identifying those .

We call this phenomenon input-similarity. As the input vector

$$\mathbf{I}(t) = [\mathbf{I}_1\ \mathbf{I}_2 \dots \mathbf{I}_{p_i}] \in R^{p_i \times 1}. \tag{15.6}$$

is usually known during the physical design, they can be represented by taking a set of 'snapshots' sampled at $\mathcal{N}$ time-points

$$\begin{bmatrix} \mathbf{I}_1(t_0) & \cdot & \cdot & \cdot & \mathbf{I}_1(t_N) \\ \vdots & & \ddots & & \vdots \\ \mathbf{I}_{p_i}(t_0) & \cdot & \cdot & \cdot & \mathbf{I}_{p_i}(t_N) \end{bmatrix} \tag{15.7}$$

in a sufficiently long period $[0, T_p]$. The sampling cycle is in a different time-scale for the thermal-power (ms) and switching-current (ns). According to the POD analysis,[1] the similarity can be mathematically described by a correlation matrix (or Grammian), estimated by a co-variance matrix:

$$\mathcal{R} = \frac{1}{\mathcal{N}} \sum_{\alpha=1}^{N} (\mathbf{I}(t_\alpha) - \bar{\mathbf{I}})(\mathbf{I}(t_\alpha) - \bar{\mathbf{I}})^T \in R^{p_i \times p_i}. \tag{15.8}$$

$\bar{\mathbf{I}}$ is a vector of mean values defined by:

$$\bar{\mathbf{I}} = \frac{1}{\mathcal{N}} \sum_{\alpha=1}^{N} \mathbf{I}(t_\alpha) \tag{15.9}$$

Usually, the input vector $\mathbf{I}(t)$ is periodic and the waveform in each period can be approximated by the piecewise-linear model.

An output similarity is defined for responses at output ports and measured by a *output correlation matrix*. To extract the output correlation matrix that is independent on the inputs, we assume that p_i inputs in the input vector $\mathbf{I}(s)$ are all the unit-impulse source $h(s)$ and define an input-port vector $\mathcal{J}(s)$ by

$$\mathcal{J} = B\mathbf{I}(s), \in R^{1 \times N}, \tag{15.10}$$

which has p_i non-zero entries with the unit-value '1'. Accordingly, the p_o output responses $y(s)$ are calculated by

$$y(s) = \mathcal{L}^T(\mathcal{G} + s\mathcal{C})^{-1}\mathcal{J}$$
$$= [y_1(s)\,y_2(s)\,...\,y_{p_o}(s)] \in R^{p_o \times 1}. \tag{15.11}$$

The according output correlation matrix is extracted in the frequency-domain. Similarly, the output signals can be represented by taking a set of 'snapshots' sampled at $\mathcal{N}$ frequency points

$$\begin{bmatrix} y_1(s_0) & \cdot & \cdot & \cdot & y_1(s_N) \\ \vdots & & \cdot & & \vdots \\ y_{p_o}(s_0) & \cdot & \cdot & \cdot & y_{p_o}(s_N) \end{bmatrix} \tag{15.12}$$

in a sufficiently wide band $[0, s_{max}]$. The s_{max} locates in a low-frequency range for the temperature and in a high-frequency range for the voltage. A co-variance matrix is defined in the frequency-domain as follows

$$R = \sum_{\alpha=1}^{\mathcal{N}} (y(s_\alpha) - \bar{y})(y(s_\alpha) - \bar{y})^T \in R^{p_o \times p_o} \tag{15.13}$$

to estimate the correlation matrix among p_o outputs.

$\bar{y}$ is a vector of mean values defined by:

$$\bar{y} = \frac{1}{\mathcal{N}} \sum_{\alpha=1}^{\mathcal{N}} \bar{y}(s_\alpha) \tag{15.14}$$

Let $\mathcal{V} = [v_1, v_2, ..., v_K]$ $(\in R^{N \times K})$ as the first K singular-value vectors of the input correlation matrix $\mathcal{R}$, and $\mathcal{W} = [w_1, w_2, ..., w_K]$ $(\in R^{N \times K})$ as the first K singular-value vectors of the output correlation matrix R. All singular-value vectors are obtained from the (SVD) of $(\mathcal{V}, \mathcal{W})$. A rank-K matrix P_i can be constructed by $P_i = VV^T$, and a rank-K matrix P_o can be constructed by $P_o = WW^T$. As shown in [Astrid *et al.* (2007)], the correlation matrix $(\mathcal{R}, R)$ is essentially the solution that minimizes the least-square between the original states $(\mathbf{I}(t), y(s))$ and their rank-K approximations $(P_i \cdot \mathbf{I}(t), P_o \cdot y(s))$. As a result, both the input signals $\mathbf{I}(t)$ and the output signals $y(s)$ can be approximated by an invariant (or dominant) subspace spanned by the orthonormalized columns of V and W, respectively:

$$\mathbf{I} = \mathcal{V}\mathbf{I}_K, y = \mathcal{W}y_K. \tag{15.15}$$

Based on (15.15), it leads to the following equivalent system equation

$$(\mathcal{G} + s\mathcal{C})\, x_K(s) = B_K \mathbf{I}_K(s), y_K(s) = \mathcal{L}_K^T x_K(s) \qquad (15.16)$$

where

$$\mathcal{L}_K^T = \mathcal{W}^T \mathcal{L}^T, B_K = B\mathcal{V}. \qquad (15.17)$$

Therefore, both the dimensions of $\mathcal{L}$ ($\in R^{N \times p_o}$) and ($\in R^{N \times p_i}$) are greatly reduced when $K << p_i$ and p_o. We call $\mathbf{I}_K$ and y_K *principal inputs and outputs* identified by *principal input-port and output-port matrices* $_K$ and $\mathcal{L}_K$, respectively.
$\mathcal{B}$

15.4.3 Dynamic Integrity and Sensitivity by Structure$\mathcal{B}$ and Parameterized Macromodel

Recall that the design parameter in our problem formulation is the via density at one track. Blindly allocating the via by searching all kinds of combinations would be computationally expensive if not impossible. Therefore, we decide the via density based on the changes at outputs, i.e., sensitivities, caused by the change of via density.

To calculate sensitivity, let's first parameterize the nominal system (15.2). The added via at one track is described by two parameters: n_j the via density and X_j the topological matrix that connects the via into the nominal system. As such, a parameterized state-space description can be obtained by

$$(\mathcal{G} + s\mathcal{C} + \sum_{j=1}^{P_o} n_j g_j + s \sum_{j=1}^{P_o} n_j c_j)x_K(n, s) = B_K \mathbf{I}_K(s),$$

$$y_K(n, s) = \mathcal{L}_K^T x_K (n, s) \qquad (15.18)$$

Similar to [Yu *et al.* (2006b, c, a, 2007a, 2008, 2009)], we expand $x(\mathbf{n},s)$ in the Taylor series with respect to n_j, and introduce a new state variable x_{ap}

$$x_{ap} = [x^{(0)}, x^{(1)}, ..., x_{P_o}^{(1)}]^T. \qquad (15.19)$$

It contains both the nominal response $x^{(0)}$ and its first-order sensitivities $[x^{(1)}_1,...,x^{(1)}_{po}]$ with respect to p_o parameters $[n_1,...,n_{po}]$. The overall response is obtained by

$$x = x^{(0)} + \sum_{j=1}^{P_o} x_j^{(1)}.$$

Substituting (15.19) into (15.18), (15.18) can be reformulated into a parameterized system with an augmented dimension by

$$(\mathcal{G}_{ap} + s\mathcal{C}_{ap})\, x_{ap} = B_{ap}\mathbf{I}_K(s), y_{ap} = \mathcal{L}_{ap}^T x_{ap}, \qquad (15.20)$$

where $\mathcal{G}_{ap}$ and $\mathcal{C}_{ap}$ show a lower-triangular-block structure and hence x_{ap} can be solved from block-backward-substitution.[12-14, 16-18]

To further compress the dimension of the state-matrices $\mathcal{G}_{ap}$ and C_{ap}, we first construct a lower-dimensioned subspace Q_{ap} from the moment expansion of (15.20), and then transform Q into the block-diagonal form Q_{ap}. After the block-orthonormalization of Q_{ap}, we apply a two-side projection to (15.20) by Q_{ap} and obtain a dimension-reduced system with preserved lower-triangular-block structure.[12-14, 16-18] We call the resulted macromodel. The accuracy of the macromodel is preserved to match the dominant moments of the original model. More importantly, due to the structure-preservation, both of the nominal response and the sensitivity with regard to the via-density change, can be calculated simultaneously. As such, we can easily embed such a structured and parameterized macromodel into the optimization flow of our TSV allocation problem.

ALGORITHM: SENSITIVITY BASED VIA ALLOCATION

---

1. **Input**: K principal input ports, K principal output ports, maximum temperature bound T_{max}, maximum voltage-bounce bound V_{max}, signal-net-congestion bound n_{max} and current-density bound n_{min}

2. *Construct* structured and parameterized macromodel;

3. *Compute* nominal voltage(V)/temperature(T) and sensitivity $\mathbf{S}_V/\mathbf{S}_T$;

4. *Check* V_{max} and T_{max} constraints for all tiles;

5. *Increase* the via density $\mathbf{n}$ according to weighted sensitivity $\mathbf{S}$ in the range of (n_{min}, n_{max});

6. *Update* the structured and parameterized macromodel;

7. *Repeat* from Step 3 until Step 4 is satisfied;

8. **Output**: Via density vector $\mathbf{n}$

Fig. 15. 4 Algorithm for the sensitivity-based via allocation with the use of macromodels.

The overall optimization flow to solve the problem formulation (15.5) is outlined in Algorithm above. Its inputs are two parts. The first is a principal system by (15.20) with the identified K principal input and output ports. The second is the user provided temperature bound T_{max}, voltage bounce bound V_{max}, signal-net congestion bound n_{max}, and current-density bound n_{min}. Then, a structured and parameterized macromodel is built once. Both nominal responses and sensitivities at K principal input-ports for each perturbed allocation-pattern are solved in one time. If the integrity constraints are not satisfied for K principal tracks, the vias density vector $\mathbf{n}$ are increased according to the sensitivity. This process repeats until the integrity constraints are satisfied. Details of this Algorithm can be found in Ref.[14]

15.5 RESULTS

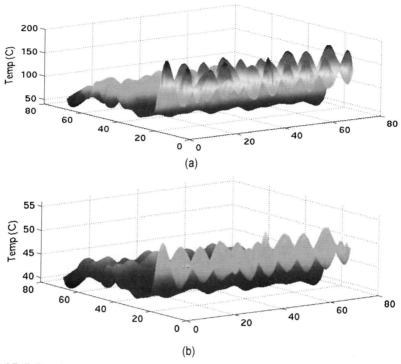

(a)

(b)

Fig. 15. 5 Steady-state temperature maps of bottom device layer (a) before allocating via, and (b) after allocating via in different temperature scales.

Experiments are implemented in C and MATLAB and run on a Sun-Fire-V250 workstation with 2G RAM. We call the separated allocation of thermal vias and power/ground vias the *sequential optimization*, and call our allocation of power/ground vias for both power and thermal integrity the *simultaneous optimization*. Moreover, the steady-state analysis is employed to calculate a static integrity.[6, 8] We use the sequential optimization with the static integrity as the baseline, in comparison to the sequential optimization with the dynamic integrity and the simultaneous optimization with the dynamic integrity proposed in this work. The electrical, thermal constants and dimensions are the same with Ref.[14] The targeted voltage violation V_t is $0.2V$ and the targeted temperature T_t is $52\,°C$. One modest 3D stacking is assumed with 2-device-layer/2-dielectric-layer. Moreover, there are 1-heat-sink and 2-P/G-plane used in the example.

Figure 15.5. further shows the steady-state temperature map across the bottom device layer. In this example, we assume that all thermal-power sources are located along one side of the device layer. The initial chip temperature at the bottom layer is 150°C, and its temperature profile at steady-state is shown in Fig. 15.5.(a). In contrast, the via-allocation results in a cooler temperature that closely approaches the targeted temperature as shown in Fig. 1.5.5.(b). Clearly, even at steady-state the temperature is still spatially variant. An accurate measure of integrity is therefore needed to consider a time and space averaged integrity at selected probing ports.

Table 15.1 The complexity of the original circuits and the reduced circuits including: the size, number of input-ports, and number of output-ports.

ckt	Total tile#	Reduced size (T,V)	Input src# (T,V)	K-input (T,V)	Output track#	K-output (T,V)
ckt 1 (2-layer)	1.9K	(30,80)	(10,20)	(10,20)	4^2	$(4^2,4^2)$
ckt 2 (2-layer)	6K	(15,48)	(100,200)	(5,8)	4^3	(6,4)
ckt 3 (2-layer)	12K	(80,160)	(300,600)	(10,16)	4^4	(8,5)
ckt 4 (2-layer)	27K	(96,180)	(1K,2K)	(12,18)	4^4	(10,8)
ckt 5 (2-layer)	52K	(96,220)	(1K,3K)	(12,20)	4^5	(12,14)

Table 15.2 Comparisons of via number and runtime for the sequential optimization with steady-state analysis, the sequential optimization with transient analysis and the simultaneous optimization with transient analysis. Two macromodels are used during the transient analysis. Macromodel-1 does not use the port-compression, and macromodel-2 uses the port-compression

ckt	Steady-state(direct)		Transient(MACRO-1)		Transient(MACRO-2)		
	runtime (s)	total via # by seq-opt	runtime (s)	total via # by seq-opt	runtime (s)	total via # by seq-opt	total via # by sim-opt
ckt 1 (2-layer)	5.4	178800	0.63	153800 (-13%)	0.63	153800 (-13%)	112800 (-36%)
ckt 2 (2-layer)	29.7	184900	0.81	159600 (-13%)	0.56	159600 (-13%)	118200 (-36%)
ckt 3 (2-layer)	182.2	218100	18.6	183800 (-16%)	4.2	184200 (-15%)	136200 (-38%)
ckt 4 (2-layer)	1269.2	234800	165.7	199000 (-15%)	10.3	199600 (-15%)	145600 (-38%)
ckt 5 (2-layer)	NA	NA	NA	NA	41.2	208600 (NA)	154200 (NA)

The details of the benchmark circuits are summarized in Table 15.1. The runtime and the number of allocated vias are compared in Table 15.1. In Table 15.2, columns 2-3 show the runtime and the number of allocated via for the baseline, and columns 4-8 show the results for the optimizations using the dynamic integrity. In detail, column 4 shows the runtime of transient analysis using macromodels without the port-compression, and column 5 shows the number of allocated vias under the sequential optimization. Column 6 shows the runtime of transient analysis using macromodels with the port-compression, and columns 7-8 show the number of allocated vias under the sequential and simultaneous optimizations, respectively.

The use of macromodels reduces the computational cost to solve power and thermal integrity and their sensitivities. Compared to the macromodel without the port-compression, the macromodeling with the port-compression reduces the overall runtime by up to 16X with similar allocation results. Compared to steady-state analysis with full-matrix analysis, our macromodel with the port-compression has a 127X smaller runtime. Additionally, the steady-state analysis can not complete the largest example in a reasonable runtime. The maximum transient-waveform difference introduced by the macromodel is about 7% when compared to the exact transient waveform.

We further compare the sequential thermal/power optimization with the simultaneous thermal/power optimization. Here both methods allocate vias with the use of dynamic integrity. Our simultaneous optimization reduces the via-cost by up to 34% when compared to the sequential optimization with static integrity, and by up to 22% when compared to the sequential optimization with dynamic integrity. This demonstrates that the reusing of power/ground vias can reduce the via cost when compared to allocating the dummy thermal vias separately from the power/ground vias.

15.6 SUMMARY

This chapter explains the need for dynamic power and thermal integrity for the high-performance 3D integration, using an example of the through-silicon-via (TSV) allocation. To cope with design complexity, an effective macromodel is employed to abstract the physical level details for the system level design. It includes the I/O compression and the structured and parameterized model order reduction, which efficiently calculate power/thermal integrity and their sensitivity with respect to via density. Compared to the design without the use of the dynamic integrity, experiments show that our approach reduces the number of TSVs by up to 38%, yet achieves the speedup by hundreds of times.

References

1. Astrid, P., Weiland, S. and Willcox, K. (2007). Missing point estimation in models described by proper orthogonal decomposition, *IEEE Trans. Autom. Control*, pp. 2237–2251.

2. Banerjee, K., Souri, S. J., Kapur, P. and Saraswat, K. C. (2001). 3D ICs: A novel chip design for improving deep submicron interconnect performance and systems-on-chip integration, *Proc. IEEE*, pp. 602–633.

3. Chiang, T.-Y., Banerjee, K. and Saraswat, K. C. (2001). Compact modeling and spice-based simulation for electrothermal analysis of multilevel ulsi interconnects, in *Proc. Int. Conf. on Computer Aided Design (ICCAD)*, pp. 165–172.

4. Cong, J., Wei, J. and Zhang, Y. (2004). A thermal-driven floorplanning algorithm for 3D ICs, in *Proc. Int. Conf. on Computer Aided Design (ICCAD)*, pp. 306–313.

5. Goplen, B. and Sapatnekar, S. (2005). Thermal via placement in 3D ICs, in *Proc. Int. Symp. on Physical Design (ISPD)*, pp. 167–174.

6. Grimme, E. J. (1997). *Krylov projection methods for model reduction (Ph. D Thesis)* (Univ. of Illinois at Urbana-Champaign).

7. Liao, W., He, L. and Lepak, K. (2005). Temperature and supply voltage aware performance and power modeling at microarchitecture level, *IEEE Trans. on Computer-Aided Design of Integrated Circuits and Systems*, pp. 1042–1053.

8. Odabasioglu, A., Celik, M. and Pileggi, L. (1998). PRIMA: Passive reduced-order interconnect macro-modeling algorithm, *IEEE Trans. on Computer-Aided Design of Integrated Circuits and Systems*, pp. 645–654.

9. Tan, C. S., Gutmann, R. J. and Reif, L. R. (2008). *Overview of Wafer-Level 3D ICs* (Springer US).

10. Teng, C. C., Cheng, Y. K., Rosenbaum, E. and Kang, S. M. (1997). iTEM: A temperature-dependent electromigration reliability diagnosis tool, *IEEE Trans. on Computer-Aided Design of Integrated Circuits and Systems*, pp. 882–893.

11. Tiwari, V., Singh, D., Rajgopal, S., Mehta, G., Patel, R. and Baez, F. (1998). Reducing power in high-performance microprocessors, in *Proc. Design Automation Conf. (DAC)*, pp. 732–737.

12. Yu, H., Chu, C. T. and He, L. (2007a). Off-chip decoupling capacitor allocation for chip package co-design, in *Proc. Design Automation Conf. (DAC)*, pp. 618–621.

13. Yu, H., Ho, J. and He, L. (2006a). Simultaneous power and thermal integrity driven via stapling in 3D ICs, in *Proc. Int. Conf. on Computer Aided Design (ICCAD)*, pp. 802–808.

14. Yu, H., Ho, J. and He, L. (2009). Allocating power ground vias in 3d ics for simultaneous power and thermal integrity, *ACM Trans. on Design Automation of Electronics Systems*.

15. Yu, H., Hu, Y., Liu, C. and He, L. (2007b). Minimal skew clock embedding considering time variant temperature variation, in *Proc. Int. Symp. on Physical Design (ISPD)*, pp. 173–180.

16. Yu, H., Shi, Y. and He, L. (2006b). Fast analysis of structured power grid by triangularization based structure preserving model order reduction, in *Proc. Design Automation Conf. (DAC)*, pp. 205–210.

17. Yu, H., Shi, Y., He, L. and Karnik, T. (2006c). Thermal via allocation for 3D ICs considering temporally and spatially variant thermal power, in *Int. Symp. on Low Power Electronics and Design (ISLPED)*, pp. 156–161.

18. Yu, H., Shi, Y., He, L. and Karnik, T. (2008). 3d ic design considering spatially and temporally variant thermal power, *IEEE Trans. on Very Large Scale Integration (VLSI) Systems*, pp. 1609–1619.

19.   Zhao, S., Roy, K. and Koh, C. K. (2002). Decoupling capacitance allocation and its application to power supply noise aware floorplanning, *IEEE Trans. on Computer-Aided Design of Integrated Circuits and Systems*, pp. 81–92.

Chapter 16

OUTLOOK

Yang, Ya Lan
IEK/ITRI

16.1 DEVELOPMENT OF 3D IC BECOMES CLEARLY

Since the demand of electronic products for semiconductor components specifications has always been focused on the pursuit of the small form factor, high integration, high performance, low cost, low power consumption and prompt time to market, the development of semiconductor technology has constantly created all sorts of new technologies under the premise of achieving these goals.

In recent years, due to the increase of consumer's demand for portable electronics of functional integration, the semiconductor technology has also been developing towards the trend of integration. Such as SoC (System on Chip) and SiP (System in Package) technologies are developed for fulfilling the requirement of integration.

According to SiP White Paper, which was published by ITRS (International Technology Roadmap for Semiconductors), the major technology development in semiconductor industry were focus on scaling down to meet Moore's Law for fulfilling the requirement of increasing density of transistor and reducing the cost.

In 2007, at Emerging Technologies Conference, which was organized by MIT, researchers pointed out that the ending of Moore's Law may come earlier due to the physical limits of scaling down for semiconductor manufacturing, as well as the difficult balance between advanced CMOS technology development and the cost required.

As the technology of scaling down has been gradually facing the challenge of physical limits, as well as enormous investment that advanced process equipment requires, the semiconductor industry has actively thought about

3D Integration for VLSI Systems
Edited by Chuan Seng Tan, Kuan-Neng Chen and Steven J. Koester
Copyright © 2012 by Pan Stanford Publishing Pte. Ltd.
www.panstanford.com

the possible technical options other than scaling down. In recent years, the market of integration via SoC and SiP technologies are booming due to the requirements of high integration, high speed, and low-power consumption driven by electronic products.

The demand for heterogeneous chips integration and the small form factor of communication applications, has promoted the development three dimensional SiP (3D SiP) to become an important technology for companies' investment in semiconductor industry, due to the shorter time for R&D (which is compared with the heterogeneous integration via SoC technology), lower cost and better integration flexibility of 3D SiP. According to the future plan of ITRS,[1] since the demand for functional integration of electronic devices has strongly increased, it will be the major focus to combine the miniaturization technology development of Moore's Law with SoC / SiP integration for semiconductor industry (Fig. 16.1).

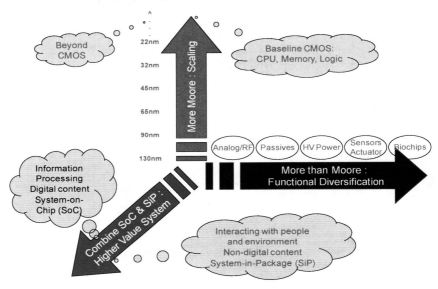

Fig. 16.1 Semiconductor technology trends.

16.2 AN OVERVIEW OF 3D IC TECHNOLOGY APPLIED STATUS

3D IC technology is now still in early stages of development. There are several 3D IC related consortiums have been formed globally and numbers of 3D IC research and development activities for different purposes have been actively proceeding. From commercial point of view for 3D IC technology, since the second half of 2008, products of CIS module, which use TSV (Through Silicon Via) as interconnection, have been launched by CMOS image sensor manufacturers. As far as other applied fields besides image sensors for applications, there are a number of manufactures related to the field, saying that planed commercial schedule shall be launched from 2010 to 2012.

From the view point of the involving companies, although there are lots of equipments and materials suppliers involving and developing 3D IC related technologies and announcing their activities and products via organizing consortiums, the IDMs (Integrated Device Manufacturers) still take the lead among the fields as for the technology and process development. As the revolution goes by, the discussions of 3D IC related technology have gradually been slowly changing from overcoming the technical problems to transferring to the exploration of the discussion of supply chain and commercialization.

Due to the financial crisis affect since 2008, the market of semiconductor was also showed a big drop during the fourth quarter in 2008 to the first quarter in 2009. And for the full year of 2009, the worldwide semiconductor market was around 17% decline compared with the market in 2008.

However, 3D IC related issues are continued to be lively discussed in the varieties of global 3D IC technology seminars without the impact of the global financial crisis. According to the mass production schedules of previous 3D IC potential applications (Fig. 16.2), which are planned by the leading devices vendors, image sensor modules with TSV technology have been mass produced since 2008.

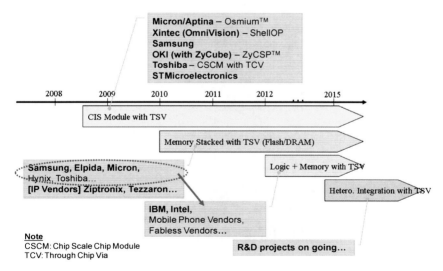

Fig. 16.2 3D IC major application categories and planned production schedules.

According to the original plan of the leading manufacturers, the application of memory module for homogeneous memory stacking with TSV technology will be mass produced from 2010 to 2011. Even under the impact of the financial crisis during 2008 to 2009, those key global memory component suppliers, such as Samsung and Micron, still strongly believe that the launching schedules of mass production will not be deferred during the economic downturn.

16.2.1 Image sensor and camera module

Wire bonding was widely applied for the image sensor packaging as the major interconnection between chip and circuit board. But it needs much more space for wire bonding, and increases the size and the thickness of packaging. For the requirement of thinning for cellular phone, lots of technology development for lens material and packaging structure make big help for improving the thickness for the package of camera module.

With the development of TSV technology, making conduction electrical vias by laser drilling not only saves the bonding space, but also reduces the space of circuit board and silicon. It means that the adoption of TSV solution will reduce cost considerably. The assessment made by Toshiba shows that the size of camera module with TSV architecture will be reduced 64% compared with traditional one with wire bonding architecture (Fig. 16.3).[2] This helps meeting the thin-oriented demand for camera phone and digital still camera applications.

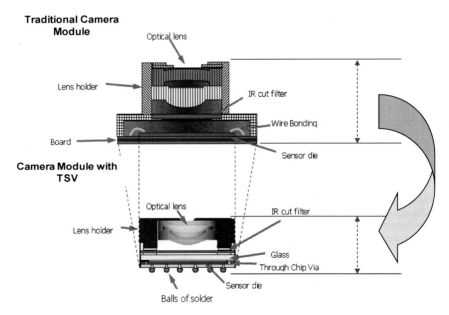

Fig. 16.3 The comparison between Toshiba CSCM camera modules and the traditional camera modules.

For current camera module application, there is only one chip for image sensing, and other components for the module are stacking lens. Due to the simple chip structure of the module, the problem of testing for adopting TSV technology is relatively easy. And because of the original package structure of image sensor is easier for adopting TSV technology, camera module became to be the earlier adopter when compare with other potential applications.

Micron, the global leading image sensor manufacturer, had invested in the development of TSV technology used by the packaging of image sensor before 2004. And Aptina, an image sensor vendor which was spun from Micron, has disclosed their first WLC (Wafer Level Camera) technology and related product in March 2008.

For the product that announced by Aptina, the interconnection technology for the camera module was TSV, and was called as Osmium™ technology by Aptina. According to the data provided by Aptina, the thickness of the camera module is only 2.5 mm, and the thickness of the camera module is decreased to 50%~70% compared with traditional camera module (Fig. 16.4). [3]

XinTec has played a major partner for OmniVision as back-end packaging segment. According to XinTec's roadmap, the TSV processes have been invested between 2006 and 2007, and transformed into the mass production stage in the second half of 2008.

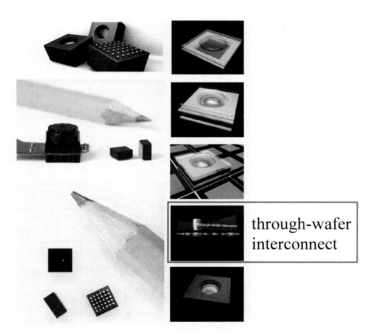

through-wafer interconnect

Fig. 16.4 Aptina's WLC module.

Samsung has started its investment in TSV technology development since 2006 and has released news of successfully applying TSV technology to memory products. As for the application of image sensor, Samsung has brought up their vertical via etching and DFR (Dry Film Resist) applied TSV technology process to replace traditional tapered via process used for CIS packaging in order to simplify the process and reduce the cost.

In Europe, STMicroelectronics has started providing the services of mass production for their TSV technology, and their ultra-thin image sensor products, VD6725, VD6853 and VD6803, which were launched later in February 2009, are also supported by TSV technology.

OKI has reached an agreement with ZyCube at the beginning of 2007, conducting the TSV technology for their CMOS and CCD image sensors named as ZyCSP™, with through-hole manufacturing process by via last process, and beginning mass production services in the second half of 2007.

In October 2007, at Japan's CEATEC exhibition, Toshiba displayed its ultra-miniaturized image sensor module-CSCM (Chip Scale Chip Module) and declared that the interconnect technology was the adoption of TSV technology. Toshiba named this technology as TCV (Through Chip Via). The products which support this technology include three modules, TCM9200MD, TCM9100MD and TCM9000MD. The mass production service had been provided since January 2008. In May 2008, Toshiba also released that the TCV technology will be applied to its more advanced packages of 2 million-pixel and the 3 million-pixel image sensor devices.

16.2.2 Memory module – NAND flash and DRAM

The application of memory can be divided into two major fields, NAND Flash and DRAM module.

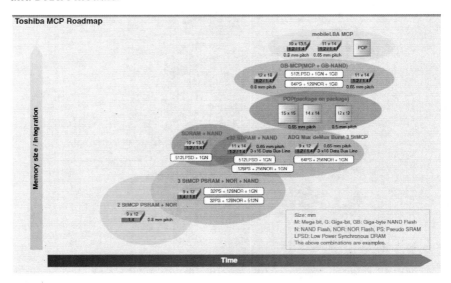

Fig. 16.5 Toshiba's MCP roadmap.

For the applications of NAND Flash module, due to the low pin count of the package for NAND Flash, the requirement for the package pitch is relatively less stringent. And for the development status and the schedule of mass production for NAND Flash module with TSV technology are slightly further than DRAM applications during the past years. In the application of NAND Flash module, the strongly demanded for high capacity memory cards and SSD (Solid State Disk) are considered to be the potential applications by using TSV technology.

Due to the rapid progress of NAND Flash component design and the fast enhance of a single chip density for NAND Flash, the need of high capacity requirements of the whole application system can be met through enhancing the density of each single NAND Flash chip.

However, the wire bonding technology can easily achieve the production of eight chips stacking and fulfill the requirement of high density for the memory module. Using TSV technology as an interconnection for NAND Flash module application can neither significantly show the efficiency nor meet the need of high memory density. Also the immaturity of process technology or non-economical scales may cause much higher costs than using traditional wire bonding technology. So there seems no urgency of using TSV technology for NAND Flash module now and in the near future.

But for the point of view for long term development, using TSV for NAND Flash module still has its chance. Since there's more and more integration of the mobile phone features in the future, as the camera function became one of the basic requirement of cellular phone, it may also sparked consumer's interest of using mobile phone for dealing image applications. After replacing some functions of camera with mobile phones, consumers may also copy the same experience of replacing video camera with mobile phones. From this point of view, the current 4GB, 8GB memory card capacity may not meet the need of memory capacity for the video related processing.

In 2010, the main memory cards supply manufacturers, Toshiba and Sony, will launch TB (Tera-byte) memory cards into the market as planed. It was predicted that the opportunity of using TSV technology is possible in the market to meet the requirement of both miniaturization in size and high-capacity memory at the same time.

According to the latest roadmap of memory packaging technology for mobile phones by Toshiba, the wire bonding technology will continue to focus on the volume production of stacked packaging during these years, especially in the PoP, which has better opportunities (Fig. 16.5). [4] But in the long run, if the cost of TSV technology can be declined to a certain degree, there should be opportunity to cut into the applications market of demand for large-capacity NAND Flash memory in the future.

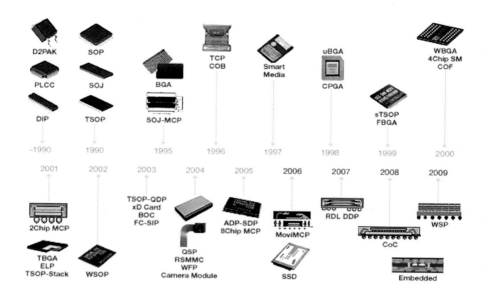

Fig. 16.6 Samsung's memory packaging technology roadmap.

The world's largest supplier of NAND Flash, Samsung, has different point of view for the adoption of 3D IC technology on NAND Flash. Back in April 2006, Samsung had disclosed its TSV technology in NAND Flash stacking — WSP (Wafer-Level Stack Process), displaying their 2Gb density, eight chips stacking NAND Flash module. One of the priorities of the demonstration was their latest TSV interconnect technology, and the advantages of thinning by using TSV technology had brought extensive discussions among the industry.

Besides the technical display and announcements, Samsung has no real mass production for NAND Flash with TSV technology during these past years. However, according to Samsung's packaging technology roadmap, 3D IC technology used by memory applications will still remain the main position of its technology development. For NAND Flash with TSV, 3D IC technology will be used by the applications for both SSD and memory card, as Samsung's plan (Fig. 16.6). [5] At ISSCC 2009 (International Solid State Circuits Conference) in San Francisco, Samsung exposed its memory applications with TSV technology to the latest masterpiece. With the increased speed of DRAM, connecting multiple memory modules via limited bus have became more challenge. In addition, since the system required for more memory capacity, the demand for larger capacity DRAM product is also rising.

For meeting the requirements of both high speed and high density for DRAM, Samsung provided two possible solutions. The first one was making the 4GB DDR3 via 56 nm process technology and improving design method for separating memory array to achieve the requirement of data rate at 1.6 Gbps with 1.2 V power supply.

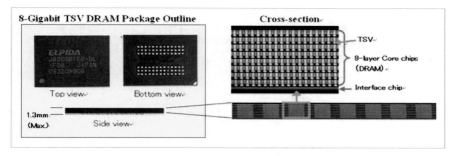

Fig. 16.7 Elpida 8GB TSV DRAM package.

Samsung provided another solution to using TSV technology for stacking 4 DRAM chips with 2GB for each chip, and an I/O chip at the bottom. The DRAM chips were stacked between the channels, through 300 TSV vias for connecting and transferring data, and meet the requirement of both high memory density and high speed data transferring.

In August 2009, Japanese DRAM maker Elpida, released its successful adoption of TSV technology for their 1Gb, eight DRAM chips stacking, which reveals replacing the filling material for the vias from Poly-Si to be copper (Fig. 16.7). [6] And due to the lower resistance feature of copper, filling with copper will greatly help reducing power consumption of high-speed DRAM in the server applications.

As the plan from Elpida's technology roadmap, published by NikkeiBP [7], the purpose of TSV technology used by DRAM module is not only for the concern of increasing the memory density, but also for providing a system level solution for heterogeneous integration which include logic, RF, MEMS and other devices in the long term (Fig. 16.8).

In addition to the positive research and development shown by a variety of memory makers, some vendors started their core business by IP (Intellectual Property) licensing and made some related standards for memory module by TSV. Currently, the 3D-IC Alliance, which was launched by Tezzaron and Ziptronix, is based on establishing 3D IC related standards.

In June 2008, 3D-IC Alliance had announced a standard named IMIS™ (Intimate Memory Interface Specification), which contains the standards about interfaces between memory and logic component stacking via using 3D IC technology. IMIS defines a vertical high-bandwidth bus between a logic and memory components. Through the vertical high-bandwidth bus, the operation performance of memory is as if the memory is embedded in a 2D SoC structure. IMIS is an open standard, which is available for the industry, and the purpose is to push 3D IC technology for commercialized quickly.

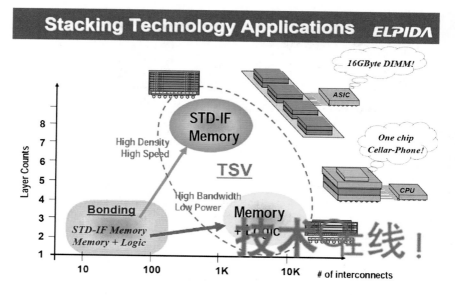

Fig. 16.8 Elpida's target applications of TSV technology.

16.2.3 Heterogeneous integration

For the application of heterogeneous integration, IBM and Intel have pointed out using TSV technology as the interconnection for the integration of logic component and stacking memory. But currently, it is till on the stage of releasing technical literature. The real commercialization products have not emerged yet.

In addition to the architecture of heterogeneous integration via 3D IC technology needed by computing related industry, Nokia had put forward their view of using TSV technology on integrating ASIC chips and memory, which was announced in RTI International meeting in 2007. But so far, traditional SiPs with wire bonding technology, such as MCP and PoP, 3D structure, are still widely used by cellular phones.

A European Union organization, IST (Information Society Technologies), had formed a project named e-CUBES (Electronic Cubes) in February, 2006. The target of the project was to construct a micro form factor and low-cost sensing system through sensors and wireless communication technologies, and to apply in intelligent air monitoring, space surveillance, health care-related wireless sensor networking and intelligent motor control system. The funding for the project was 1.2 million Euros and the implementation of the project was from February 2006 until the end of January 2009.

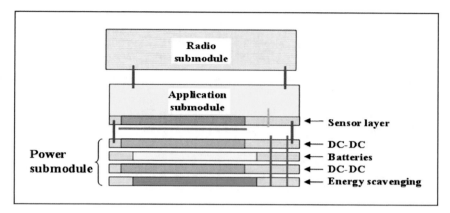

Fig. 16.9 e-Cubes program of the 3D packaging architecture.

According to the goal of e-Cubes project, the main technologies developments include various technologies related to 3D integrations, the necessary processes and thinning technologies for 3D integration, packaging technologies for 3D vertical integration, power supply and management technologies required for portable devices, design techniques for 3D SoC, verification and simulation tool developments for 3D SoC, and various technologies developments for related components such as RF components and communication technologies. The members in the project include Infineon, Alcatel Alenia Space, Honeywell Romania, Philips Research, 3D Plus, CEA, IMEC, Institute of Electron Technology, TU, and Uppsala University and other advanced European electronic industry representatives, research institutes and schools units.

In ECTC meeting in June 2008, Fraunhofer IZM pointed out that a member of the e-Cubes plan-Infineon had developed a 0.3"cube tire pressure sensor system through the result of e-Cubes research. The system combines a MEMS-tire pressure sensor and RF receiver /transceiver components and antenna in this cube. However, under the consideration of cost, the interconnection technology for components of the system is now through the stud bumps. Figure 16.9 illustrates the main structure of e-CUBES project. [8]

3D related programs and developments are also on going in Japan. NEDO started related programs with ASET for 3D IC related research and development since 2008. In the meeting of Semicon West 2009, ASET has recently exposed the 3D IC technology development project in Japan — Dream Chip program. The entire program will be carried out from 2008 until 2012. The ASET execution includes two sub-projects, "High-Density 3D-Integration Technology for Multifunctional Devices" and "Three-dimensional Reconfigurable Device Technology." Figure 16.10 illustrates the architecture of Dream Chip in Japan. [9]

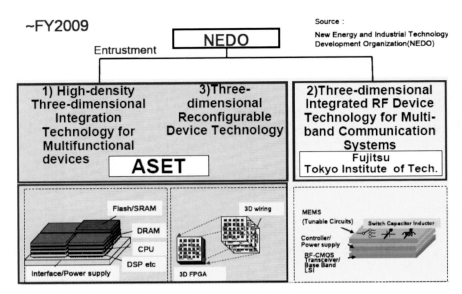

Fig. 16.10 Architecture of dream chip program in Japan.

In accordance with its relevant information, the 3D IC technology with TSV interconnection will be applied to electronic devices since 2010, with the main applications to high-end components and high-speed image processing fields. As planed, the architecture for next generation 3D SiP in Japan will include such as DRAM, Flash, SRAM memory components, RF components, DSP (Digital Signal Processor), I/O, and interposer for conducting the system. The planned projects will involving the research and developments about EDA tool design, electrical/mechanical design and testing technologies, technologies related to interposers, technologies for thermal dispassion, and stacking/bonding related fundamental technologies required by 3D integration.

16.3 SCHEDULES FROM MAIN MAKERS AND MARKET FORECAST

The potential applications of 3D IC technology have been widely discussed by the industry during the past two years. The perspectives of the application is nothing less than image sensor / camera modules, homogeneous integration of DRAM and/or NAND Flash memory stacking, and the simple heterogeneous integration include CPU/MPU logic component and memory.

Besides the simple heterogeneous integration with logic and memory components, there might be other applications such as complex heterogeneous with much more components than logic and memory, applications of opto-electronics, RF modules, MEMS related applications, and silicon interposers. As for TSV technology commercializing, MEMS and image sensor devices /

camera module have been introducing into commercialized since 2008, and TSV is applied for some specific game console with the 3D integration of processor and memory. And for other potential application categories, the market is still highly expected by the industry.

For the issue of the schedule for mass production and market rising, there are some leading companies positively plan to launch the second wave of mass production of homogeneous stacked memory products from 2010 to 2011, though there are still some other different views for the schedule and market size from the estimation of demand side.

From the market driven of the demand side, image sensor device has originally used WLP (Wafer-Level Packaging) technology. Due to the increasing demand for miniaturization, using TSV technology for interconnection helps reducing the form factor of camera module, there is still a possible existence in the market. However, since this is the early stage of technology commercialization, there is obviously a considerable gap between production yield and economic scale compared with traditional wire bonding method, and therefore the market size is limited under the harsh price / performance ratio of consumer electronics products.

Table 16.1 Status of Samsung's TSV technology and applications.

Application	Target Device	Dev. Status/ Production	Most Wanted	Driving Force	Key technology
Memory	Flash, DRAM	Done / 2008~ (Via Last) 2012~ (Via Last)	SSD, Flash Card/ DRAM Module	Size, Density	12" Infra structure Via First Middle/ Last KGD (Known-Good Die)
Analog, CIS	Analog, CIS	Done / 2008~ (8")	Analog Application CIS Module	Size, Yield	8" Infra structure Via Last Test, Lens & Module technology (CIS)
Ext.-Memory's I/O Bus Extension	GPU/ MPU, Mobile CPU	Develop / 2012~	Graphic Card PC, Works, Server, Mobile Phone	Performance, Cost	12" Infra structure Via Middle/ First Design (Process) Rule, EDA infra. / Methodology (Electrical/ Thermal/ Mechanical) Memory KGD/ Test (BIST)
Replacing L2 Cache					
Logic Die Partitioning	CPU, FPGA, ASOP	Plan / 2015~	32 nm~	Performance, Cost	

As for the application of memory module, the market condition of SSD seemed to face some obstacles under the sweeping of the global financial crisis since the end of 2008. On the other hand, under the the impact of financial crisis, manufacturers seemed to be more conservative towards the cost and the applying of TSV technology for NAND Flash.

As for the development of DRAM, due to the requirements of system products for high speed and high bandwidth for DRAM, DRAM products have gradually gone from DDR to a more high-frequency high-speed DDR2, DDR3 or DDR4 momentum development. According to specifications of DDR3 and DDR4, DDR3 will have 800 MHz, 1067 MHz, 1333 MHz, 1600 MHz and other high bandwidth specifications. At present, Micron and other leading memory makers are consistent that when the bandwidth is up to 1600 MHz or above 1333 MHz, there is need to use TSV technology to cover the large amounts of data transmission.

According to market and turn related information from Micron for DDR3 and DDR4, DDR2 has been replaced by DDR3 in the market since 2008. And it is expected to become mainstream in 2011; while DDR4 is expected to start cutting into the market since 2011. TSV applying schedule of DRAM is also expected to be implemented around 2011.

Table 16.1 shows the TSV technology-related information, which was planned by Samsung. [10] For the memory-related applications, include SSD and memory card, TSV technology for NAND Flash have been ready, and the current status is now waiting for the rise of market demand. TSV technology-related information (Table 16.1), which was planned by Samsung, shows that for the memory-related applications include SSD and memory card, TSV technology for NAND Flash have been ready, and the current status is now waiting for the rise of market demand. For the application of DRAM module, it is estimated that the mass production can be approached in 2012. As for the integration between memory and logic component, it is forecasted that there will be an outcome of the initial volume in 2012.

Table 16.2 TSV adoption by applications and tech. node.

		2009	2010	2011	2012	2013
NAND Flash	Tech. Node	2xnm	2xnm	1xnm	1xnm	1xnm
	Interconnect					
DRAM	Tech. Node	4xnm	3xnm	2xnm	2xnm	1xnm
	Interconnect		**TSV**	**TSV**	**TSV**	**TSV**
MPU	Tech. Node	32nm	32nm	22nm	22nm	15nm
	Interconnect				**TSV**	**TSV**
Logic	Tech. Node	45nm	32nm	32nm	22nm	22/15nm
	Interconnect				**TSV**	**TSV**
Foundry	Tech. Node	32nm	28nm	22nm	22nm	15nm
	Interconnect			**TSV**	**TSV**	**TSV**

In respect of process technology for 3D IC technology adoption and the development schedule, DRAM applications has entered 3xnm process technology node in 2010 and generated the demand for 3D IC technology (Table 16.2), according to the data from semiconductor equipment vendor — KLA-Tencor. [11]

According to the TSV applying schedule for applications and the technology development status by worldwide leading semiconductor companies and comparing with the market driving forces for electronic devices, 3D IC technology will be rapidly launched in the market from technology development into commercialization within 10 years in the near future. To estimate the market by packaging volume, MEMS applications and image sensors modules applications will continue to bang, and the shipping units market is expected to reach the scale of 1,565 million and 768 million in 2013.

A small trial production of memory application will be launched in 2010 the earliest. It is estimated to be 841 million units and 881 million units for NAND Flash module and DRAM module respectively in 2013. Since the packaging of memory card, memory modules and other applications of NAND Flash and DRAM are multiple memory chips stacking within a package, it is expected that the market amount of memory chips via TSV technology will be much greater than the number of the former estimation of packaging units.

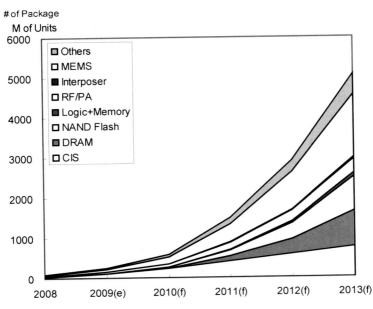

Fig. 16.11 Market forecast of 3DIC technology by applications (units of packaging).

For the market segment of heterogeneous integration with logic and memory components and the market of interposer, the two segments will be influenced by the maturity of mass production of technology, and the market is expected to remain in small application from 2012 to 2013. But with the status of maturing development of TSV technology applied for homogeneous memory stacking, it is expected that there will be an opportunity of rapid growth for the market of both heterogeneous integration and interposer segments after 2015.

References

1. ITRS 2007 edition, "ASSEMBLY AND PACKAGING" p.36, Figure AP14

2. Toshiba, http://www.toshiba.co.jp/about/press/2007_10/pr0101.htm

3. Aptina, www.aptina.com

4. Toshiba, http://www.semicon.toshiba.co.jp/docs/catalog/en/SCE0008_catalog.pdf

5. Samsung, http://www.samsung.com/global/business/semiconductor/products/fusionmemory/Products_MCP_pkginfo.html

6. Elpida, http://www.elpida.com/en/news/2009/08-27.html

7. NikkeiBP

8. e-CUBES of IST, http://ecubes.epfl.ch/public/

9. ASET, Semicon West 2009

10. Samsung, http://www.samsung.com/global/business/semiconductor/support/PackageInformation/pkg_technicalinfo_tsv.html

11. KLA-Tencor

Index